Earth Observation Data Cubes

Earth Observation Data Cubes

Special Issue Editors

Gregory Giuliani
Gilberto Camara
Brian Killough
Stuart Minchin

MDPI • Basel • Beijing • Wuhan • Barcelona • Belgrade

Special Issue Editors

Gregory Giuliani
University of Geneva
Switzerland

Gilberto Camara
Group on Earth Observations (GEO)
Switzerland

Brian Killough
National Aeronautics and Space
Administration (NASA)
USA

Stuart Minchin
Geoscience Australia
Australia

Editorial Office
MDPI
St. Alban-Anlage 66
4052 Basel, Switzerland

This is a reprint of articles from the Special Issue published online in the open access journal *Data* (ISSN 2306-5729) in 2019 (available at: https://www.mdpi.com/journal/data/special_issues/EODC).

For citation purposes, cite each article independently as indicated on the article page online and as indicated below:

LastName, A.A.; LastName, B.B.; LastName, C.C. Article Title. *Journal Name* **Year**, *Article Number*, Page Range.

ISBN 978-3-03928-092-6 (Pbk)
ISBN 978-3-03928-093-3 (PDF)

Cover image courtesy of European Space Agency.

Contents

About the Special Issue Editors

Gregory Giuliani, Dr., is Senior Lecturer in Earth Observations at the Institute for Environmental Sciences (University of Geneva, Switzerland) and Head of the Digital Earth Unit, UNEP/GRID-Geneva. He is leading the Swiss Data Cube project. He is also an ISDE Council Member and active contributor to GEO/GEOSS. His research focuses on land change science and how Earth observations can be used to monitor and assess environmental changes and support sustainable development.

Gilberto Camara, Prof. Dr., has been director of the Group on Earth Observations secretariat since July 2018. He is a leading researcher in Geoinformatics, Geographical Information Science, and Land Use Change, and has been recognized internationally for promoting free access and open source software for Earth observation data. Under his guidance as Director for Earth Observation (2001–2005) and Director General (2006–2012), Brazil's National Institute for Space Research (INPE) made significant advances in land change monitoring using remote sensing, which contributed to Brazil achieving an 80% decrease in deforestation in the Amazon rainforest, supporting the commitment pledged by Brazil's at the UNFCCC COP15. This achievement was hailed as "the biggest environmental success story in a decade" by the scientific journal Nature. In support of Brazil's pledge to the 2015 Paris Agreement, Gilberto lead a team that projected Brazil's future emissions from land use and agriculture.

Brian Killough, Dr., has been with NASA for 31 years. He currently leads the Committee on Earth Observing Satellites (CEOS) Systems Engineering Office (SEO). The SEO supports the international CEOS organization which is comprised of 60 space agencies and organizations coordinating satellite earth observation data for enhanced societal benefit. Under the leadership of Dr. Killough, the SEO has a significant role in the new Open Data Cube initiative for enhanced global use of large volumes of satellite analysis ready data.

Stuart Minchin, Dr., is currently the Chief of the Environmental Geoscience Division of Geoscience Australia. The Environmental Geoscience Division of Geoscience Australia is the center of expertise in the Australian Government for environmental earth science issues and the custodian of national environmental geoscience data, information, and knowledge. He has an extensive background in the management and modeling of environmental data and the online delivery of data, modeling, and reporting tools for improved natural resource management.

Editorial

Earth Observation Open Science: Enhancing Reproducible Science Using Data Cubes

Gregory Giuliani [1,2,*] , **Gilberto Camara** [3], **Brian Killough** [4] **and Stuart Minchin** [5]

1   Institute for Environmental Sciences, University of Geneva, Bd Carl-Vogt 66, CH-1205 Geneva, Switzerland
2   United Nation Environment Programme, Science Division, GRID-Geneva, 11 Chemin des Anémones, CH-1219 Châtelaine, Switzerland
3   Group on Earth Observations, 7bis Avenue de la Paix, Case Postale 2300, 1211 Geneva, Switzerland; gcamara@geosec.org
4   National Aeronautics and Space Administration, Langley Research Center, MS 457, Hampton, VA 23681, USA; brian.d.killough@nasa.gov
5   Geoscience Australia, GPO Box 378, Canberra, ACT 2601, Australia; stuart.minchin@ga.gov.au
*   Correspondence: gregory.giuliani@unige.ch; Tel.: +41-(0)22-379-07-09

Received: 21 November 2019; Accepted: 21 November 2019; Published: 25 November 2019

Abstract: Earth Observation Data Cubes (EODC) have emerged as a promising solution to efficiently and effectively handle Big Earth Observation (EO) Data generated by satellites and made freely and openly available from different data repositories. The aim of this Special Issue, "Earth Observation Data Cube", in *Data*, is to present the latest advances in EODC development and implementation, including innovative approaches for the exploitation of satellite EO data using multi-dimensional (e.g., spatial, temporal, spectral) approaches. This Special Issue contains 14 articles covering a wide range of topics such as Synthetic Aperture Radar (SAR), Analysis Ready Data (ARD), interoperability, thematic applications (e.g., land cover, snow cover mapping), capacity development, semantics, processing techniques, as well as national implementations and best practices. These papers made significant contributions to the advancement of a more Open and Reproducible Earth Observation Science, reducing the gap between users' expectations for decision-ready products and current Big Data analytical capabilities, and ultimately unlocking the information power of EO data by transforming them into actionable knowledge.

Keywords: open science; reproducibility; earth observations; data cube; analysis ready data; remote sensing; satellite imagery

---

Planet Earth is currently on an unsustainable pathway. Increasing pressures on natural resources induced by human activities are globally affecting the environment. Regular and continuous monitoring is necessary to assess, understand, and mitigate these environmental changes [1–3]. Consequently, timely and reliable access to data describing physical, chemical, biological, and socio-economic conditions can provide the basis for reliable and accountable scientific understanding and knowledge about the limits of our planet. This access to data can support informed decisions and evidence-based policies for the efficient use of our planet's resources [4,5].

To facilitate environmental monitoring, our planet has been under continuous observations from satellites since 1972 [6,7]. Today, remotely sensed Earth Observations (EO) data have already exceeded the petabyte-scale and increasingly are freely and openly available from different data repositories [8]. This poses a number of issues in terms of Volume (e.g., data volumes have increased by 10 in the last 5 years); Velocity (e.g., Sentinel-2 is capturing a new image of a given place every 5 days); and Variety (e.g., different type of sensors, spatial/spectral resolutions). Traditional approaches to the acquisition, management, distribution, and analysis of satellite EO data have limitations (e.g., data size, heterogeneity and complexity) that impede the massive use and analysis of Big Earth Data.

The fact that the full information potential of EO data has not yet been realized and therefore remains still underutilized is explained by various reasons: (1) it requires scientific knowledge to understand what data is needed—optical (which resolution?)—radar (which type?); (2) it is difficult to access and download the increasing volumes of data generated by satellites; (3) there is a lack of expertise and computing resources to efficiently prepare and utilize EO data; (4) the particular structure of EO data and (5) the significant effort and cost required to store and process data limit its effective use.

Addressing Big Data challenges such as Volume, Velocity and Variety, requires a change of paradigm and a move away from traditional data-centric approaches (e.g., local processing and data distribution methods to lower the barriers caused by data size and related complications in data management [9,10]. In particular, data volume and velocity will continue to grow as the demands increase for decision-support information derived from these data [11]. Using the cloud, it is now possible to move algorithms and tools to data, making large volumes of EO data available to a wide range of users, enabling them to handle and visualize data they are interested in without having to download them and consequently avoiding large-scale data transfers that can impede the efficient and effective use of EO data [12,13].

To tackle these issues and bridge the gap between users' expectations and current Big Data analytical capabilities, EO Data Cubes (EODC) have emerged as a new paradigm revolutionizing the way users can interact with EO data and providing a promising solution for the storage, organization, management, and analysis of Big EO data [14]. The main objective of EODC is to facilitate EO data usage by addressing Volume, Velocity, Variety challenges and providing access to large spatio-temporal data in an analysis-ready format [15].

Different EODC implementations are currently operational, such as Digital Earth Australia [16], the Swiss Data Cube [17], the EarthServer [18], the E-sensing platform [19] the Copernicus Data and Information Access Services (DIAS) [20] or the Google Earth Engine [21]. These initiatives are paving the way for broadening the use of EO data to larger communities of users, supporting decision-makers with timely and actionable information converted into meaningful geophysical variables and ultimately unlocking the information power of EO data.

All these developments would not have been possible without Free and Open Data policies to facilitate access to data and Open Source code to efficiently develop software solutions [22]. Open Science is a new approach to research and educational processes, which seeks to make scientific research more collaborative and transparent and to make knowledge accessible by using digital technologies and new collaborative tools [23]. Achieving reproducible knowledge requires exposing all parts of an application (e.g., code, data, executable) [24]. Therefore, Open Science is considered as an umbrella term encompassing all practices that aim to remove barriers to sharing any type of output (e.g., research data), resources (e.g., scientific publications), methods (e.g., lab notes) or tools (e.g., software). This is a practice of science to achieve more openness and to enable others to collaborate and contribute under terms that enable the reuse, redistribution and reproduction of research and its underlying data and methods [25]. In particular, with the advent of cloud computing, knowledge is easier to share [11]. Open Science is fundamental in a 21st Century where Science is embedded in societal decision-making. Increased openness and transparency are effective means to fight fake news and post-truth [26].

Despite the fact that in the EO domain various open science practices are already adopted, such as the Open Standards provided by the Open Geospatial Consortium (OGC) [27], Open Source software [28], Open Code Library (e.g., Open remote sensing http://openremotesensing.net) or the IEEE Remote Sensing Code Library (http://www.grss-ieee.org/publication-category/rscl/), data sets and algorithm evaluation standards (http://dase.grss-ieee.org), or Open Data licenses for Landsat and Sentinel data [29], EO Open Science remains underestimated and various socio-cultural, technological, political, organizational, economic and legal challenges (e.g., lack of recognition and rewards, overload for opening data, changing working procedure, missing political endorsements–strategies–policies, unclear legal frameworks) need to be addressed to adequately realize its full potential.

This Special Issue (https://www.mdpi.com/journal/data/special_issues/EODC) presents some of the most recent advancements in the use and implementation of EODC. They significantly contribute to the advancement of a more Open and Reproducible EO Science and help to reduce the gap between data and knowledge.

Most of the Open Science facets (https://www.fosteropenscience.eu/content/what-open-science-introduction) are covered by the contributions of this Special Issue. First of all, the 14 papers are accessible in *Open Access* under the terms and conditions of the Creative Commons Attribution (CCBY) license. This means that these research outputs are distributed online and freely available, removing the barriers to copying or reuse by applying an open license copyright. Together with FAIR guiding principles, this allows sharing of findings and streamlining of the creation of new data products by making them Findable, Accessible, Interoperable and Reusable [30,31].

With the different innovative solutions that are available to implement EODC, one of the major challenges is to prevent them from becoming silos of information. Interoperability is consequently an important aspect to consider. Giuliani et al. [32] demonstrated how widely adopted geospatial standards can be used to enhance the interoperability of EODC and can help in delivering and leveraging the power of EO data building, efficient discovery, access and processing services. However, to harness the information potential of satellite EO data, syntactic interoperability is not sufficient. As numerical sensory data have no semantic meaning, EO data lack semantics. Augustin et al. [33] clarify and share their definition of semantic EODC, demonstrating how they enable different possibilities for data retrieval, semantic queries based on EO data content, and semantically enabled analysis. Semantic EODC are the foundation of the EO data expert system and can facilitate deriving knowledge, as presented by Plag and Jules Plag [34].

Regarding *Open Data*, one of the main topics concerns the development of Analysis Ready Data (ARD) for Synthetic Aperture Radar (SAR) imagery. Indeed, if the provision of optical ARD is becoming common, the complexity of SAR data makes them challenging to developed procedures for the regular provision of SAR ARD. Truckenbrodt et al. [35] and Ticehurst et al. [36] assessed the feasibility of automatically producing analysis-ready radiometrically terrain-corrected (RTC) Synthetic Aperture Radar (SAR) gamma nought backscatter data from Sentinel-1. Both studies concluded that the European Space Agency (ESA) Sentinel Application Platform (SNAP) toolbox (https://step.esa.int/main/toolboxes/snap/) is a valid solution for producing Sentinel-1 ARD products. One important reward in publishing open data is the possibility to be cited. Providing a reference to data similar to scientific journal articles or conference papers is increasingly recognized as an essential practice leading to the recognition of data as important research outputs. Data citation supports (1) attribution and credit; (2) collaboration and reuse of data; (3) enables reproducibility of findings; (4) faster and efficient research progress, and (5) provides means to share data with (future) researchers. Schubert et al. [37] presented a solution for an operational service on dynamic data citation to enable the effective reuse of EO data in a collaborative and reproducible manner.

To benefit from the large volume of EO data made available with Data Cubes, recent *Open Source* developments were allowed to implement solutions in open source geoinformation and statistical software. Gebbert et al. [38] have developed spatio-temporal topological operators in the GRASS GIS software to enable the effective use of heterogenous (e.g., extent, granularity) spatio-temporal EO data. Similarly, Appel et al. [39] introduced an open source C++ library and R package for the construction and processing of on-demand data cubes from satellite image collections, and showed how it supports interactive method development workflows where data users can initially try methods on small subsamples before running analyses on high resolution and/or large areas. Finally, *Open Standards* such as the OGC Web Map Service (WMS), together with modern web browser capabilities, has enabled time-series analysis directly within a web-based application [40].

To reach the objective of facilitated and reproducible analysis of EO data, as well as empowering a large community of users to benefit from satellite EO data, *Open Notebooks* appear as promising solutions. They help to document research developments as reproducible experiments and facilitate the sharing of scientific data analysis. Electronic Lab Notebooks (ELN), such as Jupyter Notebooks, are replacing paper lab notebooks with digital versions. Kopp et al. [41] showed that such notebooks simplify access and use for end-users, enabling a wide variety of web and desktop applications. Poussin et al. [42] demonstrated the benefits of Open and Reproducible Science using a snow detection algorithm, developed in Switzerland and shared as an open notebook, to monitor snow cover evolution for the last three decades in the Gran Paradiso National Park in Italy. Furthermore, Lucas et al. [43] developed a conceptual framework to implement a Land Cover Change model, providing Australia and other countries using the Open Data Cube (ODC) environment with the opportunity to routinely generate land cover maps from Landsat or Sentinel-1/2 data, at least annually and using a consistent and internationally recognized taxonomy.

Finally, an important aspect related to any new technology lies in developing new capacities to reach large adoption, acceptance and commitment. Asmaryan et al. [44] presented how effective knowledge transfer, using *Open Educational resources*, has been achieved between Switzerland and Armenia for developing and implementing the first version of an Armenian Data Cube. This ultimately can support National Open Data Cubes to contribute to country-level development policies and practices [45].

To conclude, we believe that EODC have the potential to achieve the vision of transforming data into actionable knowledge by lowering the entry barrier to massive-use Big Earth Data analysis and therefore act as an information technology enabler. Ultimately, it can provide an effective mean to build socially robust, replicable, and reusable knowledge, to generate decision-ready products based on Open Science.

Author Contributions: G.G.: conceptualization, writing—original draft preparation; G.C.: writing—review and editing; B.K.: writing—review and editing; S.M.: writing—review and editing.

Acknowledgments: The authors gratefully acknowledge the administrative and technical support of the Data journal team. We want to thank the authors who contributed towards this Special Issue on "Earth Observations Data Cube", as well as the reviewers who provided the authors with comments and very constructive feedback.

Conflicts of Interest: The authors declare no conflict of interest.

References

1. Rockström, J.; Bai, X.; DeVries, B. Global sustainability: the challenge ahead. *Glob. Sustain.* **2018**, *1*, 1–3. [CrossRef]

2. Steffen, W.; Richardson, K.; Rockström, J.; Cornell, S.E.; Fetzer, I.; Bennett, E.M.; Biggs, R.; Carpenter, S.R.; De Vries, W.; De Wit, C.A.; et al. Planetary boundaries: Guiding human development on a changing planet. *Science* **2015**, *347*, 1259855. [CrossRef] [PubMed]

3. Biermann, F.; Bai, X.; Bondre, N.; Broadgate, W.; Chen, C.-T.A.; Dube, O.P.; Erisman, J.W.; Glaser, M.; van der Hel, S.; Lemos, M.C.; et al. Down to Earth: Contextualizing the Anthropocene. *Glob. Environ. Chang.* **2016**, *39*, 341–350. [CrossRef]

4. Giuliani, G.; Nativi, S.; Obregon, A.; Beniston, M.; Lehmann, A. Spatially enabling the Global Framework for Climate Services: Reviewing geospatial solutions to efficiently share and integrate climate data & information. *Clim. Serv.* **2017**, *8*, 44–58.

5. Lehmann, A.; Chaplin-Kramer, R.; Lacayo, M.; Giuliani, G.; Thau, D.; Koy, K.; Goldberg, G.; Sharp, R., Jr. Lifting the Information Barriers to Address Sustainability Challenges with Data from Physical Geography and Earth Observation. *Sustainability* **2017**, *9*, 858. [CrossRef]

6. Zhu, Z. Science of Landsat Analysis Ready Data. *Remote Sens.* **2019**, *11*, 2166. [CrossRef]

7. Wulder, M.A.; Loveland, T.R.; Roy, D.P.; Crawford, C.J.; Masek, J.G.; Woodcock, C.E.; Allen, R.G.; Anderson, M.C.; Belward, A.S.; Cohen, W.B.; et al. Current status of Landsat program, science, and applications. *Remote Sens. Environ.* **2019**, *225*, 127–147.

8. Woodcock, C.E.; Allen, R.; Anderson, M.; Belward, A.; Bindschadler, R.; Cohen, W.; Gao, F.; Goward, S.N.; Helder, D.; Helmer, E.; et al. Free Access to Landsat Imagery. *Science* **2008**, *320*, 1011. [CrossRef]

9. Nativi, S.; Mazzetti, P.; Craglia, M. A view-based model of data-cube to support big earth data systems interoperability. *Big Earth Data* **2017**, *1*, 75–99. [CrossRef]

10. Nativi, S.; Mazzetti, P.; Santoro, M.; Papeschi, F.; Craglia, M.; Ochiai, O. Big Data challenges in building the Global Earth Observation System of Systems. *Environ. Model. Softw.* **2015**, *68*, 1–26. [CrossRef]
11. Nativi, S.; Santoro, M.; Giuliani, G.; Mazzetti, P. Towards a knowledge base to support global change policy goals. *Int. J. Digit. Earth* **2019**, 1–29. [CrossRef]
12. Boulton, G. The challenges of a Big Data Earth. *Big Earth Data* **2018**, *4471*, 1–7. [CrossRef]
13. Guo, H. Big Earth data: A new frontier in Earth and information sciences. *Big Earth Data* **2017**, *1*, 4–20. [CrossRef]
14. Baumann, P.; Misev, D.; Merticariu, V.; Huu, B.P. Datacubes: Towards Space/Time Analysis-Ready Data. In *Service-Oriented Mapping*; Springer: Cham, Switzerland, 2019; pp. 269–299.
15. Dwyer, J.; Roy, D.; Sauer, B.; Jenkerson, C.; Zhang, H.; Lymburner, L. Analysis Ready Data: Enabling Analysis of the Landsat Archive. *Remote Sens.* **2018**, *10*, 1363.
16. Dhu, T.; Dunn, B.; Lewis, B.; Lymburner, L.; Mueller, N.; Telfer, E.; Lewis, A.; McIntyre, A.; Minchin, S.; Phillips, C. Digital earth Australia—Unlocking new value from earth observation data. *Big Earth Data* **2017**, *1*, 64–74. [CrossRef]
17. Giuliani, G.; Chatenoux, B.; De Bono, A.; Rodila, D.; Richard, J.-P.; Allenbach, K.; Dao, H.; Peduzzi, P. Building an Earth Observations Data Cube:Lessons learned from the Swiss Data Cube (SDC) on generating Analysis Ready Data (ARD). *Big Earth Data* **2017**, *1*, 100–117. [CrossRef]
18. Baumann, P.; Mazzetti, P.; Ungar, J.; Barbera, R.; Barboni, D.; Beccati, A.; Bigagli, L.; Boldrini, E.; Bruno, R.; Calanducci, A.; et al. Big Data Analytics for Earth Sciences: The EarthServer approach. *Int. J. Digit. Earth* **2016**, *9*, 3–29. [CrossRef]
19. Camara, G.; Ribeiro, G.; Vinhas, L.; Ferreira, K.R.; Cartaxo, R.; Simões, R.; Llapa, E.; Assis, L.F.; Sanchez, A. The e-Sensing architecture for big Earth observation data analysis. In Proceedings of the 2017 Conference on Big Data from Space (BiDS'17), Toulouse, France, 28–30 November 2017; pp. 1–4.
20. European Commission. The DIAS: User-friendly Access to Copernicus Data and Information. 2018. Available online: https://www.copernicus.eu/sites/default/files/Copernicus_DIAS_Factsheet_June2018.pdf (accessed on 25 November 2019).
21. Gorelick, N.; Hancher, M.; Dixon, M.; Ilyushchenko, S.; Thau, D.; Moore, R. Google Earth Engine: Planetary-scale geospatial analysis for everyone. *Remote Sens. Environ.* **2017**, *202*, 18–27. [CrossRef]
22. Ferrari, T.; Scardaci, D.; Andreozzi, S. *The Open Science Commons for the European Research Area. Earth Observation Open Science and Innovation*; Springer: Cham, Switzerland, 2018; pp. 43–67.
23. European Commission. *Open Innovation, Open Science, Open to the World — A Vision for Europe*; Directorate-General for Research and Innovation: Brussels, Belgium, 2016.
24. Peng, R.D. Reproducible Research in Computational Science. *Science* **2011**, *334*, 1226–1227. [CrossRef]
25. McKiernan, E.; Bourne, P.; Brown, C.; Buck, S.; Kenall, A.; Lin, J.; McDougall, D.; Nosek, B.A.; Ram, K.; Soderberg, C.K.; et al. How open science helps researchers succeed. *eLife* **2016**, *5*, e16800. [CrossRef]
26. Cornell, S.; Berkhout, F.; Tuinstra, W.; Tàbara, J.D.; Jäger, J.; Chabay, I.; de Wit, B.; Langlais, R.; Mills, D.; Moll, P.; et al. Opening up knowledge systems for better responses to global environmental change. *Environ. Sci. Policy* **2013**, *28*, 60–70. [CrossRef]
27. Maso, J.; Zabala, A.; Serral, I.; Pons, X. Remote Sensing Analytical Geospatial Operations Directly in the Web Browser. In Proceedings of the International Archives of the Photogrammetry, Remote Sensing and Spatial Information Sciences, Delft, The Netherlands, 1–5 October 2018; Volume XLII–4, pp. 403–410.
28. Grizonnet, M.; Michel, J.; Poughon, V.; Inglada, J.; Savinaud, M.; Cresson, R. Orfeo Toolbox: Open Source Processing of Remote Sensing Images. Open Geospat. *Data Softw. Stand.* **2017**, *2*, 15.
29. Ryan, B. The benefits from open data are immense. *Geospat. World* **2016**, 72–73.
30. Wilkinson, M.D.; Dumontier, M.; Aalbersberg, I.J.; Appleton, G.; Axton, M.; Baak, A.; Blomberg, N.; Boiten, J.W.; Santos, L.B.D.; Bourne, P.E.; et al. Comment: The fair guiding principles for scientific data management and stewardship. *Sci. Data* **2016**, *3*, 160018. [CrossRef] [PubMed]
31. Stall, S.; Yarmey, L.; Cutcher-Gershenfeld, J.; Hanson, B.; Lehnert, K.; Nosek, B.; Parsons, M.; Robinson, E.; Wyborn, L. Make scientific data FAIR. *Nature* **2019**, *570*, 27. [CrossRef] [PubMed]
32. Giuliani, G.; Masó, J.; Mazzetti, P.; Nativi, S.; Zabala, A. Paving the Way to Increased Interoperability of Earth Observations Data Cubes. *Data* **2019**, *4*, 113. [CrossRef]
33. Augustin, H.; Sudmanns, M.; Tiede, D.; Lang, S.; Baraldi, A. Semantic Earth Observation Data Cubes. *Data* **2019**, *4*, 102. [CrossRef]

34. Plag, H.-P.; Jules-Plag, S.-A. A Transformative Concept: From Data Being Passive Objects to Data Being Active Subjects. *Data* **2019**, *4*, 135. [CrossRef]

35. Truckenbrodt, J.; Freemantle, T.; Williams, C.; Jones, T.; Small, D.; Dubois, C.; Thiel, C.; Rossi, C.; Syriou, A.; Giuliani, G. Towards Sentinel-1 SAR Analysis-Ready Data: A Best Practices Assessment on Preparing Backscatter Data for the Cube. *Data* **2019**, *4*, 93. [CrossRef]

36. Ticehurst, C.; Zhou, Z.-S.; Lehmann, E.; Yuan, F.; Thankappan, M.; Rosenqvist, A.; Lewis, B.; Paget, M. Building a SAR-Enabled Data Cube Capability in Australia Using SAR Analysis Ready Data. *Data* **2019**, *4*, 100. [CrossRef]

37. Schubert, C.; Seyerl, G.; Sack, K. Dynamic Data Citation Service—Subset Tool for Operational Data Management. *Data* **2019**, *4*, 115. [CrossRef]

38. S Gebbert, S.; Leppelt, T.; Pebesma, E. A Topology Based Spatio-Temporal Map Algebra for Big Data Analysis. *Data* **2019**, *4*, 86. [CrossRef]

39. Appel, M.; Pebesma, E. On-Demand Processing of Data Cubes from Satellite Image Collections with the gdalcubes Library. *Data* **2019**, *4*, 92. [CrossRef]

40. Maso, J.; Zabala, A.; Serral, I.; Pons, X. A Portal Offering Standard Visualization and Analysis on top of an Open Data Cube for Sub-National Regions: The Catalan Data Cube Example. *Data* **2019**, *4*, 96. [CrossRef]

41. Kopp, S.; Becker, P.; Doshi, A.; Wright, D.J.; Zhang, K.; Xu, H. Achieving the Full Vision of Earth Observation Data Cubes. *Data* **2019**, *4*, 94. [CrossRef]

42. Poussin, C.; Guigoz, Y.; Palazzi, E.; Terzago, S.; Chatenoux, B.; Giuliani, G. Snow Cover Evolution in the Gran Paradiso National Park, Italian Alps, Using the Earth Observation Data Cube. *Data* **2019**, *4*, 138. [CrossRef]

43. Lucas, R.; Mueller, N.; Siggins, A.; Owers, C.; Clewley, D.; Bunting, P.; Kooymans, C.; Tissott, B.; Lewis, B.; Lymburner, L.; et al. Land Cover Mapping using Digital Earth Australia. *Data* **2019**, *4*, 143. [CrossRef]

44. Asmaryan, S.; Muradyan, V.; Tepanosyan, G.; Hovsepyan, A.; Saghatelyan, A.; Astsatryan, H.; Grigoryan, H.; Abrahamyan, R.; Guigoz, Y.; Giuliani, G. Paving the Way towards an Armenian Data Cube. *Data* **2019**, *4*, 117. [CrossRef]

45. Dhu, T.; Guiliani, G.; Juárez, J.; Kavvada, A.; Killough, B.; Merodio, P.; Minchin, S.; Ramage, S. National Open Data Cubes and Their Contribution to Country-Level Development Policies and Practices. *Data* **2019**, *4*, 144. [CrossRef]

Article

Paving the Way to Increased Interoperability of Earth Observations Data Cubes

Gregory Giuliani [1,2,*], Joan Masó [3], Paolo Mazzetti [4], Stefano Nativi [5] and Alaitz Zabala [6]

[1] Institute for Environmental Sciences, University of Geneva, enviroSPACE, Bd Carl-Vogt 66, CH-1205 Geneva, Switzerland
[2] Institute for Environmental Sciences, University of Geneva, GRID-Geneva, Bd Carl-Vogt 66, CH-1211 Geneva, Switzerland
[3] Center for Ecological Research and Forestry Applications (CREAF), Universitat Autònoma de Barcelona (UAB), 08193 Bellaterra, Barcelona, Spain
[4] National Research Council of Italy (CNR)—Institute of Atmospheric Pollution Research, Via Madonna del Piano 10, 50019 Sesto Fiorentino, Italy
[5] European Commission Joint Research Center (JRC), Via E. Fermi, 2749, 21027 Ispra, Italy
[6] Geography Department, Universitat Autònoma de Barcelona (UAB), 08193 Bellaterra, Barcelona, Spain
* Correspondence: gregory.giuliani@unige.ch; Tel.: +41-(0)22-379-07-09

Received: 14 June 2019; Accepted: 27 July 2019; Published: 30 July 2019

Abstract: Earth observations data cubes (EODCs) are a paradigm transforming the way users interact with large spatio-temporal Earth observation (EO) data. It enhances connections between data, applications and users facilitating management, access and use of analysis ready data (ARD). The ambition is allowing users to harness big EO data at a minimum cost and effort. This significant interest is illustrated by various implementations that exist. The novelty of the approach results in different innovative solutions and the lack of commonly agreed definition of EODC. Consequently, their interoperability has been recognized as a major challenge for the global change and Earth system science domains. The objective of this paper is preventing EODC from becoming silos of information; to present how interoperability can be enabled using widely-adopted geospatial standards; and to contribute to the debate of enhanced interoperability of EODC. We demonstrate how standards can be used, profiled and enriched to pave the way to increased interoperability of EODC and can help delivering and leveraging the power of EO data building, efficient discovery, access and processing services.

Keywords: Open Data Cube; remote sensing; geospatial standards; landsat; sentinel; analysis ready data

---

1. Introduction

The planet Earth is currently on an unsustainable pathway. Increasing pressures on natural resources induced by human activities are affecting the global environment. Regular and continuous monitoring is necessary to assess, understand, and mitigate these environmental changes [1–3]. Consequently, timely and reliable access to data describing physical, chemical, biological and socio-economic conditions can provide the basis for reliable and accountable scientific understanding and knowledge supporting informed decisions and evidence-based policies [4,5]. This can be done by applying the data-information-knowledge-wisdom (DIKW) paradigm [6,7]. In DIKW, information is an added-value product resulting from the comprehension of available data and their relations with physical and/or social phenomena. In turn, knowledge is generated by understanding information and elaborating on valuable patterns.

Earth Observations (EO) data, acquired remotely by satellite or in-situ by sensors, is a valid and globally consistent source of information and knowledge for monitoring the state of the planet and increasing our understanding of Earth processes [8]. EO data are essential to allow long-term global coverage and to monitor land cover changes over large areas through time [9]. With the increased number of spaceborne sensors, the planet is virtually under continuous monitoring, with satellites providing global coverage at medium-to-high spatial and spectral resolutions on a daily basis [10–12]. Furthermore, open data policies have greatly facilitated the access to satellite data, such as the United States Geological Survey (USGS) Landsat, National Aeronautics and Space Administration (NASA)'s Moderate Resolution Imaging Spectroradiometer (MODIS), the Japan Aerospace Exploration Agency (JAXA), Advanced Spaceborne Thermal Emission and Reflection Radiometer (ASTER), or European Sapce Agency (ESA) Sentinels [13,14]. However, handling such large volumes (e.g., tera to petabytes), variety (e.g., radar, optical), and velocity (e.g., new data available daily), as well as the efforts and costs required to transform EO data into meaningful information have restricted systematic analysis to monitor environmental changes [15]. Consequently, the development of large-scale analytical tools allowing effective and efficient information retrieval based on scientific questions, as well as generating decision-ready products remains a major challenge for the EO community [16].

Earth Observations Data Cubes (EODC) have recently emerged as a paradigm transforming the way users interact with large spatio-temporal EO data [17,18]. It enhances connections between data, applications, and users facilitating management, access and use of analysis ready data (ARD) [19,20]. The ambition is to allow scientists, researchers, and different businesses to harness big EO data at a minimum cost and effort [21]. This significant interest is exemplified by various implementations of platforms capable of analyzing EO data that exist, such as the Open Data Cube (ODC) [19], the EarthServer [22], the e-sensing platform [23], the JRC Earth Observation Data and Processing Platform (JEODPP) [24], the Copernicus Data and Information Access Services (DIAS) [25] or the Google Earth Engine (GEE) [26]. The novelty of the approach results in different innovative solutions and, among them, some can be considered some sort of data cube (even if there is a lack of commonly agreed definition of the EODC term), leading to interoperability issues among them precluding effective discovery, common data access and sharing processes on data stored in EODC [20,27]. Consequently, EODC interoperability has been recognized as a major challenge for the Global Change and Earth System science domains [27].

Therefore, the objectives of this paper are: (1) To better characterize EODC (e.g., differentiate between data cube and cloud-based processing facilities, such as DIAS or Google Earth Engine); (2) issue recommendations to prevent EODC from becoming silos of information; and (3) present/demonstrate how existing geospatial standards can be profiled and enriched to pave the way to increased syntactic and semantic interoperability, as well as addressing use and orchestration of EODC and can help the delivering and leveraging the power of EO data in building efficient discovery, access and processing services.

2. Earth Observation Data Cube and Analysis Ready Data Infrastructures

To better characterize EODC, six different aspects have been identified covering respective well-established data science domains, allowing to describe EODC into meaningful and manageable parts with the ultimate objective to ensure compatibility and consistency for efficient data discovery, view, access and processing [20,27].

The "faces" concept was then further elaborated, leading to the definition of six viewpoints [27] characterizing a data cube infrastructure: (1) The semantic view, covering the information stored in the content and their semantics; (2) the geometric view, covering the content in geometrical representation, in particular their discretization and digital structuring aspects; (3) the encoding view, dealing with the multi-dimension aspects, including pre-processing and analytical processing aspects; (4) the interaction/interface view, dealing with the analytical functionalities provided by the infrastructure and their accessibility via web-based Application Programming Interfaces (APIs). (5) Interconnection/platform view, dealing with the software components and services necessary to realize the cybernetics framework; (6) the composition/ecosystem view, concerning the infrastructure composability with analogous systems and governance aspects.

To enable and facilitate full interoperability of EODC, as well as leveraging the rich legacy of Business Intelligence, it is important to make sure all the views are adequately addressed and kept technology-neutral [27]. To achieve this, a crucial action consists of the identification of existing and mature models and patterns promoting the adoption of standard approaches.

To better characterize EODC, it is important to differentiate them from cloud-based processing facilities, such as DIAS, GEE or Earth on Amazon Web Services (AWS). Cloud-based EO platforms commonly provide (free and open) access to global EO datasets (available datasets are growing daily) along with powerful space and time analysis tools supporting different programming languages (e.g., JavaScript, Python and R). Recently, these online platforms have transformed the user community working with satellite EO data. They removed most of the burden for data preparation, yielding rapid results and fostering a community of contributors, which is growing fast. However, they lock users into a platform (sort of commercial) dependency, with well-known challenges. Potential identified concerns are: (1) Users do not know whether a given platform will be sustained and/or evolved in the future; (2) the provision of limited time and spatial scale for analyses; (3) the provision of cloud-based computing only (i.e., no options for hubs or local computing solutions); (4) users are requested to upload their analytical processing and even local data, while data download is discouraged or not even allowed; (5) platform providers require the right to "own" all the data utilized on the platform; (6) users get only those datasets that providers offer, limiting data interoperability (e.g., Landsat 8 or Sentinel 1 data can be missing); (7) data are often not ready to be analyzed (e.g., top-of-the-atmosphere—TOA—reflectance data).

Most of these potential drawbacks can be tackled by utilizing EODC. For example, users can install on their own computing infrastructure an open source software solution (such as the ODC), that allows for storing different type of data (e.g., Landsat, Sentinel, SPOT, MODIS, aerial and/or drone imagery, etc.). This solution provides improved control, more flexibility and scalability, both in terms of usage, and a further sense of ownership. EODC support an efficient and joint use of multiple datasets, enhancing their interoperability and complementarity. This facilitates not only data sharing but also the sharing of code, tools and algorithms. Finally, it grants the possibility to develop local and/or regional solutions that avoid commercial and internet dependence. For these reasons, more and more cloud-based EO data infrastructures are considering offering EODC services to their potential users.

3. Current Interoperability Levels of EODC

3.1. Software Systems Interoperability

Interoperability was first defined by the Institute of Electrical and Electronics Engineers (IEEE) as "the ability of two or more systems or components to exchange information and to use the information that has been exchanged" [28]. In the present digital transformation era, interoperability is a critical software system attribute, since it enables different systems interaction to support the society daily activities. The emerging technologies composing systems-of-systems have increased their importance and scope. Interoperability can be thought as the ability of software systems to interact for a specific purpose, once their differences (development platforms, data formats, culture, and legal issues) have been overcome [29]. Interoperability is not a clear-cut characteristic of a system; there exist different levels (or types) of interoperability, spanning from system integrability (including technical and syntactic interoperability levels) to system composability (including semantic, dynamic and conceptual interoperability levels). Commonly, system interoperability is achieved by pushing open standards—either de-facto or de-jure.

For the scope of this manuscript, interoperability may be defined as the ability of different data cube infrastructures to connect and communicate in a coordinated way, providing a rich experience to users. For example, interoperable data cubes should provide the necessary functions to allow users to access and analyze data with (virtually) no effort, regardless of data origin, structure and encoding.

3.2. Interoperability Contexts for EODC Software Systems

Earth Observation Data Cubes are an infrastructure managing (long) time series of observations referring to the Earth; i.e., characterized by a spatial reference system. To provide users with a rich experience, EODC interoperability must be considered in respect of the context where their use is planned. For a given EODC, it is possible to consider three different and increasing (from the composability point of view) levels of interoperability:

1. Interoperability among EODCs.
2. Interoperability of EODCs with other types of geospatial data cubes.
3. Interoperability of EODCs with general-purpose data cube infrastructures.

All the three interoperability contexts are facilitated by the past and present activities on the harmonization and mediation of EO information; i.e., the standardization process.

3.2.1. Interoperability among EODCs

For the geometric view, when characterizing an EODC [27], spatio-temporal coverages (ISO 19123, 2005) are largely recognized as the referential representation for observation of physical phenomena. Therefore, it is the general consensus on building cubes with a spatio-temporal domain. Indeed, this is an effective cube geometry for fast generation of a time-series, which is one of the most commonly used cases. However, it is worth noting that there is still heterogeneity in a number of domain dimensions (2D, 2D+T, 3D, 3D+T, and 3D+T). Besides, some commonly used cases, such as simulations, would actually need more than one temporal dimension as part of the domain. Therefore, at this level, a major challenge in EODCs interoperability concerns the harmonization of domain dimensions among different data cube implementations. Another relevant geometrical aspect is related to the metrics that is superimposed to a given data cube. In particular, this includes the coordinate reference system adopted. Harmonization of data from data cubes with significantly different spatio-temporal reference systems would require a lot of computation. This might void the processing assets stemming from the use of data cube infrastructures in respect to other (more traditional) data services.

Concerning the EODC semantic view, interoperability can leverage the on-going activities by the communities of practice, in the EO and Earth science domain, to define a set of essential variables [30–32] and variable name conventions [33]. However, semantic interoperability must be seen as ancillary to the more important pragmatic interoperability, which is the real requirement from users. data cubes are designed for efficient processing in support of specific cases of use, thus, pragmatic aspects (e.g., data resolution and fitness-for-purpose) should be considered as relevant as semantic ones. The (long) ingestion time required for efficient computation of time series may be frustrated by a time-consuming pre-processing to make data usable for a specific use-case. To work out pragmatic interoperability, the aspects related to moving from a data to an ARD system should be considered; e.g., pixel alignment, atmospheric correction.

3.2.2. Interoperability of EODCs with Other Types of Geospatial Data Cubes

Moving interoperability to the more general level of geospatial data cubes, interoperability issues increase. With the term "geospatial data cube," we consider data cubes that encode information, characterized by a spatio-temporal content, which may be represented as not making use of the coverage model. Actually, this is a common situation, in particular when socio-economic information is provided as aggregated at local, regional and national level—or with reference to any administrative boundary, in general. For example, a data cube may report a set of parameters (e.g., GDP, school enrollment, and life expectancy) by country. In this case, at least one of the dimensions (i.e., the country) has a spatial content, but it is expressed as a geographical feature and not as a coverage function.

There are also more complex cases where the geographical feature is not a dimension, such as features changing over time; e.g. the area affected by a flood, the set of protected areas in Europe, or the set of countries that are members of the United Nations. In these cases, interoperability from the geometry point of view can become very complex. Of course, the semantic viewpoint also highlights a higher complexity, since providing common semantics for different domains is still an open issue—one of multi-disciplinary interoperability.

3.2.3. Interoperability of EODCs with General-Purpose Data Cubes

Interoperability with general-purpose data cubes, where information has no explicit geographical content is even more complex. This is also a common situation in socio-economic contexts, where data (in particular statistical data) are aggregated according to non-spatio-temporal dimensions–e.g., life expectancy by job category, wealth and income by age category, etc.

In principle, most of the interoperability issues that interest data cubes have been already recognized and largely addressed by the science studying the interoperability of geospatial information systems. For that reason, some of these issues can be solved by adopting the existing standards or mediation tools.

However, from an engineering point of view, it useful to reflect on the peculiarity aspects of ARD and data cube systems: Their diversity in respect to a traditional data/information system. In particular, it is important to consider that making data cubes interoperable does not mean building a virtual data cube—like we commonly do implementing data systems federation. By simply making data cubes interoperable, it would build an information system that accesses data cubes, but that it is not necessarily a data cube itself. Data cubes are intended as systems tailored to (optimized for) specific-use cases. They were conceived to implement ARD systems. Therefore, they are required to implement interoperability at the pragmatic level. Different data cubes may be "ready" for different uses, and putting them together would likely result in a system that is not necessarily ready for a commonly defined purpose.

4. Enhancing Interoperability Using Standards

4.1. Stakeholders and Patterns

To cover all the six interoperability views, defined by Nativi et al. [27], different stakeholders must be engaged, including disciplinary experts (e.g., experts on Earth system, geospatial information, multidimensional data management, online analytical processing, HPC, and ecosystems), standardization organizations, and the users (e.g., business intelligence association, and policy makers) who must provide the use cases to be addressed by data cubes.

In developing interoperability solutions, well-accepted and innovative patterns must be considered. For example:

- Semantic interoperability:

    - Data and information typing specifications.
    - Semantic and ontological languages to be used along to enrich and disambiguate content metadata.
    - Co-design patterns.

- Geometry interoperability:

    - Geospatial information models.
    - Business intelligence and the online analytical processing multidimensional modeling.

- Encoding interoperability:

    - Well accepted file systems and formats patterns.

-     ○   Multidimensional DB.
-     ○   Big data tiling strategies.

- Interaction/interface interoperability:

    - ○   Web APIs.
    - ○   Online analytical processing (OLAP) APIs.
    - ○   Web Notebook tools.
    - ○   Well-adopted interoperability protocols (e.g. OGC, W3C, IET).

- Interconnection/platform interoperability:

    - ○   System-of-Systems (SoS) patterns.
    - ○   Software design patterns.
    - ○   Cloud computing interoperability patterns.

- Composition/ecosystems interoperability:

    - ○   Software Ecosystem (SECO) patterns.
    - ○   SoS virtual/collaborative architectures.
    - ○   SoS governance styles (e.g., directed, collaborative, acknowledged, virtual).

4.2. Documenting Data and Data Discovery

In the DIKW paradigm, the first step in the data value chain (e.g., a set of actions from data capture to decision-ready products) is known as data discovery [34]. It allows users to search, find, and evaluate suitable data that will be further used in models or other analytical workflows. Data discovery is realized through catalogs containing relevant information describing datasets (e.g., spatial resolution, spatial extent, temporal resolution) [35]. These detailed descriptions are commonly referred to as metadata [36–38]. To contribute to initiatives such as the Global Earth Observation System of Systems (GEOSS), it is required to use data description specifications (i.e., metadata standards) to document datasets and store metadata in interoperable catalogs to facilitate exchange and use by various systems [39].

Different open standard schemas have been developed to describe geospatial data [40]. The most widely used standards are developed by the International Organization for Standardization (ISO)/Technical Committee (TC), 211 Geographic Information/Geomatics, and the Open Geospatial Consortium (OGC). They concern data and service description (ISO19115-1 and ISO19119), their respective schema implementation (ISO19139-1), and the Catalog Service for the Web Interface (OGC CSW) [41,42]. With this suite of standards, users can adequately document data and provide standardized search, discovery, and query capabilities over the internet [43].

Currently, the vast majority of metadata catalogs relate to geographical data (e.g., map agency's products) and only a few of them concern EO data [44]. EO products are normally distributed by the data producers as scenes or granules (a spatial fragment of a satellite path) of data with a metadata document for each scene. In this partition of a product in space and time, most of the metadata content is identical and is repeated in each scene. That is the reason why a catalogue interface holding these metadata records will generate hundreds or thousands of hits for a thematic query. A hierarchical structure of metadata describing a product as a single unit that has multiple scenes needs to be adopted to make the catalogues useful [45]. Moreover, among the various data cube implementations, the Open Data Cube and RasDaMan/EarthServer are the most widely adopted solutions [19,22]. They arrange data in a hierarchical way, and expose data at the product level, making it visible as a single entity, but they lack metadata description and catalog interfaces impeding efficient and effective discovery mechanism.

To tackle this issue and store relevant metadata information about satellite data (e.g., acquisition, sensing, and bands) in a online metadata catalog, the XML schema ISO 19139-2 extends the original metadata schema to support additional aspects more significant for the gridded and imagery information defined in ISO19115-2, and offers an interesting possible solution [44]. Additionally, the SpatioTemporal Asset Catalog (STAC, https://stacspec.org) is under substantial development. It is an emerging metadata standard, primarily designed for remote sensing imagery, aiming at standardizing the way geospatial assets are exposed online and queried. Interestingly, there are preliminary efforts to extend ODC to use STAC files as a source of information to index data (https://github.com/radiantearth/stac-spec/tree/master/extensions/datacube).

4.3. Data Quality and Uncertainty

Nowadays, data cubes barely have data quality information in their metadata records. A couple of complimentary approaches to populate this lack of information have to be considered.

First of all, the uncertainty associated to each image can be estimated by several means. On one hand, several papers (such as [46–49]) assess the general accuracy for certain instruments, based mainly on Calibration and Validation (Cal/Val) campaigns or invariant areas. These approaches give a general uncertainty value associated to each sensor and band so that can be applied to a single product for the product-level metadata. Another refinement that can be done in this direction is to consider that this error is modulated according to the incidence angle, as the bigger this angle, the more specular effects in reflectance, and thus the higher the errors expected. Considering that effect, a different uncertainty value can be associated to each pixel of the scene; i.e. obtaining an image describing the uncertainty for each pixel of a certain image. Once this information is available, the uncertainty in any analytical operation using the imagery can be computed by propagating the original uncertainties to the final product using classical error propagation formulas and map algebra. For example, as the Normalized Difference Vegetation Index (NDVI) is computed as:

$$NDVI = \frac{IR + R}{IR - R} = \frac{M}{N}$$

(added and subtracted variables, and then divided). The uncertainty of each pixel can be computed using the two following formulas:

$$S_{NDVI} = \sqrt{(S_M^2 \times N^2 + S_N^2 \times M^2)}$$

$$S_{M \ and \ N \ (IR \pm R)} = \sqrt{(S_{IR}^2 + S_R^2)}$$

As this uncertainty propagation can be computed with map algebra, Open Data Cubes should be able to include these calculations in their routines, and thus, automatically generate the final products as well as their uncertainty.

The second approach is to assess the quality of the final product not by error propagation but by comparing the product to a known ground truth, thus validating its thematic accuracy. This is the most common approach for final products where uncertainty propagation is not directly applicable, such as land use land cover, or leaf area index (LAI) area, in which the product is validated against known values obtained by field work or other means. In this approach, the obtained quality assessment is generally documented at the dataset level (not at pixel level).

Regarding how to include data quality (both product and pixel level) in metadata, the widely selected standard is the ISO19157. ISO19157 identifies the conceptual model for describing quality elements in a geographic dataset and defines several quality elements describing different quality aspects in a dataset: Completeness, logical consistency, positional accuracy, thematic accuracy, temporal quality, usability and metaquality. Besides ISO19157, the Quality Mark-up Language (http://www.qualityml.org) is both a vocabulary and an encoding for data quality that was originally developed in GeoViQua FP7 project and extended in OGC TestBed12. This vocabulary proposes a clear encoding of quality elements (using standardized quality measures) in XML metadata documents, and can be used for describing the quality of the original products, as well as to define further quality evaluations carried out over the datasets [50].

4.4. Data Visualization and Download

Data cubes are essentially analytical frameworks. The multidimensional nature of the data cube makes it difficult to visualize, but a standard solution can be found, allowing time series to be stored in a data cube for efficient analysis, while at the same time being able to be easily visualized. Moreover, the benefit of using the data cube as the origin to create data visualizations, is that it allows creating visualizations with any combination of the data cube dimensions; for example, it can be applied to extract 2D imagery at low resolution for WMS visualization of a time series evolution, as well as 1D time profile diagrams, or x/t time slices with WCS ready to visualize. Most data cubes have data organized in ways that are optimal for these kind of operations, allowing on the fly fast visualization.

The most common way of doing optimizing as such, is by reducing the number of dimensions to less than three. Standard data visualization is generally achieved using geospatial web services, as they are particularly suitable for this purpose [51]. In that sense, OGC Web Map Service (WMS) and Web Map Tiled Service (WMTS) are common interoperable solutions to show maps from different origins together in a single image [52]. WMS and WMTS are particularly fit for accommodating multidimensional data cubes, due to their capability to define extra dimensions (on top of a two-dimensional CRS). A time parameter is defined in the WMS standards, and can be added as an extra dimension in WMTS. An additional, band parameter, can also be defined by the server to select among several bands of a EO product. This could be very convenient for an optical product (e.g., Sentinel 2), since it can be represented as a single layer and the band parameter is used to select the spectral band, while the time parameter can be used to extract a time slice of the time series. In both solutions, the server executes an internal algorithm that reduces the dimensions to the two spatial ones (on a certain CRS) and creates a 2D portrayal by applying a color palette, or by combining bands into RGB combinations. Both are characterized by requesting a simplification of the original data at a particular scale, time slice, and in a format that is easy to display. Depending on the scale requested (also known as zoom level), a small portion of the data are requested at near full resolution, or a large piece of information is requested at low resolution. Normally the client is not receiving the actual values of the datasets (e.g., EO data), only naive visualization where only colors are encoded in JPEG or PNG formats. In WMS, in trying to make the time parameter flexible, the authors introduced some variants (such as the possibility of indicating time intervals, or the nearest values) that are a bit vague and complex to implement. In that sense, the precision of the time values may introduce uncertainties in requests and responses. To prevent that, we recommend that the service enumerates the time values available in the capabilities document, and the client use them as literals and only indicated time instants in the time parameter (this recommendation coincides with the only possible use for a time extra dimension in WMTS). Another fundamental problem in WMS is that the semantics of the time parameter is unclear and could refer to the acquisition time, processing time, publication time, etc. The OGC WMS Earth Observation profile recommends using the time parameter only for the acquisition time [53]. It is always possible to create other time dimensions with other semantics if needed. We believe that by following these recommendations, the WMS time dimension is usable and modern clients are capable to overcome these restrictions by offering smarter user interfaces that, e.g., present time arrows or combine individual WMS requests in animations. A couple of previous attempts to improve the situation have been developed. On one side, ncWMS introduced several extensions in GetFeatureInfo, symbolization and vertical and temporal dimensions [54]. On the other side, a simple extension of WMS adding a binary raw data format and the move of some of the portrayal capabilities from the server to the client has also been proposed [55]. The latter option is used in the Catalan Data Cube, adding to the client much more than simple data visualization (including time profiles that take advantage of the having the data cube as the origin of the visualization), such as simple statistical calculations or pixel-based operations among layers from different sources [56].

Is it possible to build a middleware that interprets a WMS or a WMTS request, extracts the extra dimensions and translates it into specific data cube operations for cutting a re-sampling. If the response needs to be produced on the fly, it shall be fast in both extreme circumstances. Generally, fast extracting fragments from data cubes are achieved by saving the data, pre-sampled on different scales.

Regarding data download, the Web Coverage Service (WCS) standard is the right way to go. WCS 2.0 is based on the Coverage Implementation Schema (CIS) that describes the data dimensions as well as the thematic bands offered by the service in a standard language. With WCS, we can easily filter the data and extract a fragment of it, in the same number of dimensions (sub-setting) or in a reduced number (slicing). The response can be one or more common data formats (e.g., a GeoTIFF or a NetCDF). Actually, the approach could be used as standard way of exchanging products among data cubes. A data cube could act as a WCS client and progressively request the data from another data cube that implements WCS services. Moreover, a data cube that has been updated with new data could use WCS-T (transactional WCS) to automatically update other data cubes that have the same product with new time slices or completely new products.

The Web Coverage Service (WCS) standard can also be used for visualization. With the help of a modern JavaScript library (e.g., https://geotiffjs.github.io/) GeoTIFF images can also be seen in the map browser directly. Another interesting application of WCS is the extraction of a temporal profile of a point by reducing the number dimensions to one—the time. By doing so, a GetCoverage request is capable of responding to a time series (e.g., timeseriesML) that can be presented in a map client as a diagram.

4.5. Data Processing

Data cubes being analytical frameworks, computing and IT infrastructures are significant elements to enhance data flow, data transformation to information, as well as analysis and processing of the ever-increasing volume of EO data that exceed the current capacities of existing computers [12,16,57,58].

To deal with these issues of processing algorithms, sharing, and computing power, the OGC Web Processing Service (WPS) [59] and Web Coverage Processing Service (WCPS) [60], together with the high-performance and distributed computing paradigm, can be beneficial [61–63].

The WPS standard specification defines how to invoke input and output data as a web-based processing service. It provided rules on how a user/client can ask for a process execution, how a provider should publish a given processing algorithm as a service, and how inputs and outputs are managed [64]. This standard facilitates algorithms sharing in an interoperable fashion. However, it is not well adapted for raster analysis [60]. Consequently, the Web Coverage Processing Service (WCPS) specification has emerged defining a query language for processing multi-dimensional raster coverages [65]. Those two standards provide effective foundations to enhance interoperability of EO data cubes and ensure that when a user send the same request and processing algorithm can be executed on different data cubes.

Finally, to efficiently turn EO data into information and knowledge, effective processing solutions are necessary. Distributed and high-performance computing (HPC) infrastructure like clusters, grids, or clouds are adequate solutions [62,66,67]. It is now possible to benefit from the computing power of these infrastructures while using interoperable processing services in a transparent manner, hiding the complexity of these infrastructure to users [68–70]. Such integration can help leveraging the capabilities of these infrastructures and support model-as-a-Service approaches, such as the GEO Model Web [71] or data cubes [18].

4.6. Data Reproducibility

Provenance also includes the description of the algorithms used, their inputs and outputs, the computing environment where the process runs, the organization/person responsible for the product, etc. [72]. A provenance record consists of a format list of processes and data sources used to create a derived product. By documenting provenance in the metadata, traceability of scientific results is possible, and the same result can be reproduced in the same or in another environment. The new revision of 19115-2 proposes a model to record all inputs necessary to execute an analytic processes, as well as to describe the process itself. This approach is fully compatible with the WPS standard proposed in the previous section [73]. At the moment, to execute any process in the data cube, the executing environment knows everything about the job requested, and it is in the perfect position to save this information in the metadata. The data cube environment should facilitate this recording task and integrate it in the processing operations without any user intervention. If data cubes record provenance information in the same way, it could be possible use the provenance information of a result produced in one environment and reproduce the same result in another environment.

4.7. Data Integration, Semantics and Value Chains

The analysis ready data updated in real time can be the basis for elaborating Essential Climate Variables (ECV) [74]. The generalization of the essential variable concept to other areas, such us biodiversity (EBV), offers new opportunities to monitor the biodiversity, the ecosystems, and other sectors; and opens the door to generate policy related indicators (e.g., Aichi targets or the Sustainable Development Goals (SDG)) [4,32]. Many of the essential variables that can be extracted from remote sensing as high processing level products are indeed describing ECVs [75] or EBVs [76]. A semantic view is necessary for an effective usability and interoperability of data cube products [27]. Connecting data cube high level products to structured keyword dictionaries and formal ontologies is necessary. Tagging the data cube products with the essential variables concept provides a degree of formal semantics to a well-defined and accepted set of measurable class names. The adoption of a formal and common vocabulary of essential variables (EVs) can facilitate the discovery of the relevant data for a particular application. Additional metadata can make the data usable by providing concrete information on spatial resolution, periodicity, and units of measure. This information should be included in the product description of the data cube in the form of ISO 19115 keywords, or by linking to formal ontologies encoded in RDF or OWL in the Internet.

4.8. Data Ingestion from Data Cubes and from Data Providers

Currently, each remote sensing data provider is serving data in a different way. Assuming that the product we want to download is made available for free, commonly, we face an ingestion process that has two phases: Discovery of new scenes by formulating a spatiotemporal query and retrieving the individual scenes. Most data providers offer a visual interface to find the relevant data, which is good for retrieving some samples, but it is not useful to make the data ingestion process in a data cube fully automatic. Some providers offer a Web API to discover and retrieve the data. In the discovery phase, an HTTP GET or a POST request containing a spatiotemporal query and some extra parameters (e.g., the maximum cloud coverage allowed) is submitted to the server. Often the result is a file with a list of hits that includes the names of the scenes that comply with the requested constraints. The client needs to explore this file and formulate more requests; one for each scene to finally get the wanted product subset. To make the situation more complicated, the number of hits of the discovery response might be limited to a maximum number and download will only be possible once authenticated in the system. An ad-hoc script will be necessary to make the requests automatic. Assuming that we were able to get the required data, that is only the start, because there are significant differences among the composition of the scenes that a single data provider offers in terms of metadata content and data formats. There is a need for a more standardized way to document the structure of a remote sensing product and to agree on a standardized format for scene distribution that can be used to ingest the data automatically. Since this is not the case today, every product will require metadata transpositions and format transformations.

The combination of WCS and the CIS standards could open the door to define a standard profile to discover and retrieve the necessary scenes. CIS 2.0 incorporates an extended model that allows for a partition data structure that complies with the requirements of a classical remote sensing distribution in scenes or granules. WCS can incorporate security to authenticate users and will allow for formulating a single request to discover the necessary partitions that will be downloaded in a second phase.

Such approaches could be adopted not only by the remote sensing data providers but also by the data cubes, opening the door for having a protocol that a data cube could apply to harvest any other data cube has implemented the standard approach, something that is not currently possible forcing people to having to develop adaptors.

5. Examples of Enhanced Interoperability from the Swiss Data Cube and the Catalan Data Cube

5.1. Swiss Data Cube Discovery, View, Download and Process Services

To fully benefit from freely and openly available Landsat and Copernicus data archives for national environmental monitoring purposes, the Swiss Federal Office for the Environment (FOEN) is supporting the development of the Swiss Data Cube (SDC—http://www.swissdatacube.ch). The SDC is currently being developed, implemented and operated by the United Nations Environment Program (UNEP)/GRID-Geneva in partnership with the University of Geneva (UNIGE), the University of Zurich (UZH) and the Swiss Federal Institute for Forest, Snow and Landscape Research (WSL). The objectives of the SDC are twofold. First, to support the Swiss government for environmental monitoring and reporting; and second, to enable Swiss scientific institutions to take advantage of satellite EO data for research and innovation.

The SDC is built on the Open Data Cube software suite [19], and currently holds 35 years of Landsat 5, 7, and 8 (1984–2019), 5 years of Sentinel-1 (2014–2019), and 4 years of Sentinel-2 (2015–2019) of ARD over Switzerland [77,78]. This archive is updated on a daily basis with the most recent data and contains approximately 10,000 scenes, accounting for a total volume of 6 TB, and more than 200 billion observations nationwide.

Currently, one of the key challenges that SDC has to tackle to ensure its scalability, is enhancing the interoperability. Indeed, making data, metadata, and algorithms interoperable will: (1) Facilitate the interaction with the SDC from an increasing number of users; (2) allow connecting results of analysis with other datasets; (3) enhance the data value chain; and (4) ease contributions to major regional and/or international data sharing efforts, such as GEOSS.

Initial interoperability arrangements are currently under development. In the SDC, we decided to distinguish between upstream and downstream services [79,80]. The upstream tier relates to services to interact with the infrastructure (e.g., processing, view, download) while the downstream tier allows users interacting with decision-ready/value-added products. Both tiers are implementing widely adopted open standards for modeling and implementing geospatial information interoperability advanced by the OGC and ISO/TC211.

Regarding the upstream tier, the following strategy for implementing standards is being adopted:

- *Discovery:* SDC description is being done using the ISO19115-2 and ISO19139-2 standards to support gridded and imagery information. The XML schema has been deployed and exposed using the GeoNetwork metadata catalog to store all relevant information to adequately describe the SDC content (e.g., sensors, spatial resolution, temporal resolution, spectral bands). The schema plugin has been downloaded from the GeoNetwork GitHub repository: https://github.com/geonetwork/schema-plugins. Moreover, the GeoNetwork catalog allows exposing an OGC CSW interface for publishing metadata records and allowing users to query the catalog content.
- *View and Download:* To leverage the content of the SDC for visualization and download, respectively OGC WMS and WCS are under implementation. The datacube-ows component (available at: https://github.com/opendatacube/datacube-ows) implements the WMS and WCS standards allowing an interoperable access to Landsat and Sentinels data.

- *Process:* To expose analytical functionalities (e.g., algorithms) developed in the SDC using the ODC Python Application Programming Interface (API), it has been decided as using a PyWPS implementation (https://pywps.org). The main advantage is that it is also written in Python, and allows easy to expose, dedicated Python scripts as interoperable WPS services. That approach is currently under implementation and testing.

Concerning the downstream tier, it has been separated from the SDC for the reason that only final products (e.g., validated analysis results) are concerned. That facilitates the publication and sharing of good quality results through value-added/decision-ready products, while at the same time separating the usage of the Swiss Data Cube between scientific/data analysts end-users and more general end-users.

To that, a specific GeoServer instance with dedicated EO extensions and time support has been implemented. It allows users to efficiently interact with multi-dimensional (e.g., space and time), gridded, and image products generated with the SDC. It currently supports:

Discovery services

- CSW 2.0.2.
- OpeanSearch EO 1.0.

View services

- Web Map Service (WMS) with EO extension 1.1.1/1.3.0.
- Web Map Tile Service (WMTS) 1.0.0.
- Tile Map Service (TMS) 1.0.0.
- Web Map Tile Cached (WMS-C) 1.1.1.

Download services

- Web Coverage Service (WCS) with EO extension 1.0.0/1.1.0/1.1/1.1.1/2.0.1.
- Web Feature Service (WFS) 1.0.0/1.1.0/2.0.0.

To further ease user's interaction with SDC products, a web-based application called the Swiss Data Cube Viewer (http://www.swissdatacube.org/viewer/) has been developed (Figure 1). It allows visualizing, querying, and downloading time-series data generated with the SDC. This JavaScript application provides a simple, responsive template for building web mapping applications with Bootstrap, Leaflet, and typeahead.js. It provides the following functionalities:

- Visualizing and Downloading single raster product layers;
- Visualizing and Downloading time-series raster product layers;
- Generating a graph for a given pixel of a time-series raster product layer;
- Access data products in users 'client with WMS and WCS standards;
- Metadata support.

The application is entirely open-source, and the code can be download at: https://github.com/GRIDgva/SwissDataCube/tree/master/viewer

The use of OGC and ISO standards can enhance syntactic interoperability of the SDC and can help delivering and leveraging the power of EO data building efficient discovery, access and processing services provided by the SDC.

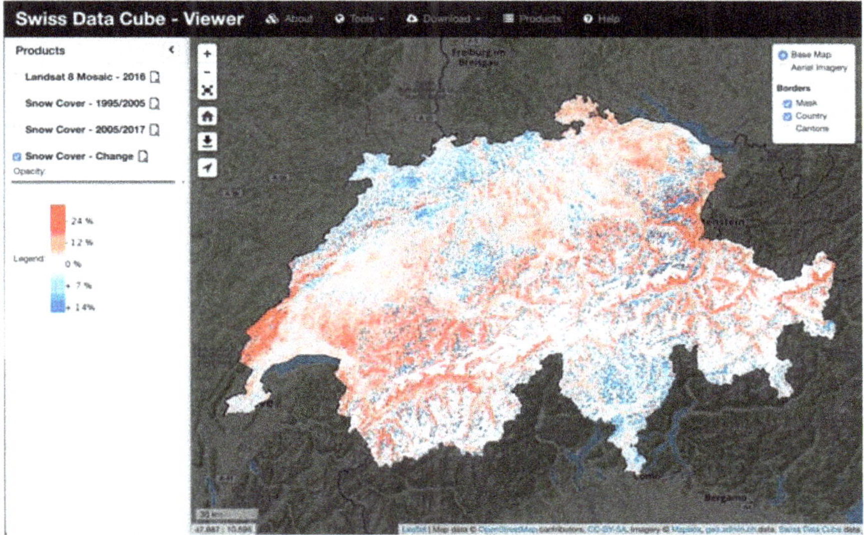

Figure 1. The Swiss Data Cube; viewer showing snow cover change over Switzerland between 1995 and 2017.

5.2. *The Catalan Data Cube View Service with Analytical Features*

The Department of Environment of the Catalan Government and Centre de Recerca Ecològica i Aplicacions Forestal (CREAF) created the SatCat data portal (http://www.opengis.uab.cat/wms/satcat) that organized the historical Landsat archive (from years 1972 to 2017) over Catalonia in a single portal, providing visualization and download functionalities based on OGC international standards [81]. The initiative is still up and running, plus continuously and manually updated; but a considerable amount of processing work is needed to keep the portal up-to-date, and to incorporate the increasing flow of the new imagery. The use of an instance of the Open Data Cube can help to automate some of the processes, thus the Catalan Data Cube (CDC) was created as a regional data cube with easily managed, modest resource requirements in mind. The CDC is currently being implemented and operated by GRUMETS research group (http://www.grumets.uab.cat/index_eng.htm—mainly composed by CREAF and Autonomous University of Barcelona), as well as the SatCat.

The CDC (http://datacube.uab.cat) is being developed, collecting the same kind of optical imagery aimed by the SatCat, but only if it is available under the ARD paradigm over Catalonia (Spain), thus it is limited to Sentinel-2 level 2A data flows coming from ESA at the moment. The CDC is built on the Open Data Cube software suite [19], with some additional Python scripts and currently holds more than 1 year of Sentinel-2 (March 2018, April 2019). The archive is updated on a monthly basis with the most recent data, and contains at the time of writing these lines, 1562 granules, forming 132 daily slides, accounting for a total volume of 1.18 TB over Catalonia.

Following the tier separation introduced in the Swiss Data Cube section, in the Catalan Data Cube we also distinguish between "upstream tier" and the "downstream tier". The upstream tier is composed of the Sentinel 2 imagery Level 2A and the downstream tier; they are elaborated on-the-fly by the web client. Regarding the upstream tier, the following strategy for implementing standards is being adopted:

- *Discovery:* Since the number of products is limited, we are using the WMS GetCapabilities response as the document that acts as a catalogue, and provides links to ISO19115-2 and ISO19139-2 metadata documents.

- *View and Download:* To leverage the content of the CDC for visualization, OGC WMS has been implemented, and for download a Web Coverage Service (WCS) is under consideration.
- *Process:* The CDC relies on the ODC processing API, but currently it is not exposing processing services at the moment. Instead, it relies on what is possible to do in the MiraMon Map Browser client side.

These services are possible as part of the MiraMon Map Server CGI suite that is encoded in C and developed on top of libraries coming from the MiraMon software.

To further ease a user's interaction with CDC products, a web-based application called the "SatCat 2.0: Catalan Data Cube" (http://datacube.uab.cat/cdc) has been developed. It allows visualizing imagery and time-series data generated with the CDC. As briefly mentioned before, this web-based application is based in the MiraMon Map Browser (https://github.com/joanma747/MiraMonMapBrowser, open source web map client), which uses an extension of OGC WMS that allows querying/retrieving data in a raw binary array format. This solution allows the client to save in memory, the actual values of each band, and then use JavaScript code to operate with the data, providing some analytical tools to the user. The SatCat 2.0 provides the following functionalities to the user (some of them are only possible by the binary arrays approach):

- Visualizing single raster product layers.
- Generating histograms or pie charts for single raster product layers.
- Modifying raster visualization by describing enhancing contrast parameters or by changing colour palettes.
- Visualizing time-series raster product layers as animations.
- Generating a graph for a given point in space of a time-series raster product layer.
- Applying spatial filters (by setting a condition in another layer; e.g., representing normalized difference vegetation index (NDVI) values only if the elevation is lower a certain value, or only for a certain land use category).
- Creating new dynamic layers by complex calculations among the bands of the different available datasets.
- Accessing products in your own client server with WMS standards.
- Accessing metadata and reading or contributing geospatial user feedback.

Figure 2 shows the list of bands made available by the Sentinel 2 sensor followed by a list of colour combinations and band indices dynamically computed by the client side. As an example, a normalized difference vegetation index (NDVI) that is dynamically computed by the client while rendering it in the view (using the values of the necessary bands of each Sentinel-2 image). Moreover, a histogram showing the frequency of the NDVI values on the image is obtained. The dynamic calculation of layers can be as simple as this vegetation index, or a complex model using several bands and layers.

The generation of animations is possible in classic WMS services (TIME parameter), but thanks to the binary arrays approach, it is possible to present plots of the temporal evolution of one or more points in the animated area and eventually detect anomalies by comparing them with the mean and variance of the visible values in the bounding box (Figure 3). For both histograms and temporal profiles, data can be copied and pasted into a spreadsheet for further analysis.

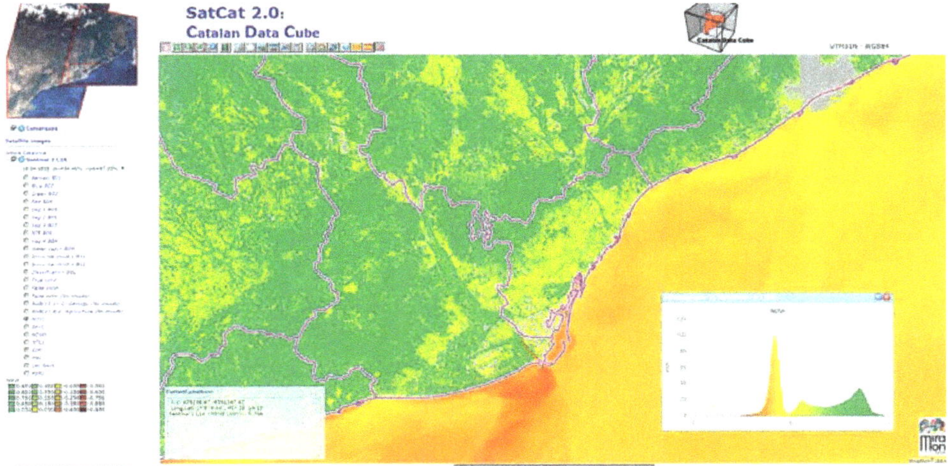

Figure 2. The Catalan Data Cube. Dynamic normalized difference vegetation index (NDVI) values and a histogram are over the view area, computed by the client using original Sentinel-2A red and infrared bands retrieved as binary arrays (centered in Barcelona and surroundings).

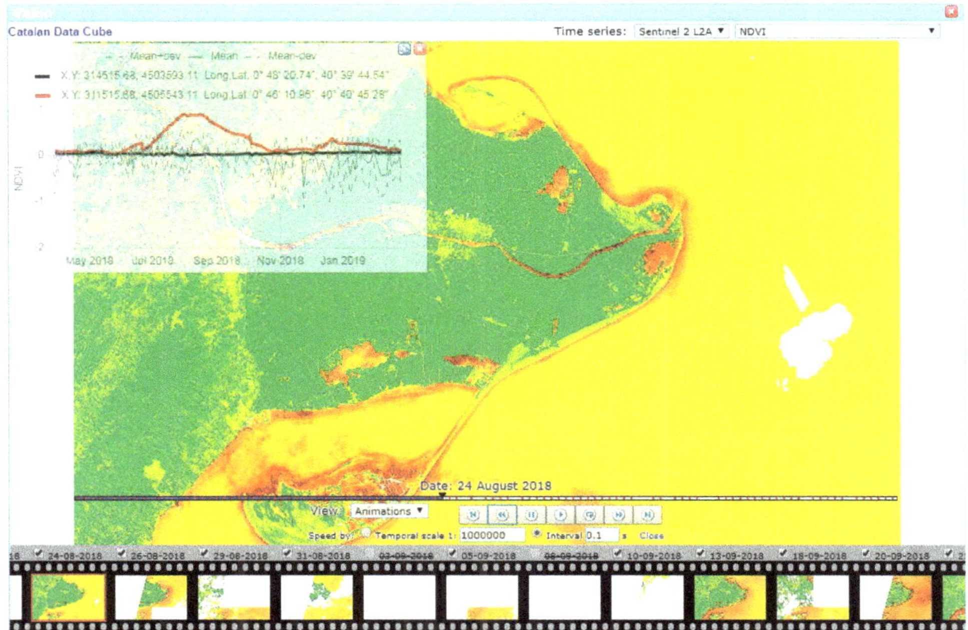

Figure 3. The Catalan Data Cube. Dynamic NDVI layer animation including, a temporal profile for sand (black) and crop (red) areas (centered in the Ebre river delta area).

6. Discussion and Perspectives

EODCs are becoming increasingly important to support Earth system science studies, enabling new global change applications. They can provide the long baseline required to determine trends, define the present, and inform the future; they can deliver a unique capability to process, interrogate, and present EO satellite data in response to environmental issues; and allow tracking changes across a given area in unprecedented details.

However, the wider use and future success of EODC will largely depend on features such as: Usability and flexibility to address various users' needs and interoperability for contributing to the digital industry revolution [27]. Consequently, interoperability can be considered as a key factor for successful and wide adoption of EODC technology and is an absolute necessity to develop and implement regional and global data cube ecosystems.

Based on the present work, we think that the following three elements need to be carefully considered:

(1)   Understanding the differences between traditional data system's interoperability (e.g., for discovery and download) and data cube system's interoperability for data analytics.
(2)   Investigation of the significance of analysis ready data for specific applications.
(3)   Explore the non-technological interoperability dimensions, such as governance and policy.

6.1. Interoperability Paradigms

In the EO domain, data interoperability has traditionally applied the "Discovery and Access" paradigm, which consists of discovering a remote dataset, downloading it to a local server, and using it locally; e.g., visualizing it or processing it to generate new data or information. In extreme synthesis, datasets have been moved though the network to be ingested in local data management platforms that support independent and monolithic applications.

With the advent of large datasets (i.e., data collected over a long-term or massive spatial data series) and the raise of virtual computing capacities, a new (and more efficient) IT paradigm emerged: The "Distributed Application" paradigm. The new approach aims at using the web as the analytics platform for building (distributed) applications, and makes use of microservices and container-based technologies. This time, datasets are not moved around, but application algorithms are deployed around (using the containerization technology) to be run where datasets are, working out a virtual collection of independent services that work together. For the EO Community, data cubes may be an important instrument to implement this paradigm—by facilitating remote data analytics. To achieve that, it is important to understand the best level of interoperability to be pursued by data cubes and which are the most effective instruments.

Advanced data cubes can be seen as databases able to organize and retrieve data on demand, and present it in the form of data structures, but also as processing facilities on top of data structures. In a data cube, data and processing are close together. Depending on the emphasis, data cubes can be used as traditional interoperability instruments (e.g., standard protocols for data discovery and access) to discover and download selections of remote sensing data, or the data cube systems can go beyond the "traditional" data systems—moving away of the "Discovery and Access" paradigm by becoming processing facilities, allowing data processing algorithms to be sent to the data cube engine and executed where the data is. Therefore, for EO data cubes, advanced interoperability tools were recently developed, such as coverage processing and query languages. Indeed, they can be used to implement system integrability. However, they are not sufficient to achieve semantic and pragmatic interoperability, which are important to achieve the "Distributed Application" paradigm that can process and use resources of more than one data cube. This interoperability shortcoming may push data cube systems to become monolithic platforms that operate in isolation only as clients of other interoperable services, but not offering interoperable interfaces by themselves.

6.2. Analysis Ready Data and Data Cubes

In general, data cubes can be seen as an analytical technological solution for taking advantage of ARD, defined by CEOS as "data that have been processed to a minimum set of requirements and organized into a form that allows immediate analysis, with a minimum of additional user effort and interoperability, both through time and with other datasets" [82]. Despite this generic definition, CEOS came up with three specifications for surface reflectance, surface temperature, and radar backscatter (http://ceos.org/ard/). Focusing on the surface reflectance products, the general concept of data readiness entails common content pre-processing, such as: Atmospheric correction, and cloud coverage masking—this may be called radiometric readiness. There is no way to create a three-dimensional (i.e., x, y, t) data product if the radiometry of the values are not homogeneous over the time dimension. The most obvious advantage for the users of remote sensing data is that when they wish to undertake large area and long-time series analyses, they no longer need to invest in computationally expensive atmospheric correction processing chains to pre-process data. ARD saves time by providing a standard solution for data preparation that should be valid for most common applications. ARD products help with saving money, because expensive processing time is only executed once by the data producer. Another kind of data readiness (at least its lower level) is the geometrical readiness, by implementing data resampling and re-projection on a common grid environment. It is not possible to build a multidimensional data cube with scenes that are not organized in a way that they become geometrically co-registered. In principle, ARD does not assume a particular application; e.g., land cover studies. For example, ARD does not assume that cloud's shadows and snow shall be removed. Instead, producers apply state of the art algorithms to detect clouds and the shadows that they create, haze, and snow; and provide an extra band that tells which pixels are clouds, which are hazy, which are shadowed, and which contain snow. With these masks, the users can filter the values that they consider unusable by themselves in the data cube. For all these reasons, it can be considered that ARD allows working at the "pragmatic-level" of interoperability. Some specific usages of remote sensing data might require a more careful consideration for preparing the product for analysis. In the H2020 ECOPOTENTIAL project, we have experienced the need for applying shadow compensation techniques for steep mountain areas that provide better results than the ones coming from ARD directly generated by the data provider. However, there is another important benefit on using ARD: Users who wish to share and compare scientific and application results can still prefer to use ARD to reduce the potential discrepancies in results, due to differences in pre-processing, incrementing interoperability and comparability of higher-level products [78]. Nevertheless, ARD procedures can also create silos: For example, Landsat collection 1 and Sentinel 2 Level 2A were created as two independent ARD products, and they cannot be used together directly. The creation of a harmonized virtual product required to define a different ARD protocol and that forces, preserving the unique features of each data source, and some compromises must be made, such as adjusting Sentinel-2 (S2)/MSI radiometry to replicate the spectral bandpasses of Landsat 8/OLI, adopt a common 30 m resolution or adopt the Sentinel 2 UTM projection and tiling system [83].

Indeed, when populating data cubes, we are forced to satisfy a set of requirements that takes into account the actual use of the data, managed by them. The choice of an EO data cube array of dimensions, coordinate reference system, or data resolution are largely optimized for a limited set of relevant uses (e.g., time series analysis or changes detection using optical data). The data cube might still be useful for other kinds of applications, but performances would be suboptimal. As a consequence, application-driven optimization affects the different aspects and levels of data cubes' interoperability. In building distributed applications, the use of heterogeneous data cubes (i.e., differing for coordinate reference system, resolutions, etc.) would provide minimal or no benefit in comparison with using general purpose (or traditional) data systems. While a set of integrable data systems is still a data system, a set of integrable (i.e., technically and syntactically interoperable) data cubes is not necessarily a data cube itself. That defining of a data Cube is such by the virtue of relevant semantic/pragmatic decisions.

6.3. Non-Technological Interoperability Dimensions

Besides all technological aspects required enabling effective interoperability of EODC, we need also to consider non-technological aspects such as governance and policy. These elements mostly relate on human and organizational aspects that are equally important from an interoperability perspective [84]. From our point of view, the following three are essential for enhancing EODC interoperability:

(1)   Currently, a commonly agreed definition and taxonomy of EODC is lacking. To our knowledge, the Data Cube Manifesto [85] is the only attempt to give a general and holistic definition, defining six requirements that must be met, in order to be considered as an EODC. This manifesto can be a good starting point to be further refined, looking at the various existing implementations, and embedding the effort in standardization processes, such as those supported by the OGC and ISO.

(2)   Efforts should be persued to support Open Data and Sharing policies. Indeed, since 2008 the entire Landsat archive has been made freely and publicly available, followed by a tremendous increase in usage, investigations, and applications [86,87]. The Landsat Open Data policy is an excellent example of how to maximize the return on large investments in satellite missions [13,86]. Withtout such a policy, develing EODC technology would not have been possible. Togerther with FAIR (Findable, Accessible, Interoperable, Re-usable) data principles [88,89], EODC can enable moving towards effective and efficient EO Open Science.

(3)   Finally, a fundamental aspect that needs to be considered is the governance. Without effective governance mechanisms and structures, it will prevent a successful implementation of EODC at national levels. Further, that will be even more important when one thinks about federated data cubes at regional and/or global levels. Governance will be the first challenge to tackle in this context. For example, in the case of the Swiss Data Cube, an incremental strategy has been developed. During the initial phase of the SDC, only one organization was involved taking care to test the data cube technology, deploy the software, ingest data, and developed initial demonstration applications. That helps fast movement and agility to closely collaborate under the mandate of the Swiss government. Now that the SDC is reaching some mature levels, new key partners in the field of EO in the country have been added to project bringing their respective expertise, and allowing consolidating the network across the country. This resulted in the signature of a Memorandum of Understanding (MoU) in June 2019 between UNEP/GRID-Geneva, the University of Geneva (UNIGE), the University of Zurich (UZH), and the Federal Institute for Forest, Snow and Landscape research (WSL). This cooperation agreement aims at fostering the use of Earth observation data for environmental monitoring on a national scale. The MoU is a pivotal instrument to clarify and implement a suitable governance structure commonly agreed by the different parties.

7. Conclusions

Addressing the interoperability challenge of EODC is essential to prevent the various EODC implementations becoming silos of information. Currently, not many efforts have been made to enhance interoperability of data cubes.

In this paper, we discuss and demonstrate how interoperability can be enabled using widely-adopted OGC and ISO geospatial standards, and how these standards can help delivering and leveraging the power of EO data building efficient discovery, access, and processing services. These standards are applied in different ways in current data cube implementations, such as the Swiss Data Cube and the Catalan Data Cube were, we have identified that OGC services mainly improve the "Discovery and Access" paradigm. An opposite paradigm of moving the processing code close to the data is facilitated by current containerization technology and rich query languages such as WCPS. The real challenge is to realize the "Distributed Application" paradigm, wherein data cubes can work together to produce analytical results.

Realizing the objective of providing EO-based information services and decision-ready products responding to users' needs requires effective and efficient mechanisms along the data value chain. EO data are essential to monitor and understand environmental changes. Consequently, it is necessary to make data and information products not in the form that it is collected, but in the form that is being used by the largest number of users possible. One step in this direction is the implantation of analysis ready data products by data providers. Being able to easily and efficiently combine EO-based data with other data sources is a crucial prerequisite to enable multi-disciplinary scientific analysis on our changing environment. Interoperable data cube services can significantly contribute to effective knowledge generation towards a more sustainable world, supporting decision and policy-makers making decisions based on evidence, and the best scientific knowledge.

Author Contributions: G.G.: Conceptualization, writing—original draft preparation; J.M.: Conceptualization, writing—original draft preparation; A.Z.: Writing—original draft preparation; S.N.: Conceptualization, writing—original draft preparation; P.M.: Conceptualization, writing—original draft preparation.

Funding: This research was funded by European Commission "Horizon 2020 Program" ERA-PLANET/ GEOEssential project, grant number 689443 and ECOPOTENTIAL project, grant number 641762. Results of this publication partly or fully rely on the Swiss Data Cube (http://www.swissdatacube.org), operated and maintained by UN Environment/GRID-Geneva, the University of Geneva, the University of Zurich and the Swiss Federal Institute for Forest, Snow and Landscape Research WSL.

Acknowledgments: The authors would like to thank the Swiss Federal Office for the Environment (FOEN) for their financial support to the Swiss Data Cube. The views expressed in the paper are those of the authors and do not necessarily reflect the views of the institutions they belong to.

Conflicts of Interest: The authors declare no conflict of interest.

References

1. Rockström, J.; Bai, X.; deVries, B. Global sustainability: The challenge ahead. *Glob. Sustain.* **2018**, *1*, e6. [CrossRef]
2. Steffen, W.; Richardson, K.; Rockström, J.; Cornell, S.E.; Fetzer, I.; Bennett, E.M.; Biggs, R.; Carpenter, S.R.; Vries, W.D.; de Wit, C.A.; et al. Planetary boundaries: Guiding human development on a changing planet. *Science* **2015**, *347*, 1259855. [CrossRef] [PubMed]
3. Biermann, F.; Bai, X.; Bondre, N.; Broadgate, W.; Chen, C.T.A.; Dube, O.P.; Erisman, W.; Glaser, M.; van der Hel, S.; Lemos, M.C.; et al. Down to Earth: Contextualizing the Anthropocene. *Glob. Environ. Chang.* **2006**, *39*, 341–350. [CrossRef]
4. Giuliani, G.; Nativi, S.; Obregon, A.; Beniston, M.; Lehmann, A. Spatially enabling the Global Framework for Climate Services: Reviewing geospatial solutions to efficiently share and integrate climate data & information. *Clim. Serv.* **2007**, *8*, 44–58. [CrossRef]
5. Lehmann, A.; Chaplin-Kramer, R.; Lacayo, M.; Giuliani, G.; Thau, D.; Koy, M.; Goldberg, G.; Sharp, R., Jr. Lifting the Information Barriers to Address Sustainability Challenges with Data from Physical Geography and Earth Observation. *Sustainability* **2017**, *9*, 858. [CrossRef]
6. Ackoff, R.L. From Data to Wisdom. In *Ackoff's Best*; John Wiley & Sons: New York, NY, USA, 1999; pp. 170–172.
7. Rowley, J. The wisdom hierarchy: Representations of the DIKW hierarchy. *J. Inf. Sci.* **2007**, *33*, 163–180. [CrossRef]
8. Giuliani, G.; Dao, H.; Bono, A.D.; Chatenoux, B.; Allenbach, K.; Laborie, P.D.; Rodila, D.; Alexandris, N.; Peduzzi, P. Live Monitoring of Earth Surface (LiMES): A framework for monitoring environmental changes from Earth Observations. *Remote Sens. Environ.* **2017**, *202*, 222–233. [CrossRef]
9. Skidmore, A.K.; Pettorelli, N.; Coops, N.C.; Geller, G.N.; Hansen, M.; Lucas, R.; Mücher, C.A.; O'Connor, B.; Paganini, M.; Pereira, H.M.; et al. Environmental science: Agree on biodiversity metrics to track from space. *Nat. News* **2015**, *523*, 403–405. [CrossRef] [PubMed]
10. Anderson, K.; Ryan, B.; Sonntag, W.; Kavvada, A.; Friedl, L. Earth observation in service of the 2030 Agenda for Sustainable Development. *Geo-Spat. Inf. Sci.* **2017**, *20*, 1–20. [CrossRef]
11. Kavvada, A.; Held, A. Analysis-Ready Earth Observation Data and the United Nations Sustainable Development Goals. In Proceedings of the IGARSS 2018 IEEE International Geoscience and Remote Sensing Symposium, Valencia, Spain, 22–27 July 2018; pp. 434–436. [CrossRef]

12. Boulton, G. The challenges of a Big Data Earth. *Big Earth Data* **2018**, *2*, 1–7. [CrossRef]
13. Ryan, B. The benefits from open data are immense. *Geospat. World* **2016**, 72–73.
14. Zhu, Z.; Wulder, M.A.; Roy, D.P.; Woodcock, C.E.; Hansen, M.C.; Radeloff, V.C.; Healey, S.P.; Schaaf, C.; Hostert, P.; Strobl, P.; et al. Benefits of the free and open Landsat data policy. *Remote Sens. Environ.* **2019**, *224*, 382–385. [CrossRef]
15. Lewis, A.; Lymburner, L.; Purss, M.B.J.; Brooke, B.; Evans, B.; Oliver, S.; Dekker, A.G.; Irons, J.R.; Minchin, S.; Mueller, N.; et al. Rapid, high-resolution detection of environmental change over continental scales from satellite data—The Earth Observation Data Cube. *Int. J. Digit. Earth* **2016**, *9*, 106–111. [CrossRef]
16. Sudmanns, S.; Lang, S.; Tiede, D. Big Earth Data: From Data to Information. *GI Forum* **2018**, *1*, 184–193. [CrossRef]
17. Baumann, P.; Misev, P.; Merticariu, V.; Huu, B.P.; Bell, B.B. Datacubes: A Technology Survey. In Proceedings of the IGARSS 2018 IEEE International Geoscience and Remote Sensing Symposium, Valencia, Spain, 22–27 July 2018; pp. 430–433. [CrossRef]
18. Baumann, P.; Rossi, A.P.; Bell, B.; Clements, O.; Evans, B.; Hoenig, H.; Hogan, P.; Kakaletris, G.; Koltsida, P.; Mantovani, S.; et al. Fostering Cross-Disciplinary Earth Science Through Datacube Analytics. In *Earth Observation Open Science and Innovation*; Springer: Cham, Switzerland, 24 January 2018; pp. 91–119. [CrossRef]
19. Killough, B. Overview of the Open Data Cube Initiative. In Proceedings of the IGARSS 2018 IEEE International Geoscience and Remote Sensing Symposium, Valencia, Spain, 22–27 July 2018; pp. 8629–8632. [CrossRef]
20. Strobl, P.; Marchetti, P.G. The Six Faces of the Data Cube. In Proceedings of the 2017 Conference on Big Data from Space, Toulouse, France, 28–30 November 2017; pp. 32–35. [CrossRef]
21. Rizvi, S.R.; Killough, B.; Cherry, A.; Gowda, S. The Ceos Data Cube Portal: A User-Friendly, Open Source Software Solution for the Distribution, Exploration, Analysis, and Visualization of Analysis Ready Data. In Proceedings of the IGARSS 2018—2018 IEEE International Geoscience and Remote Sensing Symposium, Valencia, Spain, 22–27 July 2018; pp. 8639–8642. [CrossRef]
22. Baumann, P.; Mazzetti, P.; Ungar, J.; Barbera, R.; Barboni, D.; Beccati, A.; Bigagli, L.; Boldrini, E.; Bruno, R.; Calanducci, A.; et al. Big Data Analytics for Earth Sciences: The EarthServer approach. *Int. J. Digit. Earth* **2016**, *9*, 3–29. [CrossRef]
23. Camara, G.; Assis, L.F.; Ribeiro, G.; Ferreira, K.R.; Llapa, E.; Vinhas, L. Big Earth Observation Data Analytics: Matching Requirements to System Architectures. In Proceedings of the 5th ACM SIGSPATIAL International Workshop on Analytics for Big Geospatial Data, Burlingame, CA, USA, 31 October 2016; ACM: New York, NY, USA; pp. 1–6. [CrossRef]
24. Soille, P.; Burger, A.; de Marchi, D.; Kempeneers, P.; Rodriguez, A.R.D.; Syrris, V.; Vasilev, V. A versatile data-intensive computing platform for information retrieval from big geospatial data. *Future Gener. Comput. Syst.* **2018**, *81*, 30–40. [CrossRef]
25. European Commission. The DIAS: User-friendly Access to Copernicus Data and Information. Available online: https://www.copernicus.eu/sites/default/files/Copernicus_DIAS_Factsheet_June2018.pdf (accessed on 6 July 2018).
26. Gorelick, N.; Hancher, M.; Dixon, M.; Ilyushchenko, S.; Thau, D.; Moore, R. Google Earth Engine: Planetary-scale geospatial analysis for everyone. *Remote Sens. Environ.* **2017**, *202*, 18–27. [CrossRef]
27. Nativi, S.; Mazzetti, P.; Craglia, M. A view-based model of data-cube to support big earth data systems interoperability. *Big Earth Data* **2017**, *1*, 75–99. [CrossRef]
28. Geraci, A. *IEEE Standard Computer Dictionary: A Compilation of IEEE Standard Computer Glossaries*; IEEE Std 610; IEEE Press Piscataway: Piscataway, NJ, USA, 1991; pp. 1–217. [CrossRef]
29. Motta, R.C.; Oliveira, K.M.D.; Travassos, G.H. Rethinking Interoperability in Contemporary Software Systems. In Proceedings of the 2017 IEEE/ACM Joint 5th International Workshop on Software Engineering for Systems-of-Systems and 11th Workshop on Distributed Software Development, Software Ecosystems and Systems-of-Systems (JSOS), Buenos Aires, Argentina, 23 May 2017; pp. 9–15. [CrossRef]
30. Nativi, S.; Santoro, M.; Giuliani, G.; Mazzetti, P. Towards a knowledge base to support global change policy goals. *Int. J. Digit. Earth* **2019**, 1–29. [CrossRef]
31. Lehmann, A.; Nativi, S.; Mazzetti, P.; Maso, J.; Serral, I.; Spengler, D.; Niamir, A.; McCallum, I.; Lacroix, P.; Patias, P.; et al. GEOEssential–mainstreaming workflows from data sources to environment policy indicators with essential variables. *Int. J. Digit. Earth* **2019**, 1–17. [CrossRef]

32. Masó, J.; Serral, I.; Domingo-Marimon, C.; Zabala, A. Earth observations for sustainable development goals monitoring based on essential variables and driver-pressure-state-impact-response indicators. *Int. J. Digit. Earth* **2019**, 1–19. [CrossRef]

33. Domenico, B.; Nativi, S. OGC CF-netCDF 3.0 encoding using GML Coverage Application Schema Netcdf SWG, 2 November 2015. Available online: https://docs.opengeospatial.org/is/14-100r2/14-100r2.html (accessed on 7 June 2019).

34. Miller, H.G.; Mork, P. From Data to Decisions: A Value Chain for Big Data. *It Prof.* **2013**, *15*, 57–59. [CrossRef]

35. Giuliani, G.; Guigoz, Y.; Lacroix, P.; Ray, N.; Lehmann, A. Facilitating the production of ISO-compliant metadata of geospatial datasets. *Int. J. Appl. Earth Obs. Geoinf.* **2016**, *44*, 239–243. [CrossRef]

36. Bone, C.; Ager, A.; Bunzel, K.; Tierney, L. A geospatial search engine for discovering multi-format geospatial data across the web. *Int. J. Digit. Earth* **2016**, *9*, 47–62. [CrossRef]

37. Corti, P.; Kralidis, A.T.; Lewis, B. Enhancing discovery in spatial data infrastructures using a search engine. *PeerJ. Comput. Sci.* **2018**, *4*, e152. [CrossRef]

38. Lehmann, A.; Giuliani, G.; Ray, N.; Rahman, K.; Abbaspour, K.; Nativi, S.; Craglia, M.; Cripe, D.; Quevauviller, P.; Beniston, M. Reviewing innovative Earth observation solutions for filling science-policy gaps in hydrology. *J. Hydrol.* **2014**, *518*, 267–277. [CrossRef]

39. Díaz, P.; Masó, J.; Sevillano, E.; Ninyerola, M.; Zabala, A.; Serral, I.; Pons, X. Analysis of quality metadata in the GEOSS Clearinghouse. *Int. J. Spat. Data Infrastruct. Res.* **2012**, *7*, 352–377. [CrossRef]

40. Trilles, S.; Díaz, L.; Huerta, J. Approach to Facilitating Geospatial Data and Metadata Publication Using a Standard Geoservice. *ISPRS Int. J. Geo-Inf.* **2017**, *6*, 126. [CrossRef]

41. European Commission. *INSPIRE Metadata Implementing Rules: Technical Guidelines based on EN ISO 19115 and EN ISO 19119*; European Commission Joint Research Centre: Abingdon, UK, 2010.

42. Senkler, K.; Voges, U.; Remke, A. An ISO 19115/19119 profile for OGC catalogue services CSW 2.0. Presented at the 10th EC GI & GIS Workshop, Warsaw, Poland, 23–25 June 2004; p. 9.

43. Bruha, L. Large geospatial images discovery: Metadata model and technological framework. *Geoinf. FCE CTU* **2015**, *14*, 21–36. [CrossRef]

44. Innerebner, M.; Costa, A.; Chuprikova, E.; Monsorno, R.; Ventura, B. Organizing earth observation data inside a spatial data infrastructure. *Earth Sci. Inf.* **2017**, *10*, 55–68. [CrossRef]

45. Zabala Torres, A.; Masó, J. Integrated hierarchical metadata proposal: Series, layer, entity and attribute metadata. Presented at the XXII International Cartographic Conference (ed.) Mapping Approaches into a Changing World, A Coruña, Spain, 9–15 July 2005.

46. Thome, K.J.; Helder, D.L.; Aaron, D.; Dewald, J.D. Landsat-5 TM and Landsat-7 ETM+ absolute radiometric calibration using the reflectance-based method. *IEEE Trans. Geosci. Remote Sens.* **2004**, *42*, 2777–2785. [CrossRef]

47. Markham, B.L.; Thome, K.J.; Barsi, J.A.; Kaita, E.; Helder, D.L.; Barker, J.L.; Scaramuzza, P.L. Landsat-7 ETM+ on-orbit reflective-band radiometric stability and absolute calibration. *IEEE Trans. Geosci. Remote Sens.* **2004**, *42*, 2810–2820. [CrossRef]

48. Barsi, J.A.; Markham, B.L.; Helder, D.L. Continued monitoring of Landsat reflective band calibration using pseudo-invariant calibration sites. In Proceedings of the 2012 IEEE International Geoscience and Remote Sensing Symposium, Munich, Germany, 22–27 July 2012; pp. 7007–7010. [CrossRef]

49. Mishra, N.; Helder, D.; Angal, A.; Choi, J.; Xiong, X. Absolute Calibration of Optical Satellite Sensors Using Libya 4 Pseudo Invariant Calibration Site. *Remote Sens.* **2014**, *6*, 1327–1346. [CrossRef]

50. Zabala, A.; Maso, J. *Testbed-12 Imagery Quality and Accuracy Engineering Report*; OGC 16-050; OGC: Wayland, MA, USA, 2017.

51. Vitolo, C.; Elkhatib, Y.; Reusser, D.; Macleod, C.J.A.; Buytaert, W. Web technologies for environmental Big Data. *Environ. Modell. Softw.* **2015**, *63*, 185–198. [CrossRef]

52. Hu, C.; Zhao, Y.; Li, J.; Ma, D.; Li, X. Geospatial Web Service for Remote Sensing Data Visualization. In Proceedings of the 2011 IEEE International Conference on Advanced Information Networking and Applications, Washington, DC, USA, 22–25 March 2011; pp. 594–601. [CrossRef]

53. Lankester, T.H.G. *OpenGIS Web Map Services—Profile for EO Prodcuts*; OGC: Wayland, MA, USA, 2009.

54. Blower, J.D.; Gemmell, A.L.; Griffiths, G.H.; Haines, K.; Santokhee, A.; Yang, X. A Web Map Service implementation for the visualization of multidimensional gridded environmental data. *Environ. Modell. Softw.* **2013**, *47*, 218–224. [CrossRef]

55. Maso, J.; Zabala, A.; Serral, I.; Pons, X. Remote Sensing Analytical Geospatial Operations Directly in the Web Browser. In Proceedings of the International Archives of the Photogrammetry, Remote Sensing and Spatial Information Sciences, Delft, The Netherlands, 1–5 October 2018; Volume XLII-4, pp. 403–410. [CrossRef]

56. Maso, J.; Zabala, A.; Serral, I.; Pons, X. A Portal Offering Standard Visualization and Analysis on top of an Open Data Cube for Sub-National Regions: The Catalan Data Cube Example. *Data* **2019**, *4*, 96. [CrossRef]

57. Guo, H. Big Earth data: A new frontier in Earth and information sciences. *Big Earth Data* **2017**, *1*, 4–20. [CrossRef]

58. Guo, H.; Liu, Z.; Jiang, H.; Wang, C.; Liu, J.; Liang, D. Big Earth Data: A new challenge and opportunity for Digital Earth's development. *Int. J. Digit. Earth* **2017**, *10*, 1–12. [CrossRef]

59. Open Geospatial Consortium. *OGC OWS-6 WPS Grid Processing Profile Engineering Report*; OGC: Wayland, MA, USA, 2009.

60. Baumann, P.; Misev, D.; Merticariu, V.; Huu, B.P. Datacubes: Towards Space/Time Analysis-Ready Data. In *Service-Oriented Mapping*; Springer: Cham, Switzerland, 2019; pp. 269–299. [CrossRef]

61. Nandra, C.; Bâcu, V.; Gorgan, D. Parallel Earth Data Tasks Processing on a Distributed Cloud Based Computing Architecture. In Proceedings of the 2017 21st International Conference on Control Systems and Computer Science (CSCS), Bucharest, Romania, 29–31 May 2017; pp. 677–684. [CrossRef]

62. Evans, B.; Wyborn, L.; Pugh, T.; Allen, C.; Antony, J.; Gohar, K.; Porter, D.; Smillie, J.; Trenham, C.; Wang, J.B.; et al. The NCI High Performance Computing and High Performance Data Platform to Support the Analysis of Petascale Environmental Data Collections. In *Environmental Software Systems. Infrastructures, Services and Applications, Proceedings of the 11th IFIP WG 5.11 International Symposium, ISESS 2015, Melbourne, VIC, Australia, 25–27 March 2015*; Springer: Cham, Germany, 2015; pp. 569–577. [CrossRef]

63. Purss, M.B.J.; Lewis, A.; Oliver, S.; Ip, A.; Sixsmith, J.; Evans, B.; Edberg, R.; Frankish, G.; Hurst, L.; Chan, T. Unlocking the Australian Landsat Archive—From dark data to High Performance Data infrastructures. *GeoResJ* **2015**, *6*, 135–140. [CrossRef]

64. Castronova, A.M.; Goodall, J.L.; Elag, M.M. Models as web services using the Open Geospatial Consortium (OGC) Web Processing Service (WPS) standard. *Environ. Modell. Softw.* **2013**, *41*, 72–83. [CrossRef]

65. Baumann, P. Datacube Standards and their Contribution to Analysis-Ready Data. In Proceedings of the IGARSS 2018—2018 IEEE International Geoscience and Remote Sensing Symposium, Valencia, Spain, 22–27 July 2018; pp. 2051–2053. [CrossRef]

66. Xue, Y.; Palmer-Brown, D.; Guo, H.D. The use of high-performance and high-throughput computing for the fertilization of digital earth and global change studies. *Int. J. Digit. Earth* **2011**, *4*, 185–210. [CrossRef]

67. Olasz, A.; Thai, B.N.; Kristóf, D. Development of a new framework for Distributed Processing of Geospatial Big Data. *Int. J. Spat. Data Infrastruct. Res.* **2016**, *12*, 85–111. [CrossRef]

68. Giuliani, G.; Nativi, S.; Lehmann, A.; Ray, N. WPS mediation: An approach to process geospatial data on different computing backends. *Comput. Geosci.* **2012**, *47*, 20–33. [CrossRef]

69. Mazzetti, P.; Roncella, R.; Mihon, D.; Bacu, V.; Lacroi, P.; Guigoz, Y.; Ray, N.; Giuliani, G.; Gorgan, D.; Nativi, S. Integration of data and computing infrastructures for earth science: An image mosaicking use-case. *Earth Sci. Inf.* **2016**, 1–18. [CrossRef]

70. Rodila, D.; Ray, N.; Gorgan, D. Conceptual model for environmental science applications on parallel and distributed infrastructures. *Environ. Syst. Res.* **2015**, *4*, 1–16. [CrossRef]

71. Nativi, S.; Mazzetti, P.; Geller, G.N. Environmental model access and interoperability: The GEO Model Web initiative. *Environ. Model. Softw.* **2013**, *39*, 214–228. [CrossRef]

72. Di, L.; Yue, P.; Ramapriyan, H.K.; King, R.L. Geoscience Data Provenance: An Overview. *IEEE Trans. Geosci. Remote Sens.* **2013**, *51*, 5065–5072. [CrossRef]

73. Closa, G.; Maso, J.; Zabala, A.; Pesquer, L.; Pons, X. A provenance metadata model integrating ISO Geospatial lineage and the OGC WPS: Conceptual model and implementation. *Trans. GIS* **2019**. [CrossRef]

74. Lewis, A.; Olivera, S.; Lymburner, L.; Evans, B.; Wyborn, L.; Mueller, N.; Raevksi, G.; Hooke, J.; Woodcock, R.; Sixsmith, J.; et al. The Australian Geoscience Data Cube—Foundations and lessons learned. *Remote Sens. Environ.* **2017**, *202*, 276–292. [CrossRef]

75. Wagner, W.; Dorigo, W.; de Jeu, R.; Fernandez, D.; Benveniste, J.; Haas, E.; Ertl, M. Fusion of active and passive microwave observations to create an Essential Climate Variable data record on soil moisture. *ISPRS Ann. Photogramm. Remote Sens. Spat. Inf. Sci.* **2012**, *7*, 315–321. [CrossRef]

76. Pettorelli, N.; Wegmann, M.; Skidmore, A.; Lucas, R.; Rocchini, D.; Fernandez, N.; Turak, E.; Reyers, B.; Geller, G.N.; Belward, A.; et al. Framing the concept of satellite remote sensing essential biodiversity variables: Challenges and future directions. *Remote Sens. Ecol. Conserv.* **2016**, *2*, 276–292. [CrossRef]

77. Giuliani, G.; Chatenoux, B.; De Bono, A.; Rodila, D.; Richard, J.-P.; Allenbach, K.; Dao, H.; Peduzzi, P. Building an Earth Observations Data Cube: Lessons learned from the Swiss Data Cube (SDC) on generating Analysis Ready Data (ARD). *Big Earth Data* **2017**, *1*, 1–18. [CrossRef]

78. Giuliani, G.; Chatenoux, B.; Honeck, E.; Richard, J. Towards Sentinel-2 Analysis Ready Data: A Swiss Data Cube Perspective. In Proceedings of the IGARSS 2018—2018 IEEE International Geoscience and Remote Sensing Symposium, Valencia, Spain, 22–27 July 2018; pp. 8659–8662. [CrossRef]

79. Denis, G.; Claverie, A.; Pasco, X.; Darnis, J.P.; de Maupeou, B.; Lafaye, M.; Morel, E. Towards disruptions in Earth observation? New Earth Observation systems and markets evolution: Possible scenarios and impacts. *Acta Astronaut.* **2017**, *137*, 415–433. [CrossRef]

80. European Commission. *Big Data in Earth Observation*; European Commission: Brussels, Belgium, 2017.

81. Maso, J.; Pons, X.; Zabala, A. Tuning the second-generation SDI: Theoretical aspects and real use cases. *Int. J. Geogr. Inf. Sci.* **2012**, *26*, 983–1014. [CrossRef]

82. Dwyer, J.; Roy, D.; Sauer, B.; Jenkerson, C.; Zhang, H.; Lymburner, L. Analysis Ready Data: Enabling Analysis of the Landsat Archive. *Remote Sens.* **2018**, *10*, 1363. [CrossRef]

83. Claverie, M.; Ju, J.; Masek, J.G.; Dungan, J.L.; Vermote, E.F.; Roger, J.C.; Justice, C.; Skakun, S.V. The Harmonized Landsat and Sentinel-2 surface reflectance data set. *Remote Sens. Environ.* **2018**, *219*, 145–161. [CrossRef]

84. Giuliani, G.; Lacroix, P.; Guigoz, Y.; Roncella, R.; Bigagli, L.; Santoro, M.; Lehmann, A. Bringing GEOSS Services into Practice: A Capacity Building Resource on Spatial Data Infrastructures (SDI). *Trans. GIS* **2016**, *21*, 811–824. [CrossRef]

85. Baumann, P. *The Datacube Manifesto*; European Commission: Brussels, Belgium, 2017.

86. Wulder, M.A.; Masek, J.G.; Cohen, W.B.; Loveland, T.R.; Woodcock, C.E. Opening the archive: How free data has enabled the science and monitoring promise of Landsat. *Remote Sens. Environ.* **2018**, *122*, 2–10. [CrossRef]

87. Wulder, M.A.; Loveland, T.R.; Roy, D.P.; Crawford, C.J.; Masek, J.G.; Woodcock, C.E.; Dwyer, J.; Hermosilla, T.; Hipple, J.D.; Hostert, P.; et al. Current status of Landsat program, science, and applications. *Remote Sens. Environ.* **2019**, *225*, 127–147. [CrossRef]

88. Wilkinson, M.D.; Dumontier, M.; Aalbersberg, I.J.; Appleton, G.; Axton, M.; Baak, A.; Bouwman, J.; Schultes, E.; Roos, M.; Grethe, J.S.; et al. The FAIR Guiding Principles for scientific data management and stewardship. *Sci. Data* **2016**, *3*. [CrossRef]

89. Stall, S.; Yarmey, L.; Cutcher-Gershenfeld, J.; Hanson, B.; Lehnert, K.; Nosek, B.; Wyborn, L. Make scientific data FAIR. *Nature* **2019**, *570*. [CrossRef] [PubMed]

 *data*

Article

Semantic Earth Observation Data Cubes

Hannah Augustin [1,*], **Martin Sudmanns** [1], **Dirk Tiede** [1], **Stefan Lang** [1] **and Andrea Baraldi** [2]

[1] Interfaculty Department of Geoinformatics—Z_GIS, University of Salzburg, 5020 Salzburg, Austria
[2] Italian Space Agency (ASI), 00133 Rome, Italy
* Correspondence: hannah.augustin@sbg.ac.at; Tel.: +43-662-8044-7574

Received: 15 June 2019; Accepted: 15 July 2019; Published: 17 July 2019

Abstract: There is an increasing amount of free and open Earth observation (EO) data, yet more information is not necessarily being generated from them at the same rate despite high information potential. The main challenge in the big EO analysis domain is producing information from EO data, because numerical, sensory data have no semantic meaning; they lack semantics. We are introducing the concept of a semantic EO data cube as an advancement of state-of-the-art EO data cubes. We define a semantic EO data cube as a spatio-temporal data cube containing EO data, where for each observation at least one nominal (i.e., categorical) interpretation is available and can be queried in the same instance. Here we clarify and share our definition of semantic EO data cubes, demonstrating how they enable different possibilities for data retrieval, semantic queries based on EO data content and semantically enabled analysis. Semantic EO data cubes are the foundation for EO data expert systems, where new information can be inferred automatically in a machine-based way using semantic queries that humans understand. We argue that semantic EO data cubes are better positioned to handle current and upcoming big EO data challenges than non-semantic EO data cubes, while facilitating an ever-diversifying user-base to produce their own information and harness the immense potential of big EO data.

Keywords: remote sensing; big Earth data; big EO data; information extraction; semantic enrichment; time-series

---

1. Introduction

The current Earth observation (EO) data pool is vastly different than a mere decade ago, but the main challenge remains: to produce information from data to generate knowledge [1,2]. We are surrounded by a growing ocean of EO data, but sensory data are not information and have no inherent meaning (i.e., lacking semantics) without some form of interpretation. At a minimum, this data pool is characterised by a rapidly growing data volume, accelerating data velocity (i.e., increasing data acquisition and processing speeds) and an increasingly diverse variety of sensors and products [3]. The term "data cube" is broadly understood as a multi-dimensional array organising data in a way that simplifies data storage, access and analysis compared to file-based storage and access [4]. Applying data cube technology to EO datasets attempts to address some of the challenges and opportunities rooted in these big data characteristics.

There is a growing number of implementations currently referred to as EO data cubes with the goal of lowering the barrier to store, manage, provide access to and analyse EO data in a more convenient manner. Data cubes of EO imagery typically are organised in three dimensions: latitude, longitude and time. The definitions or specifications of EO data cubes will not be discussed here but can be understood as a way of organising EO data using a logical view on them, either based on an existing archive (i.e., "indexing") or a specific, application-optimised, multi-dimensional data structure (i.e., "ingestion"). The logical view refers to the way of accessing EO data by using spatio-temporal coordinates either in an application programming interface (API) or a query language instead of file names. The main advantage of ingesting data is that the data can be stored in a query-optimised way, and specific access patterns can be realised more efficiently, such as time series analysis or spatial analysis.

Various technical solutions to create these logical views on EO data have rapidly gained traction over the past few years. The first national scale EO data cube was established in Australia [5], whose technology is now the basis of Digital Earth Australia [6] and the Open Data Cube (ODC) [7]. The free and open source ODC technology is also behind other operational EO data cubes, such as in Switzerland [8], Colombia [9], Vietnam [10], the Africa Regional Data Cube [11] and at least nine other national or regional initiatives under development [7]. Rasdaman [12], an array database system that has been around since the mid-1990s, is another leading technology behind initiatives such as EarthServer [13] and the Copernicus Data and Exploitation platform for Germany (CODE-DE, [14]). Other software implementations exist, such as the Earth System Data Cube from the European Space Agency [15] and SciDB [16].

State-of-the-art EO data cubes simplify data provision to users by facilitating data uptake and aiming to provide analysis-ready data (ARD) [4]. While there is still an ongoing discussion about how ARD are defined and specified, it is usually understood as calibrated data, and in the case of CARD4L (Committee on EO Satellites ARD for Land), even contains masks as a target requirement specification, such as for cloud and water [17,18]. The intention is to shift the burden of pre-processing from users to data providers, who are often better equipped to consistently and reliably process large volumes of high-velocity data [6,17,19]. Processing steps with a high potential level of automation can be conducted centrally where they only must be conducted once and are then available to all users. This contrasts with requiring every user to pre-process the data they would like to use on their own and improves comparability of initial data conditions between users and applications.

Web-based access to these EO data cube implementations brings users closer to the data and implements a computation platform at the data location [20]. This is a different strategy than providing EO data to users as individual, downloadable images of a pre-determined spatial extent. Data cubes make data access much more efficient and effective by providing users with data tailored more specifically to their needs, reducing unnecessary data transfer [20]. Pairing data access from EO data cubes with computational environments, (e.g., processing resources accessible using Jupyter notebooks) moves in the direction of other existing Web-based geospatial computation platforms, such as Google Earth Engine [21]. While these platforms are powerful, analyses sometimes have limited transferability to different geographic locations or points in time, or a low level of results or inferential reproducibility [22].

Even with tailored ARD access and Web-based processing capabilities, users of EO are still confronted with tons of data rather than information and the ill-posed challenge of reconstructing a scene from one or more images [23]. In this case, a scene is understood as the content of an image, whereby the result of this challenge is some form of interpretation or classification map of an image. Images suffer from data dimensionality reduction and a semantic information gap. An image is a 2D snapshot of the 4D world (i.e., three spatial and one temporal dimension), whereby all the information required to reconstruct a comprehensive and complete descriptive scene is not available from one or multiple images over time [24].

Information production from EO images still generally relies on unstructured, application-specific algorithms or increasingly popular machine learning procedures. This often results in low to no semantic interoperability between workflows, sensors or images based on the findable, accessible, interoperable and reusable (FAIR) principles [25]. The FAIR principles refer to data, and the algorithms, tools and workflows that produce them. If data-derived information is linked to the images used to generate them, provenance is maintained and accessible to users. Combining EO images with symbolic image-derived information in a collaborative, analytics environment effectively facilitates increased semantic interoperability between workflows and analyses while extending machine-actionability [26].

If the image-derived information is semantically interoperable and consistent between locations and acquisitions, semantic interoperability is established at least at the starting point of further analysis.

We are introducing the concept of a semantic EO data cube as an advancement of state-of-the-art EO data cubes. Semantic EO data cubes move beyond data storage and provision by offering basic, interoperable building blocks of image-derived information within a data cube. This enables semantic analyses that can be incorporated into simple rule-sets in domain language, and users are able to develop increasingly expressive, comprehensive rule-sets and queries. Given semantic enrichment that includes clouds, vegetation, water and "other" categories, certain semantic content-based queries covering a user-defined area of interest (AOI) in a given temporal extent are possible, such as for the most recent observations excluding clouds (e.g., user-defined cloud-free mosaic), or an observed moment in time with the maximum vegetation extent. These queries of the interpreted content of available images are independent of imposed spatial image extents and are made possible by including semantic enrichment. However, since the information is still tied to the EO images it is based on, it is also possible to search for and retrieve images based on their semantic content rather than only metadata (e.g., where and when each image was acquired). We argue that EO data cubes have the potential to offer much more than data and information product storage and access. They move towards reproducible analytical environments for user-driven information production based on EO images and allow non-expert users to use EO data in their specific context.

This paper focuses on the concept of a semantic EO data cube, assuming the basis is an EO data cube containing EO data together with a nominal interpretation for each observation. Multiple discussions and standardisation processes are currently underway to clarify what constitutes ARD and what minimum requirements constitute a data cube. However, this has little bearing on the base concepts presented here, which have implications for data access, data retrieval, semantic queries of data, semantic interoperability of different methods and results and more. We argue that semantic EO data cubes are better positioned to handle current and upcoming big EO data challenges than non-semantic EO data cubes, while facilitating an ever-diversifying user-base to produce their own information and harness the immense potential of big EO data.

2. Theoretical Framework

Concepts under the same name sometimes differ between domains. The concepts essential for our understanding of semantic EO data cubes are described for clarity, and our definition of what constitutes a semantic EO data cube is explained.

2.1. Clarifying Concepts

Data are not the same as information, and we find ourselves increasingly collecting data, yet not producing more information from them at the same rate. Information can be understood at least in two different ways: as a quantifiable measure in the sense of the information content of a message or an image (e.g., bits and bytes representing something informative [27]), or as a subjective concept, an interpretation (i.e., knowledge produced from a process) [28,29]. Information is used to generate knowledge and understanding, which might lead to wisdom [1,2].

Two terms ought to be clarified before moving forward because they are not interchangeable from our perspective, nor in the domain of computer vision: images and scenes. An EO image is broadly understood as a pixel-discretised field representing measurements of reflected radiations from Earth in different wavelengths (e.g., temperature, visible light, microwave). EO data are delivered as images or single measurements, depending on the design of a sensor. Here we refer to numerical observations represented by pixels and delivered as images. A scene, however, refers to the represented content of an image, meaning that which was observed [30].

The goal of most EO analysis is to produce actionable information to support decision-making processes. This requires transforming EO data into information, or digital numbers into subjective concepts that describe a scene. An EO image is a 2D representation of a 3D scene on Earth at a fixed moment in time, and multiple 2D images of the same 3D scene acquired over time move towards representing a snapshot of the 4D world (i.e., 3D space through time). In this context, what an image or set of images can tell you about a scene is information.

The challenge of reconstructing scenes or generating information about a scene from a mono-temporal 2D image or set of images through time underpins any classification of remotely-sensed imagery and is inherently ill-posed. It is ill-posed in the Hadamard sense because a single, unique solution may not exist, or the solution does not depend continuously on data [31,32]. The last criterion of data-dependence refers to stability, where small changes in the equation or conditions result in small changes in the solution. An ill-posed problem does not meet one or more of these criteria (e.g., there are a huge number of possible solutions when classifying imagery).

The ill-posed problem of reconstructing scenes from images stems primarily from what is known as the sensory gap [33]. For optical EO images, this gap exists between the 2D image that has been sensed (e.g., digital numbers) and the 4D world (e.g., objects, states, events, processes). This gap introduces uncertainty that inherently complicates the interpretation of images and reliable, consistent information production. One aspect of the sensory gap is the sensor transfer function, which relates to the resolvability of phenomena by the given sensor (e.g., spatial, temporal, spectral, radiometric resolution). Another aspect relates to the reduction of dimensionality inherent to images (i.e., 4D to 2D; reducing a flood event to a snapshot in time). These aspects together allow for multiple interpretations of the same or similar representations (e.g., a green pixel in a true-colour image might represent a vegetated rooftop, forest, pasture, football field or something else entirely).

In the context of EO image classification, multiple classifications are possible for any given EO image or collection of images, and many current classification methods are very sensitive to changes in input parameters or starting conditions. Certain methods even produce similar but non-identical results each time they are run on the same initial data. In the case of well-established approaches of supervised classification, different users generally use different sets of samples even if using the same data and being interested in the same categories, which consequently produces different results.

What is known as the semantic gap also contributes to difficulties in producing information from images, and it refers to the gap between something that exists and what it means, regardless of how it is observed or represented [33]. Semantics more broadly refers to a multi-domain study of meaning but influences research in many domains, such as philosophy, linguistics, technology (e.g., the semantic Web [34,35], ontology-based data access [36]) and interoperability (e.g., sharing geographic information [37], processing EO data [26]).

When we speak of semantics in EO, this refers to what an EO image represents in terms of how it is interpreted, usually by an expert. An image can be described using an unbelievable number of words and concepts, yet images do not have intrinsic meaning. Each person has their own definition or understanding of different concepts or symbols, not to mention what they find to be important in a given image or scene [38]. Images gain meaning through relations to other images and the interpretation by a viewer, which is influenced by cultural and social conventions, not to mention the viewer's intention. In the context of image databases, how users search for and interact with images creates additional meaning, especially if given an exploratory user interface [39].

Using the term semantic in relation to EO data cubes refers to how an existing EO data cube is semantically-enabled, meaning a user can interact with it using semantic concepts rather than digital numbers or reflectance values. The ability to search for and retrieve EO data using spatially-explicit semantic content-based information rather than metadata, keywords, tags, or other linked data has strong implications for changing the way EO data is queried, accessed and analysed. However, to semantically-enable an EO data cube, some level of semantics needs to be available for every observation. In the case of EO imagery, this means semantics need to be available for each representation in space and time (i.e., pixel).

2.2. Our Definition of a Semantic EO Data Cube

A semantic EO data cube or a semantics-enabled EO data cube is a data cube, where for each observation at least one nominal (i.e., categorical) interpretation is available and can be queried in the same instance. Interpreting an EO image (i.e., mapping data to symbols that represent stable concepts) results in semantic enrichment [23]. This data interpretation used in creating a semantic EO data cube may differ depending on the user and the intended purpose. Semantic variables are non-ordinal, categorical variables, but subsets of these variables may be ordinal (e.g., vegetation with sub-categories of increasing greenness or intensity) [40]. See Figure 1 (left) for a schematic illustration of a semantic EO data cube.

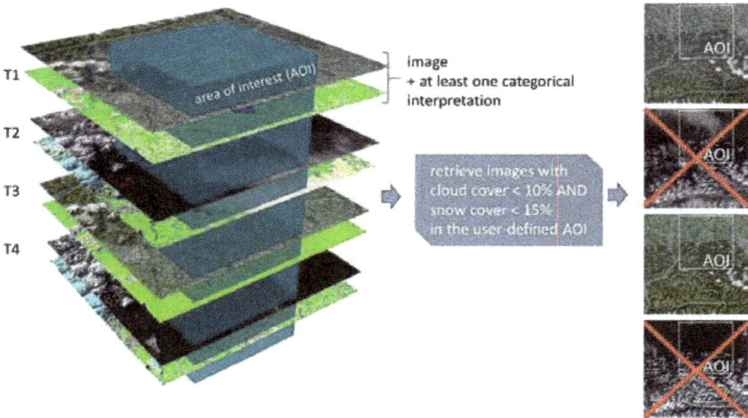

Figure 1. Schematic illustration of a semantic Earth observation (EO) data cube (left) used for an exemplary semantic content-based image retrieval (SCBIR) query. Here, a query searches for images with low cloud and low snow cover within a user-defined area of interest (AOI)-based on the associated semantic information. It retrieves images that match the semantic content-based criteria for the AOI instead of the entire image's extent. In a classic image wide query such AOI specific semantic queries are not possible.

Semantic enrichment included in a semantic EO data cube may be at a relatively low or higher semantic level. A lower semantic level means that symbols may be associated with or represent multiple semantic concepts requiring further analysis or interpretation to align with more specific concepts. The concepts in a lower level semantic enrichment can be considered semi-symbolic in that they are a first step to connecting sensory data to symbolic, semantic classes [41]. This could include information such as colour, or other ways of characterising the spatio-temporal context of each observation. A relatively high semantic level refers to explicit expert knowledge or existing ontologies. In the context of optical EO, one example of relatively high level semantic information would be land cover, such as the land cover classification system (LCCS) developed by the Food and Agriculture Organisation of the United Nations [42].

Other data and information may be combined with a semantic EO data cube to extend possible analysis, but what makes it semantically-enabled is that each observation in space over time has an interpretation. An interpretation that can be generated in an automated way with no user interaction is ideal for handling big EO data. It is also extremely beneficial if the resulting interpreted categories are transferable between different geographic locations, moments in time, images or sensors.

Only including well-known, data-derived indices for each observation (e.g., normalised difference vegetation index (NDVI)) is not sufficient to semantically-enable an EO data cube. Most of these indices are not inherently semantic, in that they still need to be interpreted to have symbolic meaning (e.g., at what NDVI is a pixel considered to contain vegetation or some other interpreted category?).

Indices can, however, contribute different quantitative insights to existing interpretations of an image in a stratified analysis (e.g., this collection of pixels is interpreted as being vegetation, but what was the average NDVI in June 2018 compared to June 2019 within this area?). While the indices can be calculated on the fly since the EO data are also present in a semantic EO data cube, it is up to the user as to whether calculating and incorporating such data-derived indices in a data cube reduces computational resources or has other benefits for further analyses.

Including additional data or information that is not directly derived from EO data does not semantically enable an EO data cube but might enable new query possibilities of EO data in space and time. Such data or information could concern the geographic area (e.g., digital elevation model (DEM)), socio-economic data, or masks of various kinds (e.g., urban area or forest mask). All of these data and information sources are not derived directly from the EO data such that they: (1) do not add information about each EO image's content, but rather the scene content or other characteristics pertaining to the time they were acquired; and (2) may no longer be true for the moment in time an EO image was captured (e.g., a DEM acquired before an earthquake). A DEM, for example, could be used as a spatial selection criterion, even if not specifically related to the semantic content of each image (e.g., selecting observations above a given elevation for alpine areas). Another example would be including an annual forest mask used by environmental regulatory bodies, but that annual mask may not be true even for the EO data available for that given year contained in the data cube.

In semantic EO data cubes it is crucial that EO data be stored with data-derived information for each acquisition. A data cube containing only data-derived interpretations could be considered semantic, but EO data have too much potential to be constrained to a single interpretation, especially since there is no single correct interpretation of image content. World ontologies are infinite. Multiple different perspectives and interpretations need to be possible to close the semantic gap [38], and users should be allowed to generate their own interpretations within a semantic EO data cube should those available not be suitable for their needs. The loss of connection to original EO data constrains semantics to the available interpretation, eliminates access to the source of the data-derived information important for provenance and limits further analysis. Some users might benefit from incorporating reflectance values from specific bands (e.g., calculating an index), using the semantic information to generate composite images through time, or generating different information based on the data to augment existing semantic enrichment.

The focus of semantic EO data cubes is to facilitate ad hoc, flexible information generation from data, that might have potential to lead to knowledge. Semantic EO data cubes combine concepts from EO, image processing, geoinformatics, computer vision, image retrieval and understanding, semantics, ontologies and more. Similar to how the semantic Web can be considered an extension of the Web [34], semantic EO data cubes offer a solution to combining EO data with meaning. This ultimately better enables people and computers to work together to access, retrieve and analyse EO data and data-derived information in a semantically-enabled and machine-readable way.

3. Examples from Existing Semantic EO Data Cubes

Three applied examples of semantic EO data cubes are presented, and each of them uses the same relatively low-level, generic, data-derived semantic enrichment as the basis for each of the semantic EO data cubes. This general-purpose semantic enrichment is application- and user-independent and thus can support multiple application domains. The semantic enrichment used in the following examples is automatically generated (i.e., without any user-defined parameterisation or training data) by the Satellite Image Automatic Mapper™ (SIAM™). This software is an expert system that employs a per-pixel physical spectral model-based decision-tree to images calibrated to at least top-of-atmosphere reflectance in order to accomplish automatic, near real-time multi-spectral discretisation based on a priori knowledge [43]. The decision tree maps each observation located within a multi-spectral reflectance hypercube to one multi-spectral colour name, which is stable and sensor agnostic. It is sensor-agnostic in that data calibrated to at least top-of-atmosphere reflectance by optical sensors can be used to generate semantic enrichment comparable between sensors (e.g., Sentinel-2, Landsat). SIAM™'s output has been independently validated at a continental scale by [44].

This colour naming results in a discrete and finite vocabulary referring to hyper-polyhedra within a multi-spectral feature space, whereby the colour names create a vocabulary that is a mutually exclusive and totally exhaustive partitioning of the multi-spectral reflectance hypercube. These colour names have semantic associations using a knowledge-based approach and thus are considered semi-symbolic (i.e., semi-concepts). More broadly, this vocabulary of colour names can be thought of as stable, sensor-agnostic visual "letters" that can be used to build "words" (i.e., symbolic concepts) that have a higher semantic level using knowledge-based rules. The output may be considered sufficient for generating CARD4L masks as specified in the product family specification [18], but also offers building blocks for a complete scene classification map.

In the following examples, these data-derived information building blocks (i.e., semi-concepts) are based on Landsat 8 or Sentinel-2 images and are stored using either Open Data Cube or rasdaman technology to create semantic EO data cubes. While the semi-concepts themselves are inferior in semantics to land cover classes, they are reproducible, transferable between images and geographic locations, and each colour has a semantic association. These implementations serve as the foundation for semantic content-based image retrieval (SCBIR) (Section 3.1) or other semantic queries (Sections 3.2 and 3.3). Spectral-based semi-concepts can serve as the basis for more expressive, automated scene classification, queries and analysis within each of these prototypical semantic EO data cubes using knowledge-based rules (see Section 4.4).

3.1. Semantic Content-Based Image Retrieval

The example of operational SCBIR has been prototypically implemented within a semantic EO data cube based on Landsat 8 data and the rasdaman array database system as an underlying data cube technology [45]. While this prototypical implementation (see Figure 1) did not cover a large database, it is designed for scalability by relying on parameter free, fully automated and multisensory enabled semantic enrichment, as well as on a data cube technology proven to be scalable to PB sizes [13].

Unlike a traditional content-based image retrieval system, a SCBIR system is expected to cope with spatially-explicit (i.e., area of interest (AOI)-based), temporal, semantic queries (e.g., "retrieve all images in the database where the AOI does not contain clouds or snow"). Very few SCBIR system prototypes targeting EO images have been presented in the literature [46,47]. None of them is available in operating mode to date.

The implementation of SCBIR is urgently needed in today's big EO archives to overcome the limitations of currently implemented image data retrieval methods using image metadata (e.g., acquisition date, sensor, pre-processing level) and image wide statistics like average cloud cover. The latter is especially a problem because the average cloud cover statistic is one of the most used pre-selection criteria for image retrieval of big EO data but is an average over an entire image. Spatially-explicit AOI-based querying that makes use of the semantic information of each pixel in a data cube could help in making use of hidden or "dark" data in big EO databases. This could, for example, lead to retrieving more cloud-free time series or improving cloud free mosaic composition, utilising data contained in images with low average cloud cover.

A SCBIR query is visualised in Figure 1 based on the prototypical implementation. A query based on the semantic information for low cloud cover combined with low snow cover in the selected AOI would only retrieve 2 of the 4 sample images in this example, making query results better posed for following analyses. While our definition of a semantic EO data cube does not prescribe any particular level of semantic enrichment, SCBIR queries beyond cloud/snow cover are possible depending on the available image interpretation, e.g., searches for images where flooding occurred, containing a low tidal range, or where a peak in vegetation coverage occurs.

3.2. Flood Extent in Somalia Based on Landsat 8 Imagery

One of the first implementations of a semantic data cube was a study to extract surface water dynamics and the maximum flood extent as an indicator for flood risk using a dense temporal stack of 78 Landsat 8 images [48]. By using water observations of three years, areas are delineated that are prone to being flooded, as illustrated in Figure 2. In this study, the array database system rasdaman was used to instantiate a semantic EO data cube with pre-processed Landsat imagery and semantic enrichment generated with SIAM™, which can be accessed by using a self-programmed Web frontend, visually supporting the design of semantic queries. In this system the analyses are automatically translated database queries, which increase reproducibility, readability and comprehensibility for a human operator and can be conducted within a few minutes. The study showed how a generic semantic EO data cube can be used for on-the-fly information production using a very simple ruleset.

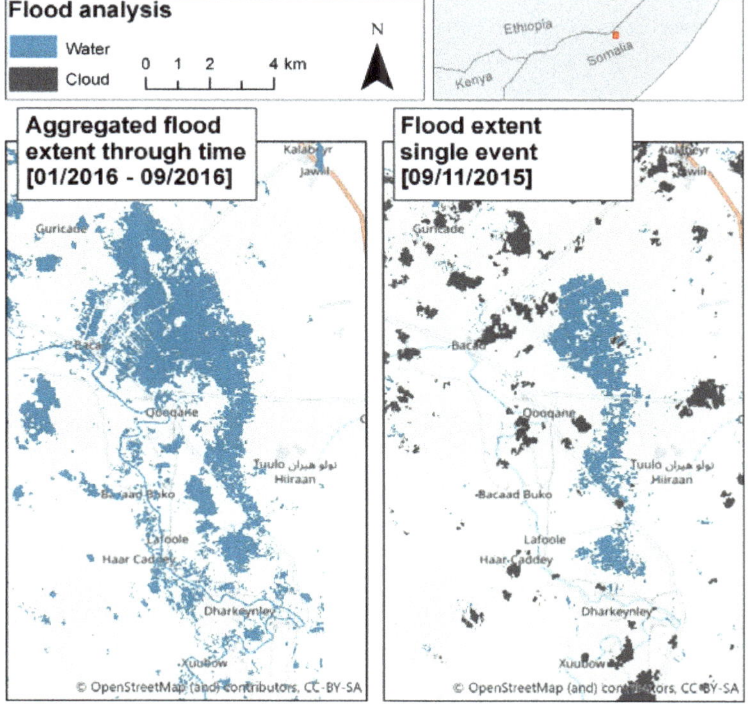

Figure 2. A flood mask generated from 78 semantically enriched Landsat 8 images over 9 months in Somalia (left) as an indicator for flood risk is compared to a single event analysis following a reported flood event in the year before (right). Both maps are the result of basic user queries using the semantic information only, without the use of additional parameters or calculations on the original data sets. Originally published as CC-BY-ND by [48], modified.

3.3. Semantic EO Data Cube along the Turkish/Syrian Border

The potential of semantic EO data cubes is demonstrated here using a proof-of-concept implementation based on ODC technology, described in detail by [49]. In this case EO refers to satellite-based remote sensing data produced by the Copernicus programme's Sentinel-2 satellites. All available Sentinel-2 data (i.e., ca. 1000 images to date) covering over 30,000 km^2 along the north-western Syrian border to Turkey (latitudes 36.01°–37.05°N; longitudes 35.67°–39.11°E) are continuously incorporated in an automated way including being mapped into semi-symbolic colour names by SIAM™. The example output generated here demonstrates that traditional statistical model-based algorithms may be replaced by querying symbolic information, starting from semi-symbolic colour names with semantic associations that are not bound to a specific theme or application within a semantic EO data cube.

In March 2019 flash flooding was reported in various parts of Syria [50]. The worst flooding was reported in Idlib province, which is just south of the western most part of the study area (see Figure 3). While optical imagery is often hindered by cloud cover in rain events, a query for water-like pixels around the time of intense precipitation shows that certain flooded areas have been observed by Sentinel-2 satellites. A normalised observed surface water occurrence (SWO) over time is calculated for two spatio-temporal extents of interest, namely 15 March to 15 April for the entire study area in 2018 and 2019 (see Figure 4). The calculation of the result for each spatio-temporal extent took around 10 minutes to complete using the same hardware and software as described by [49]. The algorithm, described in pseudocode in Figure 5, can be applied to any semantic concept that exists in the semantic EO data cube. This is demonstrated in Figure 6 where the same algorithm was applied to the semantic EO data cube, but for vegetation-like pixels rather than water-like pixels for the same two spatio-temporal extents.

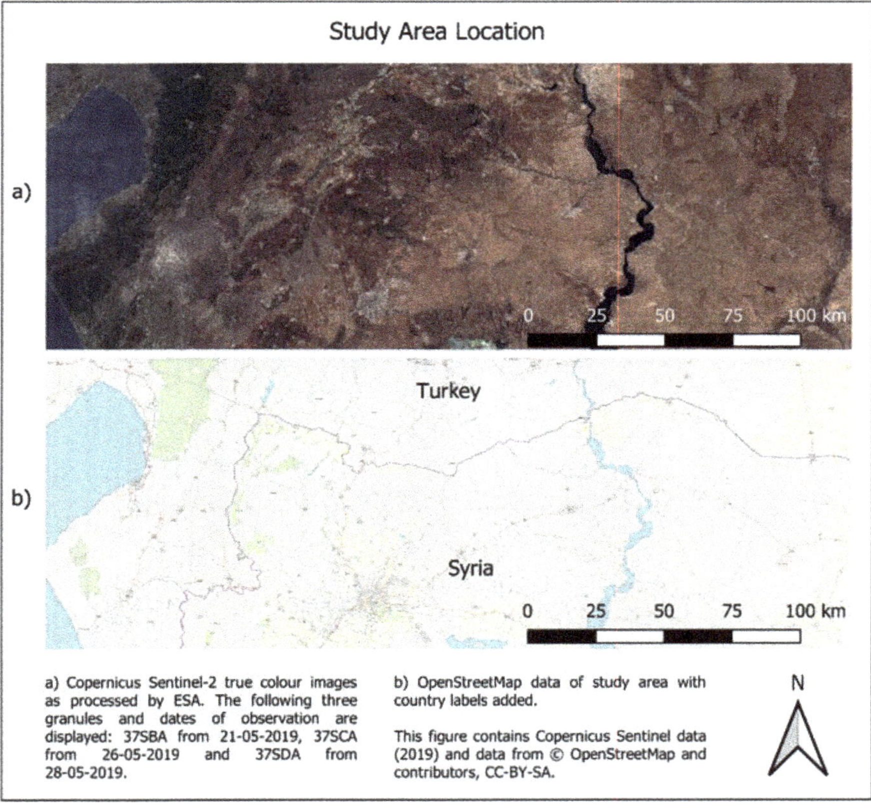

Figure 3. The spatial extent of the semantic EO data cube comprises three Sentinel-2 granules. (**a**) displays the true colour Sentinel-2 images as processed by the European Space Agency (ESA); (**b**) shows the area as represented in OpenStreetMap.

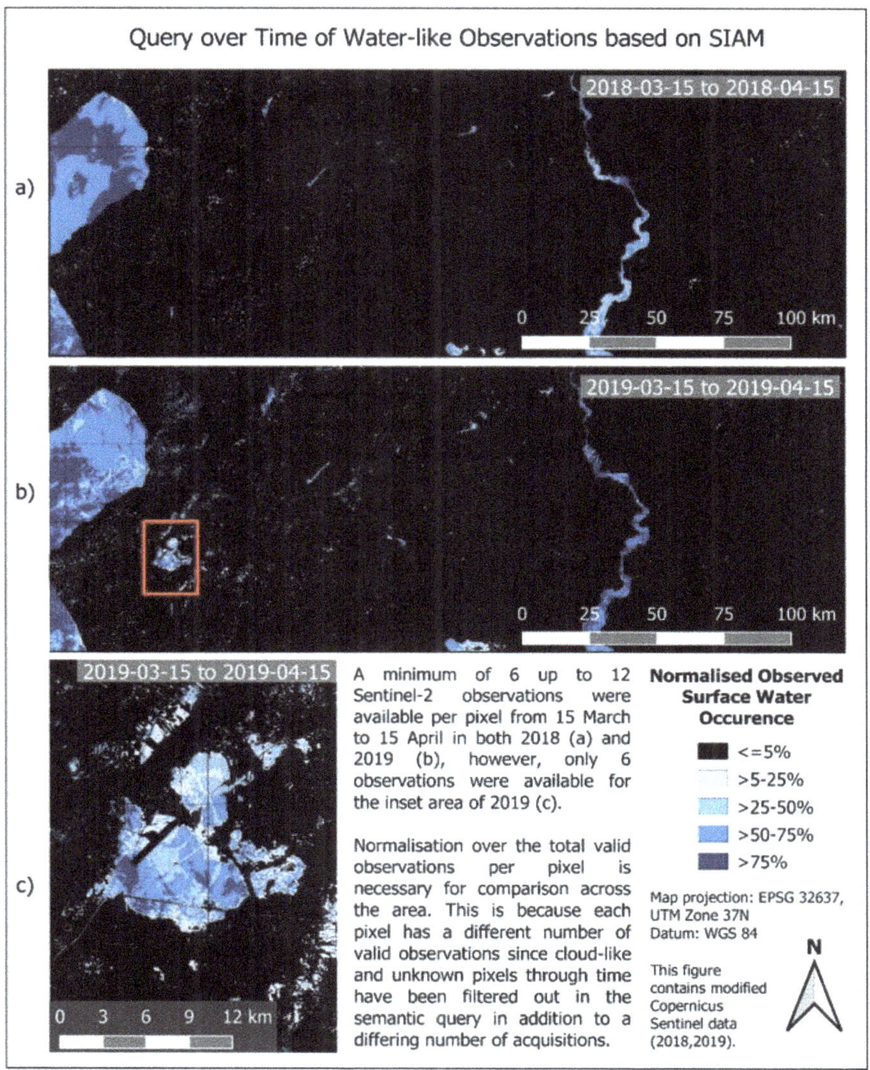

Figure 4. This figure displays the results of the semantic query for water-like observations for two spatio-temporal extents of interest. (**a**) Query for water-like observations from 15 March to 15 April 2018. (**b**) Query for water-like observations from 15 March to 15 April 2019. (**c**) Close-up of an area where water-like observations were present in 2019 but not in 2018.

```
1    connect to data cube
2    query data cube:
3        what: ingested tiles of SIAM's 33 semantic semi-concept granularity for Sentinel-2 data
4        where: covering [area]
5        when: acquired from [start] to [end]
6        return 3D array of per-pixel SIAM categories over time
7    split SIAM categories over time into smaller cells along spatial axes for processing
8    for each cell:
9        load into memory
10       create "water" array of same shape with all values set to 0
11       ask semantic query:
12           if water-like (>=21 and <=24), change 0 to 1
13       if invalid, ice/snow, clouds or unknown (NaN or ==25 or ==29 or ==33), change 0 to 255
14       perform time-series analysis (input: 3D "water" array from semantic query):
15           change values 255 to NaN
16           calculate "total water observations" array:
17               sum values over time (water == 1; else == 0)
18           calculate "total clean observations" array:
19               create Boolean array where values not NaN are True
20               sum values over time (True == 1; False == 0)
21           calculate "normalised occurrence" array:
22               divide "total water observations" by "total clean observations"
23           change NaN to 0 in all arrays
24           return three resulting 2D arrays:
25               "normalised occurrence"
26               "total water observations"
27               "total clean observations"
28       create GeoTiff of each returned array
29   mosaic all cell-based GeoTiffs back together for each respective analysis
```

Figure 5. Pseudocode describing how the normalised observed surface water occurrence (SWO) over time is calculated based on semi-concepts, in addition to two other outputs necessary for its calculation. The array of "total clean observations" provides the number of observations over time per-pixel after excluding cloud-like, snow-like and unknown pixels in the spatio-temporal extent of interest. Snow-like are excluded in this case based on the knowledge that there is generally no snow within the spatio-temporal extent of interest. "Total water observations" refers to the number of observations over time per-pixel that water-like spectral profiles were observed. It is the ratio between these two outputs (i.e., total divided by clean observations per-pixel) that results in the normalised observed SWO.

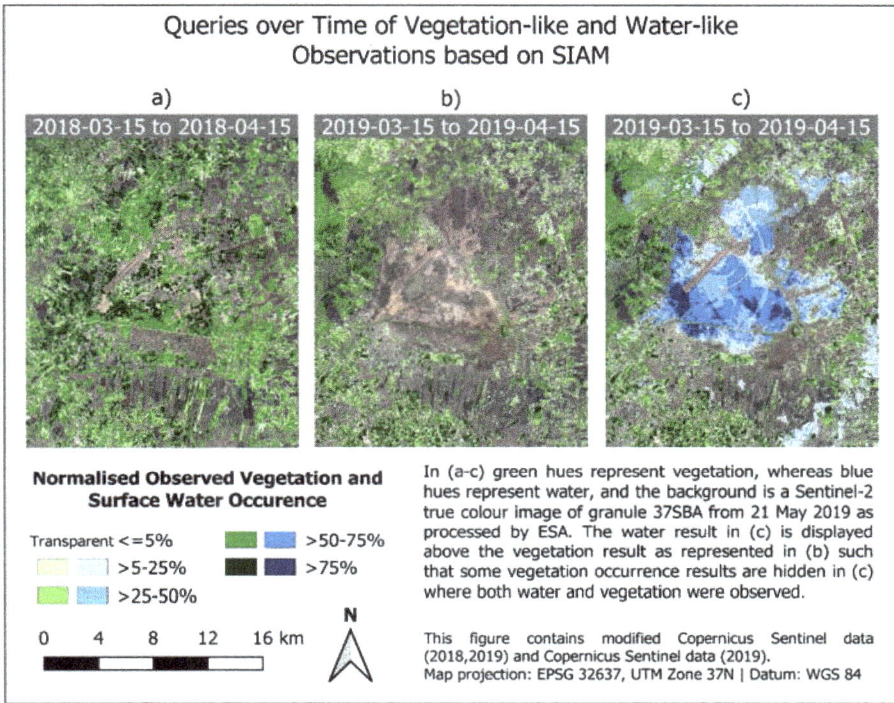

Figure 6. This figure displays the results of a different semantic query for the same two spatio-temporal extents of interest used in the query of water-like observations seen in Figure 5. (**a**) Normalised observed vegetation occurrence from 15 March to 15 April 2018. (**b**) Normalised observed vegetation occurrence from 15 March to 15 April 2019. (**c**) Normalised observed SWO from 15 March to 15 April 2019 overlaid above normalised observed vegetation occurrence as represented in (**b**).

4. Discussion and Outlook

Semantic EO data cubes are interdisciplinary in their conceptualisation, combining concepts related to image retrieval, computer vision, human cognition, semantics, world ontologies, remote sensing and more. The applied examples presented in Section 3 are brought into context of semantic EO data cubes, according to the definition and concepts provided in Section 2. Semantic EO data cubes also have the potential to be a foundational element in image understanding systems, which is discussed briefly in Section 4.4 and is a focus of on-going research.

4.1. Improvements to Data and Image Retrieval

Combining semantic enrichment with EO images has implications for EO archives, databases and the ways in which users can search for and select images [45,51]. EO data cubes already enable users to retrieve data independent of the image's spatial extent. The best-case scenario is when images processed to ARD specifications are used as the basis of an EO data cube and not just any images or quality indicators. Semantic EO data cubes enable users to search for and retrieve EO data in their spatio-temporal extent of interest based on their content, rather than image-wide statistics.

Since data and semantic enrichment are both available, SCBIR can improve ARD provision to users by expanding the possibilities that users have to retrieve images that meet their requirements. Currently a user may be interested in an area that occupies only 10% of an image. If this section of the image is cloud-free but the rest of the image is not, this image will not be returned when searching for low average image-wide cloud coverage statistics (Figure 1). Semantic EO data cubes can provide average cloud coverage information about a user-defined AOI that could be used for data retrieval instead of aggregated image-wide metadata or statistics. Not only can users retrieve data for a given spatio-temporal extent that has low cloud cover, but based on any other category available. This means that such queries theoretically could also include searches for images containing a certain percentage of water, snow or vegetation given reliable semantic enrichment at that semantic level.

Including semantic enrichment with EO data can also serve to improve automated user-defined image composites or mosaics. The classic example is creating a cloud-free composite for a given spatio-temporal extent. As long as the semantic enrichment offers some information about cloud cover, users can retrieve cloud-free pixels for their spatio-temporal extent of interest without having to run a complex algorithm or rely on pixel-based statistics over time. A user could search for the most recent cloud-free pixels within a given spatio-temporal extent (e.g., May 2019) based on semantics instead of statistics, whereby the result could look something like Figure 3a.

SCBIR and semantically-enabled best pixel selection is even more important in the big EO era so that the data best suited for the analysis can be efficiently and effectively retrieved from huge archives in an automated way. An overview of different capabilities between file-based hubs or archives (e.g., Copernicus Open Access Hub), non-semantic EO data cubes and semantic EO data cubes is provided in Table 1.

Table 1. Feature matrix for different approaches of storing and analysing EO images.

Feature	File-Based EO Image Hubs	Non-Semantic Data Cubes	Semantic EO Data Cubes
• Image download	X	X	X
• Metadata-based search	X	X	X
• Image-wide processing	X	X	X
• AOI-based processing	-	X	X
• Fast access to imagery	-	X	X
• Time series analysis (statistical)	-	X	X
• Time series analysis (semantic)	-	-	X
• SCBIR	-	-	X
• Content-based best pixel selection for cloud-free composites	-	-	X^1
• Generic approach with re-usable and sharable tools	-	-	X^1

[1] Depending on the implementation level.

4.2. Semantic Content-Based Queries

The presence of a categorical interpretation for each observation allows users to pose semantic queries in EO data cubes. Semantic queries are queries about the world that exist and "make sense" regardless of whether images or data exist. They move beyond answering questions related to image retrieval (e.g., "Which data in my area of interest have less than 10% cloud cover?") towards queries about the world (e.g., "Where and when have glaciers in the Alps grown over the last decade?"). These queries may or may not be able to be answered based on available EO data. The query space is only limited by the semantic level of enrichment and any additional information or knowledge that is available (e.g., DEM, image-derived indices).

Semantic EO data cubes enable information retrieval and semantically-enabled analysis while allowing users to better explore what is possible with available EO data in an ad hoc way beyond the confines of specific applications. There is a difference between requiring a user to know what application-specific information they want to produce from EO data, and trying to answer the question, "what is possible with these data?" [52]. For example, flooding in Turkey and Syria was known to have occurred in the spatio-temporal extent in 2019 used in queries shown in Figures 5 and 6, but it was unclear whether optical Sentinel-2 imagery was able to capture any of it, and if so, where. Applying a query for water or water-like pixels aggregated over time, such as shown in Figure 2, is a spatially-explicit way to help answer that question. Additionally, such a query might be even more powerful if the user has limited spatially-explicit precipitation or temperature information and is unaware of any flooding that may or may not have occurred in a given area at any point in time.

It is important to emphasise that any analysis of EO data is only relevant for the snapshots in time that are available. Information derived from them may also not necessarily be valid for much of the time between acquisitions. For example, just because flooding is not observed or detected using Sentinel-2 data does not mean that flooding did not occur in a given spatio-temporal extent. Even big EO data with a high temporal sampling rate must always be interpreted keeping this in mind and is best when combined with additional information or domain knowledge.

Including semantic enrichment for each image enables semantic queries to be applied to EO data and derived information without requiring complex algorithms to process all data for a geographic area or given timespan. Even though the semantic level of the interpretations may vary amongst implementations, algorithms can access the reflectance values already associated with an interpretation that can be referenced later in the workflow, if necessary. Data-derived content-based information is available for each existing observation and can be read in a machine-based way using categories that users understand.

Working with symbolic categories instead of reflectance values means that users can work with queries that are readily understandable if the vocabulary of a community is being used, or a standard set of classes such as LCCS or similar. However, using categories means an unfortunately non-reversible data reduction, or reduction of the feature space in comparison to a multitude of bands with a higher bit depth (e.g., 48 categories stored as 8-bit data in comparison to 13 bands of 12-bit data, such as for Sentinel-2). This data reduction benefits query performance, in particular, but needs to be taken into account for every analysis. Based on our definition of a semantic EO data cube, the original data is available and accessible should users require them.

Having the original data available with categories also creates new possibilities for other applications, such as stratifying data analysis based on semantic enrichment. This could be relevant for improving sampling for machine-learning algorithms based on the frequency and distribution of certain categories through space and time. For example, samples could be stratified based on the occurrence of spectrally similar pixels by category within a study area in an attempt to mitigate sampling bias. Other analysis can also benefit from stratification based on category, such as topographic correction (e.g., certain categories will be darker in terrain shadow than others, and clouds are unaffected), or calculating indices (e.g., first querying for vegetation before calculating NDVI to avoid having to set a threshold to distinguish vegetation with the index alone).

4.3. Automated, Generic Semantic Enrichment for Big EO Data

Semantic EO data cubes are most powerful when combined with semantically rich yet generic interpretations because semantics differ between domains, applications, users and the targeted purpose of analysis. Closing the semantic gap when generating information from EO data is very difficult and goes beyond the focus of this paper (refer to [44] as a starting point on this topic), but even the simplest semantic enrichment better positions EO data cubes for analysis than ones containing no semantics at all. Any data-derived semantic information can be used as the basis of a semantic EO data cube, but generic semantic enrichment is highly extendible. It allows multiple domains to simultaneously benefit from EO data and derived information without having to reprocess huge amounts of data for every analysis. Workflows utilising the same generic, data-derived building blocks for analysis also supports increased semantic interoperability. However, big EO data necessitates data-derived interpretations that can be generated without user parameterisation (i.e., automated), are reliable and acceptable in quality and with reasonable processing times [20].

The semantic enrichment generated by SIAM™ and used in the applied examples was chosen because it is fast, fully automated, scalable to handle big EO data, sensor-agnostic and comparable between images captured at different locations and times. The limited semantic depth can be partly compensated through the availability of the temporal dimension in dense time series because the concepts are particularly stable (i.e., robust to changes in input data and imaging sensor specifications). Semantic categories that are sensor agnostic means that users can compare the semantic content of different images acquired by different sensors using the same semantic concepts. Higher level semantics can improve information generation but are generally limited to a specific theme or application. This may be beneficial in some cases and those interested in generating information can decide what is necessary for them before processing massive amounts of EO data to create a semantic EO data cube.

The examples presented in Sections 3.2 and 3.3 both queried water-like pixels based on the low-level generic semantic enrichment available over time. Even with a semi-symbolic level of semantic enrichment, queries for water-like observations could be conducted for a single acquisition or aggregated over multiple acquisitions (Figures 2 and 5). Query results shown in Section 3.3 took an additional step of excluding cloud-like and snow-like pixels and normalising the results over time for increased comparability given spatio-temporal heterogeneity of available data. The same query for two different spatio-temporal extents as shown in Figure 5 were generated within 10 minutes on relatively limited computing resources as documented by [49]. Especially in situations where timely information generation is critical, such generic implementations may be particularly useful. They can also serve in finding spatio-temporal locations interesting for further analysis using available data. Figure 6 demonstrates two semantic queries on two spatio-temporal extents based on the same semantic EO data cube, showcasing the benefit of being able to conduct various semantic queries using generic semantic enrichment.

Many other surface water occurrence algorithms and analysis for EO data exist but cannot necessarily be conducted ad hoc for user-defined spatio-temporal AOIs, are more computationally expensive, and results are not necessarily able to be queried. For example, work conducted at the European Commission's Joint Research Centre by [53] has generated various high-resolution global surface water information layers. These results provide valuable information based on EO data, but cannot be queried for content, are separate from the data that they were derived from and are limited to pre-defined temporal extents (e.g., annually). The surface water information generated by [54] or [55] for each EO observation and used in their surface water dynamics analysis could be the basis for a semantic EO data cube, but it would be semantically limited to the concept of water and does not seem to be continuously updated with newly available data (i.e., images acquired up to now) in an automated way. These implementations provide static layers, and are not currently posed to provide more dynamic, near-real-time or continuously updated results such as information about the maximum observed water extent in 2019 as it happens based on cloud-free/clean pixels.

In Figure 5 it is visible that large, permanent water bodies sometimes returned less than 100% of normalised observed surface water occurrence. This has to do with the semantic query not taking pixels associated with haze or very thin clouds into consideration, which are not necessarily water-like nor cloud-like. Queries can be improved, and more complex knowledge-based rules implemented. These proof-of-concept results demonstrate that even queries low in complexity based on low-level semantic enrichment can produce higher-level information that might be useful in certain scenarios.

4.4. Towards an Image Understanding System

While our definition does not specify applications and implementations of semantic EO data cubes, a prominent use-case is as part of an application-independent expert system, where the semantic EO data cube serves as a fact base. In an expert system, users connect rules stored within a knowledge base to a fact base to infer new information. In such a set-up, the knowledge base is continuously-augmented with rules based in domain knowledge. This allows using already existing encoded expert knowledge or having users contribute their own knowledge. An overall architecture such as proposed by [45] consists of an image understanding sub-system in addition to the semantic EO data cube, which both makes use of already existing interpretations and feeds the fact base with newly derived, true information.

A prototype of an expert-system-based architecture is currently under development for Austria, where a semantic EO data cube serves as a backbone for user-generated semantic queries [56]. This system combines a fully automated semantic enrichment of Sentinel-2 images up to basic land cover types with a semantic EO data cube and Web interface for human-like queries based on semantic models of the spatio-temporal 4D physical-world domain. Although still under development, first results are promising and show that users are able to formulate even complex queries using the semantic pre-processing as simple building blocks to derive information at a higher semantic level than the initial building blocks.

5. Conclusions

The aim of this paper was to define what a semantic EO data cube is and what they make possible in terms of image retrieval, analysis and information production potential. Lots of EO data are being collected, yet proportionally less are being used to produce information, many domains are underserved in relation to what EO could offer, and users of EO data need to have a high level of technical competence to produce information from EO data.

By combining EO data with an interpretation for each observation of a scene, semantic EO data cubes allow users to run queries on big EO data and time-series that were not previously possible and provide imaged-derived information building blocks for analysis that are more meaningful than measured surface reflectance. Semantic enrichment enables semantic content-based image retrieval, allowing users to retrieve specific observations based on what they contain rather than image-wide statistics. Semantic queries (i.e., queries that exist independent of EO images) can be run on EO data that are at least at the semantic level of enrichment or higher without having to necessarily run complex, application-specific algorithms for each analysis. Including semantics in an EO data cube also establishes a minimal level of semantic interoperability for different analyses conducted within the same semantic EO data cube or a different implementation using the same semantic enrichment. This has implications for improving reproducibility of methods and results, especially when applying the same methods based on the same semantic enrichment to different spatio-temporal extents.

Semantic EO data cubes go beyond state-of-the art EO data cubes by managing image-derived information together with data accessible for querying, and thus serve as initial building blocks for semantic queries. Instead of attempting to answer a specific question using EO data, semantic EO data cubes move towards exploring what questions can possibly be answered using the EO data available for a given spatio-temporal extent of interest. Analysis is only limited by the semantic enrichment included and can be extended using transparently coded rule-sets or additional information and knowledge to produce information with a higher semantic level.

We believe that semantic EO data cubes are better positioned to serve big EO data than existing EO data cube implementations, especially when containing ARD and generic, sensor-agnostic semantic enrichment that can be automatically generated in a scalable way. The potential of semantic EO data cubes is just beginning to be explored, but hopefully it is evident that there is plenty of potential yet to be discovered. Semantic EO data cubes are the foundation for big EO data expert systems, where new information can be inferred automatically in a machine-based way using semantic queries that humans understand.

Author Contributions: All authors were involved in the conceptualisation of this paper. Software used for generating semantic enrichment in the applied examples, SIAM™, was developed by A.B. Example 3.1 was provided by D.T., 3.2 by M.S. and D.T. and example 3.3 was provided by H.A. Original draft preparation was predominantly conducted by H.A. with review and editing prior to submission by D.T., M.S., S.L. and H.A.

Funding: This research has received funding from the Austrian Research Promotion Agency (FFG) under the Austrian Space Application Programme (ASAP) within the project Sen2Cube.at (project no. 866016) and from the Austrian Science Fund (FWF) through the Doctoral College GIScience (DK W1237-N23).

Acknowledgments: We would like to thank Christian Werner for his contributions in discussions about the various concepts included in this paper.

Conflicts of Interest: The authors declare no conflict of interest. The funders had no role in the design of the study; in the collection, analyses, or interpretation of data; in the writing of the manuscript, or in the decision to publish the results.

References

1. Rowley, J. The Wisdom Hierarchy: Representations of the DIKW Hierarchy. *J. Inf. Sci.* **2007**, *33*, 163–180. [CrossRef]
2. Ackoff, R.L. From data to wisdom. *J. Appl. Syst. Anal.* **1989**, *16*, 3–9.
3. Laney, D. 3-D data management: Controlling data volume, velocity and variety. In *Application Delivery Strategies*; META Group Inc.: Stamford, CT, USA, 2001.
4. Baumann, P. The Datacube Manifesto. Available online: http://www.earthserver.eu/tech/datacube-manifesto (accessed on 30 January 2018).
5. Lewis, A.; Oliver, S.; Lymburner, L.; Evans, B.; Wyborn, L.; Mueller, N.; Raevksi, G.; Hooke, J.; Woodcock, R.; Sixsmith, J.; et al. The Australian Geoscience Data Cube—Foundations and lessons learned. *Remote Sens. Environ.* **2017**, *202*, 276–292. [CrossRef]
6. Dhu, T.; Dunn, B.; Lewis, B.; Lymburner, L.; Mueller, N.; Telfer, E.; Lewis, A.; McIntyre, A.; Minchin, S.; Phillips, C. Digital earth Australia—Unlocking new value from earth observation data. *Big Earth Data* **2017**, *1*, 64–74. [CrossRef]
7. Killough, B. Overview of the Open Data Cube Initiative. In Proceedings of the 2018 IEEE International Geoscience and Remote Sensing Symposium (IGARSS), Valencia, Spain, 22–27 July 2018; pp. 8629–8632.
8. Giuliani, G.; Chatenoux, B.; Bono, A.D.; Rodila, D.; Richard, J.-P.; Allenbach, K.; Dao, H.; Peduzzi, P. Building an Earth Observations Data Cube: Lessons learned from the Swiss Data Cube (SDC) on generating Analysis Ready Data (ARD). *Big Earth Data* **2017**, *1*, 100–117. [CrossRef]
9. Ariza-Porras, C.; Bravo, G.; Villamizar, M.; Moreno, A.; Castro, H.; Galindo, G.; Cabera, E.; Valbuena, S.; Lozano, P. CDCol: A Geoscience Data Cube that Meets Colombian Needs. In Proceedings of the Advances in Computing, Cali, Colombia, 19–22 September 2017; Springer: Cham, Switzerland, 2017; pp. 87–99.
10. Cottom, T.S. An Examination of Vietnam and Space. *Space Policy* **2019**, *47*, 78–84. [CrossRef]
11. Group on Earth Observations (GEO). Digital Earth Africa: Project Overview. Available online: https://www.ga.gov.au/__data/assets/pdf_file/0008/73376/Digital-Earth-Africa.pdf (accessed on 28 May 2019).
12. Baumann, P.; Furtado, P.; Ritsch, R.; Widmann, N. The RasDaMan approach to multidimensional database management. In Proceedings of the 1997 ACM symposium on Applied computing—SAC '97, San Jose, CA, USA, 28 February–2 March 1997; ACM Press: San Jose, CA, USA, 1997; pp. 166–173.
13. Baumann, P.; Mazzetti, P.; Ungar, J.; Barbera, R.; Barboni, D.; Beccati, A.; Bigagli, L.; Boldrini, E.; Bruno, R.; Calanducci, A.; et al. Big Data Analytics for Earth Sciences: The EarthServer approach. *Int. J. Digit. Earth* **2016**, *9*, 3–29. [CrossRef]
14. Storch, T.; Reck, C.; Holzwarth, S.; Keuck, V. Code-De—the German Operational Environment for Accessing and Processing Copernicus Sentinel Products. In Proceedings of the 2018 IEEE International Geoscience and Remote Sensing Symposium (IGARSS), Valencia, Spain, 22–27 July 2018; pp. 6520–6523.
15. Gans, F.; Mahecha, M.D.; Reichstein, M.; Brandt, G.; Fomferra, N.; Permana, H.; Brockmann, C. The Earth in a Box: A light-weight data cube approach to empower the study of land-surface processes and interactions. *EGU Gen. Assem. Conf. Abstr.* **2018**, *20*, 9841.
16. Appel, M.; Lahn, F.; Buytaert, W.; Pebesma, E. Open and scalable analytics of large Earth observation datasets: From scenes to multidimensional arrays using SciDB and GDAL. *ISPRS J. Photogramm. Remote Sens.* **2018**, *138*, 47–56. [CrossRef]
17. Lewis, A.; Lacey, J.; Mecklenburg, S.; Ross, J.; Siqueira, A.; Killough, B.; Szantoi, Z.; Tadono, T.; Rosenavist, A.; Goryl, P.; et al. CEOS Analysis Ready Data for Land (CARD4L) Overview. In Proceedings of the 2018 IEEE International Geoscience and Remote Sensing Symposium (IGARSS), Valencia, Spain, 22–27 July 2018; pp. 7407–7410.

18. Committee on Earth Observation Satellites CEOS Analysis Ready Data for Land (CARD4L): Product Family Specification Optical Surface Reflectance (CARD4L-OSR) Version 4.0. Available online: http://ceos.org/ard/files/CARD4L_Product_Specification_Surface_Reflectance_v4.0.pdf (accessed on 15 June 2019).

19. Dwyer, J.L.; Roy, D.P.; Sauer, B.; Jenkerson, C.B.; Zhang, H.K.; Lymburner, L. Analysis Ready Data: Enabling Analysis of the Landsat Archive. *Remote Sens.* **2018**, *10*, 1363.

20. Sudmanns, M.; Tiede, D.; Lang, S.; Bergstedt, H.; Trost, G.; Augustin, H.; Baraldi, A.; Blaschke, T. Big Earth data: Disruptive changes in Earth observation data management and analysis? *Int. J. Digit. Earth* **2019**, 1–19. [CrossRef]

21. Pagani, G.A.; Trani, L. Data Cube and Cloud Resources as Platform for Seamless Geospatial Computation. In Proceedings of the 15th ACM International Conference on Computing Frontiers, Ischia, Italy, 8–10 May 2018; ACM: New York, NY, USA, 2018; pp. 293–298.

22. Goodman, S.N.; Fanelli, D.; Ioannidis, J.P.A. What does research reproducibility mean? *Sci. Transl. Med.* **2016**, *8*, 341ps12. [CrossRef]

23. Baraldi, A.; Tiede, D. AutoCloud+, a "Universal" Physical and Statistical Model-Based 2D Spatial Topology-Preserving Software for Cloud/Cloud–Shadow Detection in Multi-Sensor Single-Date Earth Observation Multi-Spectral Imagery—Part 1: Systematic ESA EO Level 2 Product Generation at the Ground Segment as Broad Context. *ISPRS Int. J. Geo-Inf.* **2018**, *7*, 457.

24. Matsuyama, T.; Hwang, V.S.-S. *SIGMA: A Knowledge-Based Aerial Image Understanding System*; Plenum Press: New York, NY, USA; London, UK, 1990.

25. Wilkinson, M.D.; Dumontier, M.; Aalbersberg, I.J.; Appleton, G.; Axton, M.; Baak, A.; Blomberg, N.; Boiten, J.-W.; da Silva Santos, L.B.; Bourne, P.E.; et al. The FAIR Guiding Principles for scientific data management and stewardship. *Sci. Data* **2016**, *3*, 160018. [CrossRef] [PubMed]

26. Sudmanns, M.; Tiede, D.; Lang, S.; Baraldi, A. Semantic and syntactic interoperability in online processing of big Earth observation data. *Int. J. Digit. Earth* **2018**, *11*, 95–112. [CrossRef] [PubMed]

27. Shannon, C.E. A Mathematical Theory of Communication. *Bell Syst. Tech. J.* **1948**, *27*, 379–423. [CrossRef]

28. Buckland, M.K. Information as thing. *J. Am. Soc. Inf. Sci.* **1991**, *42*, 351–360. [CrossRef]

29. Capurro, R.; Hjørland, B. The concept of information. *Annu. Rev. Inf. Sci. Technol.* **2003**, *37*, 343–411. [CrossRef]

30. Nazif, A.M.; Levine, M.D. Low Level Image Segmentation: An Expert System. *IEEE Trans. Pattern Anal. Mach. Intell.* **1984**, *PAMI-6*, 555–577. [CrossRef]

31. Hadamard, J. Sur les problemes aux derivees partielles et leur signification physique. *Princet. Univ. Bull.* **1902**, *13*, 49–52.

32. Bertero, M.; Poggio, T.A.; Torre, V. Ill-posed problems in early vision. *Proc. IEEE* **1988**, *76*, 869–889. [CrossRef]

33. Smeulders, A.W.M.; Worring, M.; Santini, S.; Gupta, A.; Jain, R. Content-based image retrieval at the end of the early years. *IEEE Trans. Pattern Anal. Mach. Intell.* **2000**, *22*, 1349–1380. [CrossRef]

34. Berners-Lee, T.; Hendler, J.; Lassila, O. The Semantic Web. *Sci. Am.* **2001**, *284*, 28–37. [CrossRef]

35. Heflin, J.; Hendler, J. A portrait of the Semantic Web in action. *IEEE Intell. Syst.* **2001**, *16*, 54–59. [CrossRef]

36. Poggi, A.; Lembo, D.; Calvanese, D.; De Giacomo, G.; Lenzerini, M.; Rosati, R. Linking Data to Ontologies. In *Journal on Data Semantics X*; Spaccapietra, S., Ed.; Springer: Berlin/Heidelberg, Germany, 2008; pp. 133–173.

37. Harvey, F.; Kuhn, W.; Pundt, H.; Bishr, Y.; Riedemann, C. Semantic interoperability: A central issue for sharing geographic information. *Ann. Reg. Sci.* **1999**, *33*, 213–232. [CrossRef]

38. Bahmanyar, R.; de Oca, A.M.M.; Datcu, M. The Semantic Gap: An Exploration of User and Computer Perspectives in Earth Observation Images. *IEEE Geosci. Remote Sens. Lett.* **2015**, *12*, 2046–2050. [CrossRef]

39. Santini, S.; Gupta, A.; Jain, R. Emergent semantics through interaction in image databases. *IEEE Trans. Knowl. Data Eng.* **2001**, *13*, 337–351. [CrossRef]

40. Baraldi, A. Vision Goes Symbolic Without Loss of Information Within the Preattentive Vision Phase: The Need to Shift the Learning Paradigm from Machine-Learning (from Examples) to Machine-Teaching (by Rules) at the First Stage of a Two-Stage Hybrid Remote Sensing. In *Earth Observation*; Rustamov, R., Ed.; IntechOpen: London, UK, 2012.

41. Baraldi, A.; Boschetti, L. Operational Automatic Remote Sensing Image Understanding Systems: Beyond Geographic Object-Based and Object-Oriented Image Analysis (GEOBIA/GEOOIA). Part 1: Introduction. *Remote Sens.* **2012**, *4*, 2694–2735. [CrossRef]

42. Di Gregorio, A.; Henry, M.; Donegan, E.; Finegold, Y.; Latham, J.; Jonckheere, I.; Cumani, R. *Land Cover Classification System: Advanced Database Gateway*; Software Version 3; FAO: Rome, Italy, 2016.

43. Baraldi, A.; Durieux, L.; Simonetti, D.; Conchedda, G.; Holecz, F.; Blonda, P. Automatic Spectral-Rule-Based Preliminary Classification of Radiometrically Calibrated SPOT-4/-5/IRS, AVHRR/MSG, AATSR, IKONOS/QuickBird/OrbView/GeoEye, and DMC/SPOT-1/-2 Imagery—Part I: System Design and Implementation. *IEEE Trans. Geosci. Remote Sens.* **2010**, *48*, 1299–1325. [CrossRef]

44. Baraldi, A.; Humber, M.L.; Tiede, D.; Lang, S. GEO-CEOS stage 4 validation of the Satellite Image Automatic Mapper lightweight computer program for ESA Earth observation level 2 product generation—Part 1: Theory. *Cogent Geosci.* **2018**, *4*, 1–46. [CrossRef]

45. Tiede, D.; Baraldi, A.; Sudmanns, M.; Belgiu, M.; Lang, S. Architecture and prototypical implementation of a semantic querying system for big Earth observation image bases. *Eur. J. Remote Sens.* **2017**, *50*, 452–463. [CrossRef]

46. Dumitru, C.O.; Cui, S.; Schwarz, G.; Datcu, M. Information Content of Very-High-Resolution SAR Images: Semantics, Geospatial Context, and Ontologies. *IEEE J. Sel. Top. Appl. Earth Obs. Remote Sens.* **2015**, *8*, 1635–1650. [CrossRef]

47. Li, Y.; Bretschneider, T. Semantics-based satellite image retrieval using low-level features. In Proceedings of the 2004 IEEE International Geoscience and Remote Sensing Symposium (IGARSS), Anchorage, AK, USA, 20–24 September 2004; Volume 7, pp. 4406–4409.

48. Sudmanns, M.; Tiede, D.; Wendt, L.; Baraldi, A. Automatic Ex-post Flood Assessment Using Long Time Series of Optical Earth Observation Images. *Gi_Forum* **2017**, *1*, 217–227. [CrossRef]

49. Augustin, H.; Sudmanns, M.; Tiede, D.; Baraldi, A. A Semantic Earth Observation Data Cube for Monitoring Environmental Changes during the Syrian Conflict. *Gi_Forum* **2018**, *1*, 214–227. [CrossRef]

50. Flood List News Iraq and Syria—Flooding Hits Syria Refugee Camps, Displaces Thousands Near Tigris River in Iraq—FloodList. Available online: http://floodlist.com/asia/iraq-and-syria-floods-march-april-2019 (accessed on 30 May 2019).

51. Tiede, D.; Baraldi, A.; Sudmanns, M.; Belgiu, M.; Lang, S. ImageQuerying—Earth Observation Image Content Extraction & Querying across Time and Space. In Proceedings of the 2016 Conference on Big Data from Space (BiDS'16), Santa Cruz de Tenerife, Spain, 15–17 March 2016; pp. 192–195.

52. Willcocks, L.P.; Mingers, J. *Social Theory and Philosophy for Information Systems*; John Wiley & Sons Ltd.: Chichester, UK, 2004; ISBN 978-0-470-85117-3.

53. Pekel, J.-F.; Cottam, A.; Gorelick, N.; Belward, A.S. High-resolution mapping of global surface water and its long-term changes. *Nature* **2016**, *540*, 418. [CrossRef]

54. Tulbure, M.G.; Broich, M.; Stehman, S.V.; Kommareddy, A. Surface water extent dynamics from three decades of seasonally continuous Landsat time series at subcontinental scale in a semi-arid region. *Remote Sens. Environ.* **2016**, *178*, 142–157. [CrossRef]

55. Mueller, N.; Lewis, A.; Roberts, D.; Ring, S.; Melrose, R.; Sixsmith, J.; Lymburner, L.; McIntyre, A.; Tan, P.; Curnow, S.; et al. Water observations from space: Mapping surface water from 25 years of Landsat imagery across Australia. *Remote Sens. Environ.* **2016**, *174*, 341–352. [CrossRef]

56. Tiede, D.; Sudmanns, M.; Augustin, H.; Lang, S.; Baraldi, A. Sentinel-2 Semantic Data Information Cube Austria. In Proceedings of the 2019 Big Data from Space (BiDS'19), Munich, Germany, 19–21 February 2019; Soille, P., Loekken, S., Albani, S., Eds.; Publications Office of the European Union: Brussels, Belgium, 2019; pp. 65–68.

Concept Paper

A Transformative Concept: From Data Being Passive Objects to Data Being Active Subjects

Hans-Peter Plag [1,2,4,*] and Shelley-Ann Jules-Plag [3,4]

[1] Department of Ocean, Earth, and Atmospheric Science, Old Dominion University, Norfolk, VA 23529, USA
[2] Mitigation and Adaptation Research Institute, Old Dominion University, Norfolk, VA 23529, USA
[3] Engineering, Management and System Engineering, Old Dominion University, Norfolk, VA 23529, USA; julesplag@gmail.com
[4] Tiwah UG, 53547 Rossbach, Germany
* Correspondence: hpplag@tiwah.com

Received: 28 July 2019; Accepted: 29 September 2019; Published: 2 October 2019

Abstract: The exploitation of potential societal benefits of Earth observations is hampered by users having to engage in often tedious processes to discover data and extract information and knowledge. A concept is introduced for a transition from the current perception of data as passive objects (DPO) to a new perception of data as active subjects (DAS). This transition would greatly increase data usage and exploitation, and support the extraction of knowledge from data products. Enabling the data subjects to actively reach out to potential users would revolutionize data dissemination and sharing and facilitate collaboration in user communities. The three core elements of the transformative DAS concept are: (1) "intelligent semantic data agents" (ISDAs) that have the capabilities to communicate with their human and digital environment. Each ISDA provides a voice to the data product it represents. It has comprehensive knowledge of the represented product including quality, uncertainties, access conditions, previous uses, user feedbacks, etc., and it can engage in transactions with users. (2) A knowledge base that constructs extensive graphs presenting a comprehensive picture of communities of people, applications, models, tools, and resources and provides tools for the analysis of these graphs. (3) An interaction platform that links the ISDAs to the human environment and facilitates transaction including discovery of products, access to products and derived knowledge, modifications and use of products, and the exchange of feedback on the usage. This platform documents the transactions in a secure way maintaining full provenance.

Keywords: data discovery; metadata; knowledge base; graph data; intelligent semantic agents

---

1. Introduction

The current conceptual approach for discovery of Earth observation (EO) data and derived products is to a large extent based on a perception of data as passive objects. Extracting information and creating new knowledge from data often requires a high level of expertise. Users have to engage in often tedious search processes to discover data. Missing metadata reduce the chance to match data to requirements and determine applicability. Utilizing the data for research most often involves lengthy processes to access products and translate them into a format suitable for the purpose. For decision support, the high level of expertise required to extract information from data is a major obstacle. Feedback on the usability of data for different applications is mostly not collected and not available to users searching for data and knowledge. Semantic issues hamper discoverability and reduce usability of the data and products. Users who would benefit from collaborations often discover potential collaborators by chance. Linking of users with similar interests happens in social networks disconnected from data discovery and access tools. As a result, exploitation of Earth observations (EOs) in Earth sciences is at a level much lower than desirable and feasible. The use of products and knowledge derived from Earth observations (EOs) for decision and policy making is also hampered by the level of expertise required to extract relevant information from data products and by the limited discoverability.

Currently, the challenges to the discovery, access and use of the increasingly comprehensive Earth observation (EO) data greatly limit the exploitation of the potential societal benefits of this global resource. In fact, the value of Earth observation (EO) as a 'public good' depends mainly of the conditions of access to that good [1]. At the same time, humanity is facing growing global threats, see, e.g., [2,3]. Humanity's quest for sustainable development expressed in the United Nations' Agenda 2030 [4] is hampered by a lack of information on the biosphere and humansphere, and much of this information could be extracted from Earth observations (EOs) [5]. Developing the interventions that can facilitate progress towards the seventeen Sustainable Development Goals (SDGs) set in the Agenda 2030 and monitoring progress toward the associated Targets requires comprehensive input from Earth observation (EO) communities, see, e.g., [6,7]. Sustainable development as defined in the Agenda 2030, as well as, developing sustainability in general requires a scientific paradigm shift toward systems thinking [8] and this transition has to be informed by comprehensive integrated Earth observation (EO) data. The current description of globally connected systemic and catastrophic risks captures poorly the role of human-environment interactions [9], and this creates a bias towards solutions that often ignore the new realities of the Anthropocene [10]. Understanding "Anthropocene risks", i.e., risks that emerge from human-driven processes, interact with global social-ecological connectivity, and exhibit complex, cross-scale relationships [10], requires full and easy access to information that can be derived from Earth observation (EO) data and tools for the extraction. The large human-caused changes in the planetary physiology carry the risk of unexpected new phenomena with potentially global consequences and threats [9]. Examples are the emerging threats of sargassum blooms [11], the potential existence of a tipping points for a trajectory towards a "hothouse climate" [12], the possibility of ocean anoxic events [9], and the potential overload of the ocean with carbon [13].

Assessments of risks in general and "Anthropocene risks" in particular very often show a tendency to assume that the large risks are more likely in the far future [14]. For example, a potential state shift in the biosphere [15], reaching tipping points for a hothouse trajectory [12], or the overload of the ocean with carbon [13], etc., are all very often considered as a possibility in the far future, thus ignoring that there are potential hidden risks that could trigger such catastrophic events in the near future. Assessing risks, developing interventions to address the threats today and having early warnings concerning hidden risks also need full access to comprehensive Earth observations (EOs) to address the many knowledge gaps regarding catastrophic risks and to inform interdisciplinary and transdisciplinary mapping and tracking of the multitude of factors that could contribute to global catastrophic risks [16]. In the light of the challenges modern society is facing and the enormous value easy access to comprehensive and integrated Earth observation (EO) data and derived information would have for addressing these challenges, it seems imperative to transform the current relationship between data and users [5]. Thus, the goal of utilizing the societal benefits of Earth observations (EOs) has to be a major design criterion for systems that manage and provide access to such data.

1.1. Meeting Societal Data and Knowledge Needs

Over several decades, Earth observation (EO) communities have made efforts to increase the realization of the societal benefits of Earth observation (EO). The Integrated Global Observing Strategy (IGOS) initiated by the G7 in 1984 as a framework for Earth observations (EOs) was developed with the goal to identify what was essential to be observed in order to document comprehensively the changes that are happening on the planet [17]. The Integrated Global Observing Strategy Partnership (IGOS-P) was established in 1998 with the mandate to ensure that Earth observations (EOs) would respond to societal needs. This partnership brought together major organizations in the scientific and Earth observation (EO) fields and engaged in efforts to first identify what needs to be monitored and then to facilitate the implementation of corresponding observing systems. IGOS-P used a well-defined theme approach to define the overall strategy, with the themes being motivated by real-world challenges [18]. The resulting IGOS-P theme reports documented very well the outcomes of the first step defining from observational needs for societally relevant themes, see, e.g., [19–23]. However, IGOS-P was less successful in the second step.

Already the Agenda 21 [24], which wasa result of the World Summit in Rio in 1992, emphasized the need for coordinated Earth observations and for the creation of knowledge that would support decisions for sustainable development. The World Summit on Sustainable Development in Johannesburg in 2002 reconfirmed the need for coordinated Earth observations, and this led in 2003 to the initiation of the ad hoc Group on Earth Observations (GEO) with the task to develop in eighteen months an implementation plan for the Global Earth Observation System of Systems (GEOSS). The outcome of this activity resulted in 2005 in the establishment of Group on Earth Observations (GEO). The vision of GEO is a future where decisions can be informed by Earth observations. Considering the spectrum of challenges and threats to our global civilization, this is no longer a nice-to-achieve vision; it is a necessity for survival. For GEO, the tool for making progress towards this vision is Global Earth Observation System of Systems (GEOSS). Initially, GEOSS was intended to be integrated into an end-to-end feedback loop with GEOSS providing data and information in support of decision making and users providing feedback on information needs for the further development of GEOSS (Figure 1). Importantly, this initial concept included for GEOSS the task of integrating Earth observation (EO) data with other data and the use of Earth system models to generate the information and knowledge required by societal decision makers.

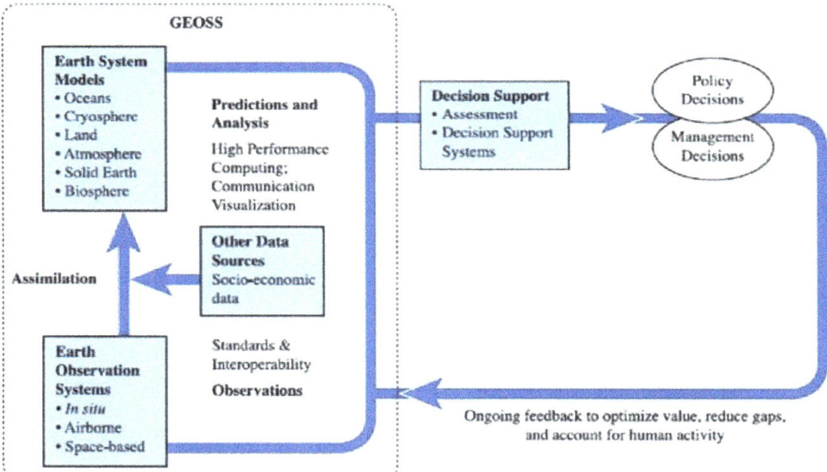

Figure 1. The initial concept for global Earth observation system of systems (GEOSS) emphasized its aim to inform decision making through an end-to-end feedback loop of data and knowledge supporting decision making and feedback from users informing the development of GEOSS. GEOSS was intended to integrate Earth observation (Earth observation (EO)) data with other data and Earth system models to provide the information needed for decision and policy making [25].

In the first ten years of GEO, considerable efforts were made on the feedback part of the loop to improve the knowledge of societal needs in support of defining EO priorities, both in communities of practice that mostly originated in IGOS-P themes, see, e.g., [26], and dedicated efforts to gain overviews of observational requirements derived from societal needs, see, e.g., [27–30]. For the development of GEOSS, the main effort was on improving data discoverability, availability, and accessibility, while the integration with other data and models had much lower priority. As a result, GEOSS up to today serves best expert communities who have the capacity to search, access, and process the data. Efforts to combine data with a knowledge base remain at an early conceptual state. As recent as 2019, a new concept paper has been accepted by the GEO Executive Committee that proposes the development of a GEOSS Knowledge Hub mainly for expert communities as a framework for transforming Earth observation (EO) data to knowledge for decision making [31]. On the other hand, participatory workshops bringing Earth observation (EO) and science communities together with societal stakeholders again and again reveal that there is a lack of capacity outside relatively small expert communities for the extraction of information from Earth observations (EOs), see, e.g., [32].

Considerable efforts have been made to measure the potential and actual societal benefits of Earth observations (EOs). For example, from 2009 to 2011, a community effort led by NASA aimed at an assessment of societal benefits of Earth observations (EOs) as a basis for the prioritization of Earth observation (EO) systems [27,33]. For several years, the GEO Work Programme included a Fundamental Task on Societal Benefits organizing a sequence of workshops addressing the assessment of societal benefits of Earth observations (EOs). NASA has set up the "VALUABLES" collaboration to measure how satellite information benefits people and the environment when it is used to make decisions [34]. However, very often the results of these assessments are published in reports and not easily available in digital format to link benefit-based knowledge needs to observational requirements.

New societal knowledge needs emerged in 2015 with the United Nations' adoption of the 2030 Agenda for Sustainable Development [4], the adoption of the Sendai Framework for Disaster Risk Reduction 2015–2030 [35] by the United Nations, and the Paris Climate Agreement reached under the United Nations Framework Convention on Climate Change (UNFCCC). GEO has responded to the emergence of these agreements by including the support for the UN 2030 Agenda for Sustainable Development, the Paris Climate Agreement, and the Sendai Framework for Disaster Risk Reduction in the global priorities. Likewise, several United Nations agencies give the support of these frameworks high priority. Among others, the urgent need for a transformative digital ecosystem for the environment is emphasized by [5] to ensure that progress towards sustainability is informed by data.

Considering the example of the 2030 Agenda, the development and validation of interventions to reach the many targets associated with the seventeen Sustainable Development Goals (SDGs) pose wicked problems to society. Wicked problems are social or cultural problems that are difficult or impossible to solve because of incomplete and often contradictory knowledge, the large number of people and opinions involved, the heavy economic burden associated with progress towards a solution, and the interconnected nature of each problem with many other problems [36]. All of this applies to the Sustainable Development Goals (SDGs). In particular, knowledge on how to make progress towards the Sustainable Development Goals (SDGs) is incomplete and contradictory, reaching the SDGs even on a local level involves the whole of society, making progress requires a rethinking of economy [37], and the goals are strongly interconnected, see, e.g., [38–40]. Moreover, there are many interactions between the individual goals that are variable across different economic, social, and cultural settings [7].

Monitoring progress towards the targets associated with the Sustainable Development Goals (SDGs) requires metrics defined by a set of indicators, and developing indicators that provide useful quantitative metrics is a long process involving the scientific community, see, e.g., [41,42]. The United Nations Statistical Commission (UNSC) created the Inter-Agency and Expert Group on SDG Indicators (IAEG-SDGs) with the aim to develop a manageable indicator framework. Based on a proposal of the IAEG-SDGs, an initial framework with a total of 232 global indicators was adopted in 2017 by the United Nations General Assembly as a voluntary and country-led endeavor to monitor progress towards the SDG Targets. According to the level of data availability and methodological development, the SDG Indicators have been grouped in three different Tiers: From Tier I, for the ones having an established methodology and widely available data, to Tiers II and III, for those not having data available or no methodology established, respectively. As of 11 May 2018, the updated tier classification contains 93 Tier I indicators, 72 Tier II indicators, and 62 Tier III indicators [43]. However, actually being able to quantify these indicators for individual countries poses an insurmountable challenge to small countries like the Small Island Developing States (SIDS) and those countries with very limited economic resources. Many of the indicators depend very much on Earth observations (EOs) and an integration of Earth observations (EOs) with other socioeconomic data and models [6,7,44,45].

Many efforts have focused on archiving and publishing datasets. An example is the World Data Center PANGAEA [46], which is a member of the ICSU World Data System. PANGAEA provides services for archiving, publishing, and re-usage of data [46]. Most of the datasets are open access, and a search engine provides a high level of discoverability. However, being a repository, the dataset are passive objects and extracting information from a dataset requires accessing the data and using expertise in the analyses of the data. The datasets are structured under a set of themes and sub-themes, which limits transdisciplinary approaches.

Efforts are also being made to utilize relationships between datasets and products to increase data discoverability and utilization. For example, the Linked Open Data Cloud (LODC) captures the relationships between an increasing number of datasets [47]. As of March 2019, the dataset contains 1239 datasets with 16,147 links. More datasets can be registered manually and links can be recorded. The LODC generates domain specific sub-clouds. Users can interactively explore the cloud to retrieve information of specific datasets or explore the relationships captured in the links. The full LODC is available for analyses. However, links to other objects such as applications, user types, processing tools, etc., are not comprehensively captured and feedback on the datasets is not solicited.

Recommender systems that would promote datasets and products to potential users are very limited in the Earth observation (EO) community. However, recommender systems are increasingly used for the promotion of commercial products. Commercial retailers increasingly use advanced algorithms including big data analyses, deep learning, deep search, and crowd-sourcing to bring their products to potential customers. In the early use of the Web, customers often had to carry out lengthy searches over limited domains to discover the products and services they were looking for, a conceptual approach that is denoted here as Customers Discover Products (CDP). The recent development in the commercial domain constitutes a transition to a conceptual approach where a framework enables products to discover potential customers, a concept denoted here as Products Discover Customers (PDC). Customers of, e.g., Amazon are informed when new books and other products appear on the market that might be of interest for them based on previous searches or purchases. Recommender systems have been developed and deployed in supermarkets to aid customers in decisions of what to choose from the large variety of products, see, e.g., [48]. Web advertisements are targeted to likely recipients based on social media behavior or Web searches. In Products Discover Customers (PDC), data from social media are increasingly collected and analyzed to explore connections among people and between people and products to propose and facilitate new connections. Extensive feedback on products and services is collected from customers and users and made available to inform decisions of other customers and users. In some cases, attempts are made to stimulate feedbacks with rewards, e.g., when hotels have very low numbers of reviews, Hotels.com offers coupons for special nights in return for reviews, and feedbackrewards.com manages for companies customer feedback programs using rewards for stimulating feedback [49].

Recent artificial intelligence (AI) developments have opened the door for intelligent software agents, see, e.g., [50,51]. Theoretical concepts have been developed to capture connections between societal agents, products, tools, activities, and transactions, and to construct graph data describing the chains and networks between these elements.

1.2. From Passive Data Objects to Active Data Subjects

The ability to design intelligent software agents that can represent a data product and provide comprehensive information derived from this product, combined with the ability to construct extensive graph data provides a basis for a transition in the Earth observation (EO) domain from the perception of Data as Passive Objects (DPO) to a perception of Data as Active Subjects (DAS). The DAS concept has the overarching goal to greatly increase data usage and exploitation. It has the potential to revolutionize data discovery, sharing, dissemination and usage and by doing so greatly enhance the exploitation of Earth observations (EOs) for research and the realization of societal benefits. In contrast with the current DPO concept, in which datasets are passive and isolated in repositories, the DAS approach pairs datasets with intelligent software data agents that can connect and interact with other software and human agents. These software data agents are comparable to human agents who provide links between people (such as actors, musicians, etc.) and potential jobs. Similar to those human agents for people, the software data agents have full knowledge about the dataset(s) they represent, including among others comprehensive metadata as well as information on usability and applicability, and they have the ability to discover potential applications and users for their datasets(s).

The subject does the action. The object is the center of action. In the DPO perception, e.g., researcher X analyzed the global temperature data to quantified global warming. In the DAS perception, the global temperature dataset Y would inform that global heating has reached 0.1 °C per decade. In the first case, the temperature data is the object. In the second case, the data is the subject and this subject informs about knowledge it could extract from its data.

Another example would be a minister in a government who is in need to quantify one of the indicators for the SDGs. In the DPO world, the minister could have to engage a team of experts to discover and collect the relevant data, use appropriate processing tools, and, following a best practice, generate the quantitative indicator. In this case, all data used would be objects and even the indicator would be an object. However, in the DAS world, there would be a software agent representing this indicator, and this agent could inform the minister of the quantitative development of the indicator in the minister's country. This would be of great value particularly for the smaller and less resourceful countries such as the SIDS, see, e.g., [52].

Having active data-based subjects, these subjects also could have the capability to promote their data and knowledge to societal human agents who would benefit from this. Today, the dominating concept for data distribution is one of Users Discover Data (UDD). Within the Data as Active Subjects (DAS), a transition to a new concept of Data Discover Users (DDU) would be possible. This would be comparable to the ongoing transition in the commercial world mentioned above from Customers Discover Products (CDP) to Products Discover Customers (PDC).

1.3. Structure of The Paper

In the next section, the DAS concept is outlined in more detail. After an overview, three subsequent subsections discuss the three core elements of this concept, i.e., the Intelligent Semantic Data Agents (ISDAs) that are representing datasets, products and services (Section 2.2), the knowledge base that creates and provides access to extensive graph data (Section 2.3), and the interaction platform on which human users and ISDAs interact (Section 2.4). Section 3.1 explores the potential of DAS not only in terms of increased data exploitation but also in terms of capacity building, decision and policy making, and realization of societal benefits of Earth observations (EOs) and derived knowledge. Section 3.2 outlines a case study for the validation of the concept, and Section 3.3 provides thoughts on the implementation and identifies challenges for the implementation of DAS. Section 4 summarizes the main conclusions.

2. The DAS Concept

2.1. Overview

The overarching design criterion for the DAS concept (Figure 2) is the goal of enabling data products to actively respond to information and knowledge needs of societal users and to reach out to those who may benefit from knowing about a data product and having access to the product or information derived from the product. To some extent, this change in perception of data objects is comparable to the one from considering cars as passive objects that are driven by humans to cars as active subjects that provide transportation to humans and other objects as needed. In the same way as autonomous cars may lead to a Gestalt shift [53] in how we perceive transportation, the transition to perceiving data as active subjects could lead to a Gestalt shift in how we perceive knowledge derived from data.

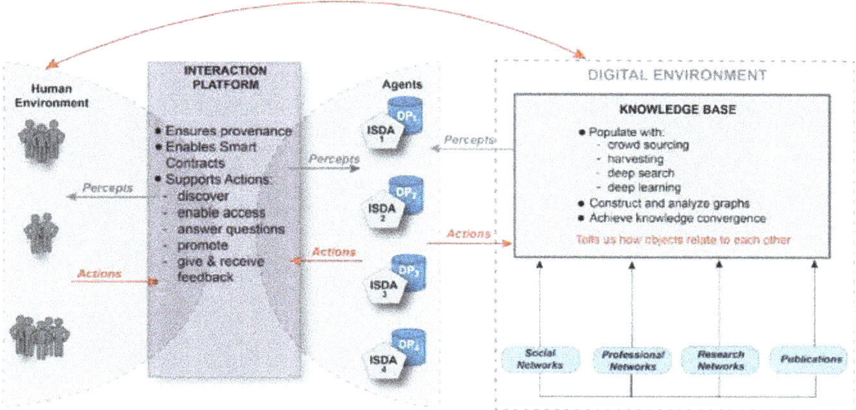

Figure 2. In the data as active subjects (DAS) concept, each intelligent semantic data agent (ISDA) represents a data product (DP). The ISDAs utilize the graph data in a knowledge base to discover applications and users that could benefit from their data products. They interact with those users, or users that contact them, to provide knowledge or manage access to data. All interactions that impact the data are recorded to ensure provenance. The knowledge base generates graph data based on information obtained through crowd sourcing or extracted from social and research networks and publications.

The DAS concept introduced here hinges on three core elements (Figure 2):

1. Intelligent Semantic Data Agents (ISDAs) that are software agents that represent data products. They have the goal to serve potential users and to increase the exploitation of the societal benefits of the data product they represent. To achieve this, an ISDA has comprehensive knowledge about the data product it represents including quality, uncertainties, access conditions, previous uses, user feedbacks, etc. These non-human software agents have the semantic capabilities to communicate with potential users in the human environment and comprehensive graph data in the knowledge base. The ISDAs also have semantic and pragmatic descriptors that allow them to meaningfully interconnected with software agents of other datasets through complex and dynamic relations. These relations are continuously updated as users interact with the data agents and provide feedback on the data.

2. A knowledge base that can construct and analyze extensive graphs presenting a comprehensive picture of the elements in a community of people, applications, models, tools, and resources. Earth observation (EO) data is mostly polyglot spatial data representing properties at points, lines, or polygones in space and their changes over time (Figure 3). Graph data captures the connections between objects and can consist, e.g., of property graphs linking persons, network graphs linking locations, semantic graphs linking language elements in ontologies, and more generalized graphs linking diverse objects such as data sets, information needs, and societal agents. Polyglot data are helpful in answering questions such as "how did land cover change over time at this point?" Graph data can answer questions such as "which researcher could benefit from land cover data?" The knowledge base will focus on graph data providing links between, e.g., knowledge needs and data types, user types and applications, publications and datasets, processing tools and datasets. None of the objects linked in the graph data resides in the knowledge base.

3. An interaction platform to negotiate and execute "contracts" under which users gain access to knowledge extracted from data, access data, modify data, use data and provide feedback on their usage, and to document these interactions in a secure and reliable way maintaining full provenance.

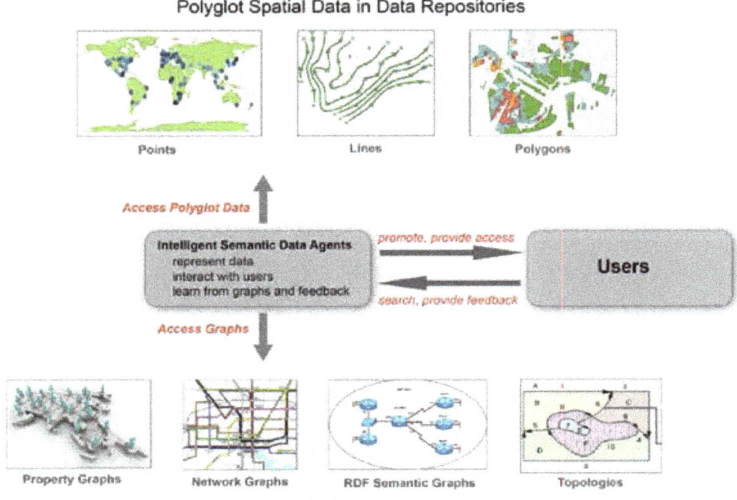

Figure 3. In the DAS concept, graph data capturing the properties and connections in diverse networks (people, applications, models, datasets) are used by data agents representing data to match users and data both on request (searches) and through promotion. The data agents "learn" from user feedback and dynamically adjust to changes in the graphs.

In the DAS concept, datasets and products derived from Earth observations (EOs) are associated with the Intelligent Semantic Data Agents (ISDAs) that can communicate semantic information in response to queries including access conditions, derived knowledge, quality, uncertainties, guidance on applicability, and user feedback. Conceptually, these ISDAs utilize the graph data in the knowledge base to explore the user landscape in search for users that might have interest in the data (Figure 2). They can interact with users as well as other ISDAs. An ISDA will also have knowledge about tools that can make use of the data or derive other products from the data. The sharing of this knowledge with users facilitates rapid capacity building in the use of the data and broadens the range of scientific applications of the data represented by the ISDA. Thus, the DAS concept provides remedies to many of the current issues associated with a perception of passive data objects paired with passive metadata that often are maintained separately from the actual data. All interactions with a data agent are either integrated into the agent as an innate part or recorded in the provenance system.

The knowledge base in the DAS concept uses deep searches, big data analyses and crowd sourcing to map for specific use cases the user landscape in the communities engaged in research and applications and to identify their knowledge and information needs. Based on deep searches and deep learning, graphs of user types, what they do, their tools, and their potential needs are constructed from publications, social networks, social media communications, and observation inventories. The graph data are analyzed to enable the ISDAs to promote their data products to users with potentially matching interests and needs.

The ISDAs utilize the interaction platform for communication and interactions with users. This platform provides a system that tracks interactions with users, ensures provenance and increase reproducibility of research that is based on the represented data. The matching of users and data products takes place on this interaction platform, which will ensure provenance. The interactions are handled with an approach similar to smart contracts. Searches and feedbacks are analyzed by the knowledge base to update graphs and by the ISDAs to add intelligence to the ISDAs and to enable them to identify new potential use cases for the data they represent.

2.2. Intelligent Semantic Data Agents

The introduction of the software Intelligent Semantic Data Agents (ISDAs) (Figure 2) is a concept that has the potential to revolutionize the interaction of users and data. The principle idea is comparable to the human agent of, e.g., a movie star, who has the task to promote the actor and to negotiate new engagements for the actor. Ideally, the human agent has all relevant information about the actor, including past engagements, preferred partners, limitations, and preferences, and fully understands the capabilities of the actor. Similarly, an ISDA has all relevant information about a dataset, including comprehensive provenance, related datasets, models and applications to be used by users, user types that might be interested, applicability and limitations, quality and uncertainties, and more. The ISDA has the task to promote the dataset actively to potential users (thus making progress toward the Data Discover Users (DDU) concept), to respond to queries, to inform about the dataset, to provide derived information (e.g., selected statistics, subsets, etc.), receive feedback from users, and to learn from user interactions to be better prepared for future users.

From a semantic point of view, the knowledge base will formulate the semantics of the domain, such that each data product has a meaning attached to it. However, it will go beyond the semantics of datasets to a pragmatic approach, in which a data product is represented by an agent that is aware of the data product's meaning and is capable of learning potential use cases of the data product. Thus, data products will be represented by agents (the ISDAs) that can act on knowledge within the knowledge base and generate new knowledge.

Data products present in the graphs of the knowledge base will be represented by ISDAs that act on their behalf. The ISDAs are purposive software agents whose aim is to facilitate the interaction between users and the data product. In particular, an ISDA will be able to respond to questions about its data product, provide access to parts or all of the data product, and solicit feedback on the data product. Initially, the ISDAs will be goal-based agents [50,51] but they will have to evolve into learning agents. The ISDAs can request specific analytics from the knowledge base to discover potential users and to enter into communication with them. In particular, it can find users with the skills and interest to use the data or who might need these data to corroborate a published study, even if these potential users did not know of the existence of the data. The ISDAs will be able to use the social media and contact information of users in the knowledge base to enter in communication with them. A core research question on the path to implementation is how rich the data description will have to be to enable these capabilities.

The ISDAs are capable of executing complex transaction patterns with users, such as granting access, executing custom queries to aggregate, truncate, convert, randomly sample data, and provide references or meta-data. For that, the agents will adopt a transaction processing framework to manage its interactions with other agents and users [54].The concept of rough set [55–58] can be considered as a capability of the ISDAs.

The ISDAs will be able to grow from initial "seeds" with very limited capabilities into fully developed "adult" agents that have access to all the information related to the dataset, including all uses, experiences, feedbacks. Thus, the agents gain in knowledge as the knowledge base becomes more complete. A deep-learning algorithm will be used to further enrich the information available to an ISDA about the represented dataset so that it can link to users with potentially matching interests and needs and inform users about products of potential interest to them, including the data sharing and access conditions.

The ISDAs will also benefit from a generalization of the concept of digital object identifier which comprehensively identifies a dataset including the relevant metadata, the ISDA, and derived datasets in a consistent identification scheme. Having the main identifier pointing to the ISDA instead of the dataset itself will ensure that a user who aims to access the data always will have access to the full history of transformations and applications of the data.

2.3. The Knowledge Base

The knowledge base is envisioned as an extended version of the existing Socio-Economic and Environmental Information Needs Knowledge Base (SEE-IN KB), which has the main function to construct and analyze graph data capturing the connections between datasets, products, applications, user types, and other elements in scientific communities and society at large. To the extent permissible under privacy and personal data protection regulations (such as the European General Data Protection Regulation (GDPR)), individual persons can be integrated into graph data. This knowledge base provides the graph data and analytical tools to connect users and facilitate collaborations.

Graph data consists of two basic elements: The nodes (or vertices), and the links (or edges) between these nodes. Both the nodes and links are objects that are characterized by a set of properties. Each link is associated with two nodes. Links can be directional with head and tail nodes or bidirectional. In the Socio-Economic and Environmental Information Needs Knowledge Base (SEE-IN KB), the nodes are not limited in terms of what objects can constitute a node. For example, nodes can be as diverse as a specific person, a group or type of humans (e.g., a user type), a dataset, an information need, a societal goal, a modeling software, or a specific observation sensor. The set of properties for each class of nodes and links is dynamic and can be extended as more information about an object becomes available. Importantly, each node and link has a unique identifier.

The knowledge base uses big data analysis techniques to map the user landscape in the communities engaged in research and applications and identify their knowledge and information needs. It generates graph data that describe user types and their potential needs based on publications and social media communications and links them to tools and datasets. In utilizing published information on persons, such as paper authorship and owners of data and processing tools, it will be important to ensure compliance to privacy and personal data protection regulations, such as the General Data Protection Regulation (GDPR). Individual persons can be integrated as nodes into the graphs. During the development of the Global Earth Observation System of Systems (GEOSS) User Requirements Registry (URR), which initially only captured user types, users of the User Requirements Registry (URR) repeatedly requested the possibility to link themselves to user types and establish a social network of users within the User Requirements Registry (URR) [30]. It is expected that similar requests are made for the knowledge base. The knowledge base also maps the Earth observation (EO) landscape in terms of available datasets, products, and processing tools. The research communities are being mapped in terms of research topics, needs, and challenges, as well as the tools available to process and analyze data and to use data for modeling and simulation. An important source for mapping research communities is the comprehensive publication and citation data compiled in rapidly expanding research knowledge hubs. Increasingly, journals require information on data and tools used for the research published in a paper, see, e.g., [59]. This information can be exploited to inform the construction of graph data and to increase the knowledge and skills of the ISDAs. The development of the graph data also is based on deep searches and deep learning from scholarly and other publications, social networks, etc. In particular, the knowledge base will employ parallel crawlers to inform the construction of graph data.

The knowledge base requires the capability to provide the information needed to bring data and products to potential users. This capability has to be based on the full spectrum of graph theory. This includes the detection of components and communities applying, e.g., the deep search algorithms depth-first search (DFS) [60] and Kosaraju, see, e.g., [61], and the concept of weakly connected components, label propagation, and spacification [62]. Evaluating community structures can focus on conductance, modularity, and clustering coefficients [63], and this provides a basis to identify collaboration potentials between research groups and individuals. Ranking and walking along graphs provides a basis for prioritization as well as discovery of relevant nodes in support of data promotion and can be based on algorithms applying pageranks and different centralities, see, e.g., [64], random walking and sampling. Path-finding facilitates the identification of users who's requirements could be a match for a dataset, applying, e.g., Dijkstra's [65] and Bellman-Ford's [61] algorithms. Importantly, detection of unreliable or fake information [66,67] has to be integrated into the graph development processes.

The Socio-Economic and Environmental Information Needs Knowledge Base (SEE-IN KB) provides extensive search and feedback utilities and the analysis of both searches and feedbacks with deep learning methods can further improve the capability to add intelligence to the Intelligent Semantic Data Agents (ISDAs). Crowd-sourcing opportunities can be used to gather both primary graph data and feedback on data and the performance of the ISDAs. The lexicon (ontology) contained in the Socio-Economic and Environmental Information Needs Knowledge Base (SEE-IN KB) as the primary source for all semantic aspects will grow based on deep learning from other registries and from user interactions. The Socio-Economic and Environmental Information Needs Knowledge Base (SEE-IN KB) provides access to a large set of user needs (originally collected in the Global Earth Observation System of Systems (GEOSS) User Requirements Registry (URR) [68,69]) and observational requirements (partly harvested from OSCAR, see http://www.wmo-sat.info/oscar). The Socio-Economic and Environmental Information Needs Knowledge Base (SEE-IN KB) explores existing and new data repository in an effort to link Earth observations (EOs) and the global community of potential users.

Big data analytics on the graph data in an extended version of the Socio-Economic and Environmental Information Needs Knowledge Base (SEE-IN KB) is at the core of the DAS concept. In the current DPO concept datasets are passive and isolated in repositories. In contrast to this, the DAS approach will create the graph data of a "Web of things" where each dataset will be represented by a node with semantic and pragmatic descriptors, and meaningfully interconnected with the other entities (other datasets, users, models, instruments, etc.) through complex and dynamic relations, which will be updated as users and ISDAs interact with the graph data and provide feedback.

The graph data requires a generic model for metadata (referred to below as metamodel) that enables the networked representation of a population of entities and their mutual relations. Since the system is open-ended, and the final extent of all datasets that may be added is not known at inception, it would be illusive to attempt to create a fixed and comprehensive ontology that would encompass every future addition of datasets in the knowledge base.

A dataset provides a partial, biased, and time-bounded description of an object of interest in the real world. This means that the dataset expresses a reference in a semiotic relationship that involves the real world object as a referent, and the specific form of the data as symbol. The data provider and data users relate with the dataset both at a semantic level to uncover the meaning expressed in it, and also at a pragmatic level to achieve some practical ends, communicative or otherwise. In this sense, datasets seem to be more complex objects to manipulate and recommend automatically than products on Amazon or videos of Youtube. Even the individuation of the real-world object to which the data is pointing is subject to the researcher's interests and underlying theories or a user's preconceptions and world view. Similarly, the characteristics of the object represented by the dataset depend on the technical means of observation, on the methodology adopted, and on the level of fidelity decided by the data provider.

Other aspects to be covered in the DAS approach involve the origin of the data (what actors made it available), how it was obtained, for instance, whether the measurement is punctual or longitudinal, whether the data originated from a model (and what kind of model), a survey, observations (what kind of sensor), and what use-cases the data can support. The Socio-Economic and Environmental Information Needs Knowledge Base (SEE-IN KB) will also have to enforce integrity rules through mechanisms like reputation management, voting, and read/copy/write access rules, to make sure that datasets are not tampered with, and that single source of truth principles are maintained for every given data entity.

An important step towards the implementation of the DAS concept is the introduction of an extensible metamodel that covers these aspects of the graph data, so that the Intelligent Semantic Data Agents (ISDAs) initiated by data providers may represent their associated datasets as precisely as possible, that advanced search capabilities may be implemented, and that the big data algorithms have a rich basis upon which to analyze a continuously growing knowledge base, and ultimately bring the data to those data users who need it.

Besides the graph-data metamodel, an important ingredient for the DAS concept is the introduction of advanced machine learning algorithms to bring the data to potential users. Broadly speaking, machine learning refers to capability of a computer program to learn a knowledge-intensive task while improving its performance on the task as it gains more experience [70]. The task at hand is the suggestion of datasets and potential collaborators to a set of users. The performance corresponds to the practical value of the suggested datasets to the users, while the experience is derived from the feedback obtained from users regarding the quality of the suggestions. The machine learning algorithms will take advantage of the underlying structure of the graph data, the similarity between datasets, and the similarity between users as obtained from social media and scholarly publications. The machine learning techniques that can be used to achieve this include clustering, collaborative filtering, case-based reasoning, and deep learning.

Clustering is a computing task in which a set of objects is segmented in subsets such that the objects in one cluster are more similar to each other than the objects out of the cluster [70]. Clustering can be used to create categories of datasets on the one hand, and categories of users and applications on the other hand. The clustering of datasets can be performed by applying the highly connected subgraph algorithm [71] on the graph data. Datasets will be found in the same cluster if they are highly connected in the graph data, which would mean that the datasets within one cluster will share relevant variables and methodological features. The similarity metric of the clustering algorithm will be continuously adapted based on the feedback received from users. Thus, as the algorithm gains in experience, the clustering of the datasets will result in groups more and more homogeneous, thereby enabling more customized suggestions. Since the graph links have different semantics, the same dataset element will potentially belong to multiple clusters, for instance geographic clusters, data fidelity clusters, topical clusters, etc.

Using social network data (such as Facebook posts or Twitter hashtags), parsed publications, research knowledge hubs with citation data, newspaper articles (particularly those discussing science-related topics), co-citation analysis, as well as past patterns of dataset search and use, it will be possible to similarly cluster the users into multiple groups based on their scientific disciplines, their application domains of interest, their geographic area of focus, etc. Here again, as the algorithm learns more about the relevant properties that users share, they will be placed in clusters that become more and more specific, so that the recommendations will become more accurate.

Collaborative filtering uses the ratings and feedback provided by users of a product to recommend the same product to users with a high level of similarity. A commonly used similarity metric is the Pearson correlation [72] or the vector cosine-based similarity [73]. In this approach, crowd-sourced user feedback is exploited to provide better suggestions. This method may be inadequate at the beginning when user feedback data is sparse, but improves exponentially as user data becomes more widespread [74]. Collaborative filtering works well in combination with the clustering method described before, since, initially, recommendations may be forwarded to users in the same cluster, as they share some similarity.

In case-based reasoning, properties of datasets and of users entities are utilized to match users and products. The cases encode knowledge such as "users sufficiently similar with user u and who accessed dataset with property x also used dataset with property y." As such, case-based reasoning will exploit the results of the clustering algorithms. Case-based reasoning algorithms are often based on decision trees [75] and have some major benefits: They are suitable for non-formalized knowledge domains, they are robust and easy to maintain, and they allow for incremental improvement. However, just as with collaborative filtering, the approach becomes computationally inefficient when the domain is too dynamic and when the number of cases becomes very large [76].

To remedy these shortcomings, deep learning, based on restricted Boltzmann machines [77] are emerging as very promising techniques for data intensive learning tasks, owing to the availability of parallelized computational resources. These techniques use successive layers of neural networks and perform computations of increasing levels of abstraction to discover a hierarchy of features, from low-level features to higher level ones [78], i.e., a bottom-up approach. Deep-learning algorithms have been successfully applied to computer vision and language processing and have only recently begun to be used in commercial recommender systems [79]. As shown in [80], deep-learning algorithm can be used to learn about the attitudes of a user toward a dataset from the review text of dataset posted by users and the features of the product itself, and thereby match datasets with types of users to maximize the utility of a dataset for a certain type of user.

2.4. Interaction Platform

The interaction platform is the space in which users and ISDAs interact with each other (Figure 2) and where a track record of these interactions is being kept. Users of the platform can take on the role of data provider, who want to make datasets available to a community of users, and data users who may be scientists who need some data in the context of their research or other social agents (individuals, governmental bodies, NGOs) who may have interest in knowledge derived from the data to answer practical questions relevant to their problems.

Experience and events should be captured in schemes that provide a complete history of a given dataset. While such a scheme for the recording of the transactions could be based on blockchains, there are concerns that this would be far too demanding in terms of energy, see, e.g., [81]. Blockchain is an emerging interaction paradigm for transmission and storage of information without centralized control. It is a secure and distributed database that is hosted locally by the human or software agents engaged in a transaction. It contains the history of all transactions performed by these agents, without a centralized intermediary, thereby allowing each participant to independently verify validity of a chain of interactions. Furthermore, blockchains can be made public or limit access to only users with specified credentials.

The first blockchain was introduced by Bitcoin [82], but its use as an architectural model for secure user interaction has now expanded beyond the domain of digital currencies [83]. User transactions are structured in blocks. Each block is validated by an algorithmic key or "proof-of-work." Once a block is validated, it is timestamped and added to the chain of blocks and becomes publicly visible to the members of the network. The decentralized, transparent and robust nature of blockchain makes it particularly well adapted for a distributed and intelligent data search system. However, the choice of whether to use one of the existing blockchains (for a discussion of potential candidates, see, e.g., [84]) or to develop a new blockchain dedicated to data and knowledge-related transactions would be a difficult one. In addition, there are concerns that the trust in blockchains is not fully justified [85]. An important application of blockchains is to provide provenance particularly with respect to transfers of ownership in something. This comes with a very high use of resources. In fact, a white paper developed by the World Economic Forum states that the energy consumed in the blockchain network is unsustainable [81]. Energy consumption can be reduced significantly depending on the consensus algorithms used [86], and replacing the "proof-of-work" algorithms by "proof-of-stake" or "proof-of-authority" results in drastically reduce energy consumption decoupled from the number of users engaged in a blockchain [87]. For the access to data, tools to process the data, information derived from data, and knowledge created using the data, the ownership in general remains with the orginator, and only the rights to access, processing, use and further distribution are points of negotiation. For this purpose, provenance may be achieved without blockchains. However, a distributed ledger that validates and records transactions between several ISDAs as well as between ISDAs and human agents seems to be mandatory for the interaction platform.

For the management of interactions between agents (data agents, models, persons, repositories, etc.), a concept similar to that of "smart contracts" could be developed. These "smart contracts" would automatically perform delegated terms of a contract without user intervention. The traceability of blockchains or a similar distributed ledger would allow the capture of events and user experiences as blockchain-based schemes to provide a complete history of datasets addition, access, purchase, updates, etc. To the extent possible, protocols would facilitate, verify, or enforce the negotiation or performance of a "contract" between a user and the ISDA representing the data product. With this concept, many aspects of the transactions could be made partially or fully self-executing, self-enforcing, or both. Conceptually, this approach provides security superior to traditional more open transactions. The "smart contract" concept seamlessly interfaces with a distributed ledger.

However, as noted above, blockchains are very demanding in terms of computational resources and energy, and a careful assessment of the trade-off between the amount of resources needed and the level of security, perseverance, and documentation achieved needs to be carried out to inform the design of the interaction platform.

3. Discussion

3.1. Current Status and New Contributions

Many Earth observation (EO) communities have made considerable efforts to improve data discoverability and accessibility. In particular, Group on Earth Observations (GEO) has made a significant contribution serving users of data with means to discover data, see, e.g., [88]. In many scientific communities, efforts have been made towards the integration of data and modeling tools. A particular focus has been on the development of data models that support interdisciplinary and cross-disciplinary data integration, see, e.g., [89]. Harmonization of metadata across thematic areas and beyond poses a major challenge, see, e.g., [90]. Brokering of data and metadata for a large number of datasets is often at the core of efforts to overcome this challenge, see, e.g., [88,91]. The need for new transformative approaches is acknowledged, see, e.g., [5,92].

For the development of Earth observation (EO) systems with high scientific and societal benefits, comprehensive knowledge of information needs is mandatory. Over the last few decades, there have been abundant efforts at national and international levels to assess user needs that constitute requirements for Earth observations (EOs). Examples are the reports produced by IGOS-P themes, see, e.g., [19–21,23,93], and the reports that resulted from the GEO task US-09-01a, see, e.g., [94]. In most cases, mapping of user landscapes was based on limited surveys, user forums, or literature reviews by experts with emphasis mostly on one or another methodology. Surveys of users often resulted in limited responses, and the main input was provided by expert groups and communities (see, e.g., [33] and the references therein). The output of most of these efforts consists mainly in written reports with no functionality for further machine and algorithm-based analyses. While these reports have a high value, exploitation is low. Repositories of observational requirements (such as OSCAR, see http://www.wmo-sat.info/oscar) are mostly limited to relational databases and in most cases lack a linkage of the observational requirements to societal users and their decision and policy making processes. In most cases, feedback capabilities are limited or absent and users have limited opportunities to comment on and augment the information in the repositories. The construction and analysis of graphs is not supported in these approaches. However, implementing DAS can build on these initiatives and utilize the resulting reports and repositories in the construction of graphs.

The Global Earth Observation System of Systems (GEOSS) User Requirements Registry (URR) aimed to construct graphs that represented user types, applications, observational requirements, and needs in terms of research, technology, infrastructure, and capacity [69,95,96]. These graphs captured the connectivity between instances in one group as well as cross-group interdependencies.

The experience with the URR shows that users wanted the graphs to be extended to include far more groups, such as models, tools, people, data, knowledge, decision and policy making, etc. [69]. This user-based request was one of the main motivations for the transition of the URR to the Socio-Economic and Environmental Information Needs Knowledge Base (SEE-IN KB).

Identifying and describing comprehensively all characteristics of data relevant for the matching of potential users still presents a major challenge to the provision of sufficient metadata. Different communities have different views and understandings of a given characteristics, and this severely hampers harmonization (see, e.g., the discussion on data quality in [97]).

Abundant efforts have been made to map the user landscapes for Earth observations (EOs). For example, numerous efforts have been made to characterize those users engaged in water sustainability that depend on information derived from Earth observations (EOs), see, e.g., [94]. However, the full picture of the user landscape has not been captured, at least not in a form that could easily be analyzed by algorithms to discover unexploited linkages and unmatched needs. In the past, focus has been too much on writing reports and articles and not on getting the information on needs and requirements in a form available for machine-based analyses. The reports (e.g., the set of reports produced by US-09-01; see [33]) often disappear in shelfs and are not really used in guiding the development of observing systems and knowledge services or in linking users and data.

Recent developments in unstructured databases allow for a far more flexible approach to data that represents a system of graphs. Advances in big data analysis and the availability of abundant information in social media, research networks, social communication channels, governmental and non-governmental Web sites, and online publications enables the machine-based construction of complex graphs that include, among others, also the decision and policy making processes and agents that depend on evidence and knowledge derived from Earth observations (EOs). Likewise, improvements in the presentation and analysis of graph data open new avenues for comprehensive user assessments and the detailed mapping of user landscapes. Importantly, the theory for the analysis of graph data is fully developed (see Section 2.3) and provides a powerful tool for those who need to explore the landscape in order to identify and engage with users, discover gaps, and improve the services they provide to better meet the needs of the users. Utilizing these recent developments, efforts have been made to utilize large Web-based knowledge sources to develop new avenues for access to data sources. For example, knowledge has been extracted from Wikipedia to link this knowledge to data by [98]. Other efforts aim at unifying the access to knowledge, see, e.g., [99]. The Linked Open Data Cloud (LODC) provides an opportunity to publish data and integrate it into a graph connecting data across many domains [47].

Despite the many efforts to improve access and usability of Earth observations (EOs), to increase knowledge of information needs, and to link users better to available data and knowledge resources, the current techniques available to Earth scientists and other users to discover and access data are still at a very low level with respect to comprehensive discovery, easy access, options for feedback, etc. The separation of passive metadata from the actual data often leads to incomplete metadata with crucial information missing. This has major impacts on provenance and reproducibility of research. Data citation is also impacted by incomplete metadata, see, e.g., [100]. What appears necessary is a fundamental transformation, a "Gestalt shift", in the view of how data and users should interact [5]. The DAS concept could provide for this transformation.

The overall DAS concept is fully developed (see Figure 2). The knowledge base builds on the Socio-Economic and Environmental Information Needs Knowledge Base (SEE-IN KB). The SEE-IN KB is being developed as a knowledge base to construct, store, present, and analyze complex systems of graphs. It is populated with graphs fully capturing the stakeholder landscapes for societally relevant themes. It provides the means to explore the graphs to discover connectivity and to identify gaps in terms of unmatched linkages. The current version of the Socio-Economic and Environmental Information Needs Knowledge Base (SEE-IN KB) is at a prototype level with respect to storing graph-data. In most approaches to graph data, the concept of triples is used, where a triple consists of two nodes (a subject and an object) and a link or predicate connecting these two nodes (e.g., the Resource Description Framework (RDF), see [101]). In a number of approaches, the nodes carry information on the links (in and out links) they are attached to (an example is the "Oracle Big Data Spatial and Graph" package; see, e.g., [102]). The Socio-Economic and Environmental Information Needs Knowledge Base (SEE-IN KB) does not include information on links with the nodes. Moreover, links are not necessarily directional. This generalized graph data model provides on the one hand for more flexibility and on the other hand requires more analytical skills to extract relationships from the graph data.

The artificial intelligence (AI)-based construction of graphs applying deep search and deep learning methodology is at the beginning. The currently available knowledge base design specification and architecture description needs to be further detailed and discussed with relevant communities, including the providers of GEOSS infrastructure. Comments from experts communities, including the private sector, will be crucial for the full conceptual development.

The concept for ISDAs is developed in terms of functionality and desired capabilities. Conceptually, the ISDAs are fully software agents that represent specific datasets and have the authority to answer queries from potential users and to negotiate with interested users conditions of data access and use. The ISDAs could be designed similar to Web servers giving access to information about extended metadata and contents of datasets, as well as derived attributes. Among others, an ISDA can provide the full dataset it represents or subsets of it in a user-requested format, can give access to tools used to process the data, and answer questions that require certain processing of the data. The ISDAs can access and query the graph data in the knowledge base to discover potential users and contact these users with promotional information about their dataset. They also can collect feedback from users of their datasets and provide this feedback to the knowledge base. The machinery the ISDAs could initially work on is the Web. For example, a dedicated main domain could ensure easily recognizable URLs and enhanced browsers could facilitate the communication between ISDAs and humans. In a later stage, a new framework for the world of the ISDAs could be created. The ecosystem of the ISDAs would be a core part of the digital ecosystem for the environment and the planet envisioned by [5,92].

The main advantage of the DAS concept is the fact that ISDAs are local agents associated with the data products where these data products exist. Thus, the need to publish data in archives or repositories, to develop large catalogs of datasets, etc., would be much reduced or disappear.

The interaction platform is conceptually developed in terms of the software and human actors and the documentation of interactions to ensure provenance. It will provide a matching and recording framework, where users and ISDAs can interact in promotion of datasets and in transactions that can lead to data modifications (e.g., for data providers) and use of data (for users). The platform could utilize the blockchain concept to ensure provenance, but a key question to be researched is the trade-off between the amount of resources needed and the level of security, perseverance, and documentation achieved needs to be carried out to inform the design of the interaction platform (see Section 2.4).

3.2. Validation Through Case Studies

Detailed case studies addressing societal problems in a transdisciplinary approach could provide validation of the DAS concept. Initially, focus should be on broad scientific communities that depend heavily on Earth observations (EOs) and are researching societally relevant problems. Most of the problems related to sustainable development or developing sustainability are wicked problems [36] or super-wicked problems [103], and for most of these problems transdisciplinary collaborative approaches are most suited to address the problem [104].

Problems that appear to be ideal candidates for such case studies are within the Food-Water-Energy Nexus (FWEN). The FWEN provides an excellent example of interactions in a complex system of systems [7,38,44] with many potentially severe societal consequences [105]. In particular, a water crisis has been identified as a global catastrophic risk, see, e.g., [106]. Earth observations (EOs) are crucial to address the FWEN comprehensively, see, e.g., [107,108], and to make progress towards the SDGs. The FWEN links sustainability of water use to almost all of humanity's activities. Achieving the 2030 Agenda for Sustainable Development [4] is conditioned by addressing the FWEN and making progress towards global food, water, and energy sustainability. The Sustainable Development Goals (SDGs) 2 (no hunger), 6 (clean water), and 7 (affordable and clean energy) are directly interdependent, while almost all other SDGs are impacted or are impacting the sustainability within these three domains. This makes the landscape of users depending on knowledge of the state and trends in the planetary physiology including the water, nitrogen and phosphorus cycles a very complex one. Diagnosing the time and spatial patterns of problems and co-developing and validating solutions for food, water and energy-related problems constitutes a suite of wicked problems. Addressing these issues requires access to comprehensive data, and the need for increased cross-domain data sharing has been emphasized within the relevant domains e.g., [109]. Likewise, building capacity to use the available cross-domain knowledge for decision and policy making and management of the relevant cycles in the planetary physiology is a complex task that needs to use many different avenues to engage with users in their activities. Comprehensive knowledge of the landscape of stakeholder, decision makers, and knowledge providers engaged in sustainability in a form that supports matchmaking, collaboration, and participatory activities is a prerequisite for identifying problems as well as providing evidence and knowledge to those who need this, and to build capacity.

The goal of such case studies would be to improve the understanding of the relationship between the FWEN and modern global change, including modern climate change, changes in the nitrogen and phosphorus cycles, and loss of biodiversity, and to develop transformative interventions that could change the trajectory of the underlying system towards desirable futures. The knowledge base would be used to construct the graph data relevant for research and user communities related to these challenges and to construct a data Web of relevant datasets. ISDAs for these datasets would be trained and would interact with researchers in the participating communities to discover and access data products. The ISDAs would also promote data products to potential users. Feedback collected from those participating in the use cases would provide a basis to validate and improve the DAS components. The communities that ideally should be involved in this validation include, among others, the Group on Earth Observations (GEO) Initiative "Earth Observations in Service of the 2030 Agenda for Sustainable Development" (http://eo4sdg.org/), the GEO Water Cycle Community of Practice (http://www.earthobservations.org/wa_igwco.shtml), the Future Earth Sustainable Water Future Programme (https://water-future.org/), and the Sustainable Water-Energy-Food Nexus Working Group (http://water-future.org/working_groups/sustainable-w-e-f-nexus-working-group/).

3.3. Considerations For Implementation

To ensure broad acceptance and support for the transition from the DPO perception to a DAS perception, the design and implementation of the DAS concept should be further developed in a participatory modeling. The planning of a versatile, secure, efficient, and active system linking observations and users for the benefit of society constitutes a wicked problem, and participatory modeling could be the first step in a collaborative approach to this problem. Group on Earth Observations (GEO) could utilize its convening power to bring a wide range of stakeholders together for such a participatory modeling. Again, the FWEN and related SDGs could be the societal challenge for this participatory modeling effort to focus on.

As a result of this effort, the design specifications for the DAS concept would be further detailed, including a detailed description of the functionality. The architecture will have to consider distributed cloud-based elements and will most likely require modifications of the current graph data model. The current graph data model separates the graph information from the objects. In many other graph software implementation, objects carry part of the graph information, and it will have to be researched whether a complete separation of objects and links is desirable and feasible within legal constraints. The specification will include the description of the methodology for the construction, presentation and analysis of graph data as well as the functionality for user feedback collections. For the latter, potential legal constraints will have to be assessed to ensure that the collection of user information is conform with legal requirements.

The concept for the ISDAs as representatives of datasets and products has to ensure that the ISDAs have semantic capabilities. A core research question to address is how rich the data description available to the ISDAs will have to be to enable these capabilities. The development of a genuine knowledge model that enables AI to reason and search is a necessity for the implementation of the DAS concept.

The specification of a communication protocols for the ISDAs is an important step towards implementation. The methodology for self-learning ISDAs can be based on deep learning methodology to increase their knowledge relevant to the data they represent as well as the potential and actual applications and users of the data. To some extent, the ISDAs could utilize crawlers to collect relevant information. It is anticipated that ISDAs will be initiated as minimal seeds and then grow into more adult ISDAs. A research question relates to the minimum capability of the seeds necessary for them to grow. Among others, the ISDAs will need limited data processing capabilities to extract rough datasets or statistical or average properties, and they will have "magnifying glasses" to allow users to zoom into large datasets. They also should have the capability to provide data in a format requested be a user. Thus, a user would not have to know anything about the details of how the data are actually stored in the original data archive.

The generic design specification and architecture of the virtual interaction platform for ISDAs and users requires careful considerations. In terms of interactions, the platform will support the capabilities of the ISDAs to respond to user queries, identify users and needs and to promote data accordingly or to suggest collaborations between users to users. For this, the ISDAs will need to utilize and analyze the graph data available in the knowledge base to assess where their data would be beneficial. The ISDAs will be able to provide access to data in various ways. Actual transactions could be recorded in a scheme derived from blockchains to ensure provenance of both the original data and derived products,

The knowledge base is currently implemented as an extension of the already existing Socio-Economic and Environmental Information Needs Knowledge Base (SEE-IN KB). The SEE-IN KB contains considerable graph data for several research areas including water cycle, geohazards, health, and air quality. The data model of the Socio-Economic and Environmental Information Needs Knowledge Base (SEE-IN KB) is specifically designed for graph data, and a methodology for graph construction based on deep search and deep learning approaches is being implemented.

The GEOSS Common Infrastructure provides access to a large number of datasets. It will be important to ensure that the three core elements of DAS can communicate with the GEOSS Common Infrastructure (GCI) to train ISDAs for relevant datasets and to allow access to the knowledge base, ISDAs and interaction platform through the GCI.

4. Conclusions

The amount, quality, and diversity of Earth observations (EOs) is rapidly increasing but exploitation of this extremely valuable resource is hampered by limited discoverability, lack of information on applicability, and insufficient capacity in extracting relevant information from this resource for knowledge creation. Most efforts to improve in all these aspects are incremental improvements of existing concepts. At the same time, as outlined in Section 1, humanity in the Anthropocene is challenged increasingly with global catastrophic risks while aiming for more sustainability. Assessing and addressing these risks requires comprehensive information on the biosphere, the humansphere and the impacts of the humansphere and technosphere on the biosphere.

In this situation, a transformational paradigm shift in the relationship between data and users is required. The transition from the DPO to a DAS perception could facilitate this "Gestalt shift" and would have far reaching transformational consequences. In particular, it is expected that this transition would provide novel ways of integrating data into transdisciplinary approaches to wicked problems discussed, e.g., by [110]. Implementing the United Nations' 2030 Agenda for Sustainable Development [4] poses many wicked problems to society, and most of the seventeen Sustainable Development Goals (SDGs) detailed in the agenda have all the additional properties of super-wicked problems identified by [103]. In particular, for most of the Sustainable Development Goals (SDGs), there is no central authority for the implementation, time is running out, and those who are causing the challenge are now attempting to solve the problem. For the validation of the DAS concept, use cases can be built around selected wicked problems associated with the implementation of the Sustainable Development Goals (SDGs).

The implementation of the DAS concept requires a major community effort and GEO could use its convening power to bring together selected communities for pilot projects aiming at the further development and validation of the DAS concept. A specific use cases of interest would be the Food-Water-Energy Nexus (FWEN) and the related Sustainable Development Goals (SDGs) 2 (no hunger), 6 (clean water), and 7 (clean energy). A DAS-related use case would aim at understanding the relationship between the FWEN and modern global change, including modern climate change, changes in the nitrogen and phosphorus cycles, and loss of biodiversity.

Author Contributions: The authors contributed equally to all sections.

Funding: The authors would like to acknowledge the European Union "Horizon 2020 Program" that funded the ConnectinGEO (Grant Agreement no. 641538) projects. Part of the work for one author (Plag) was conducted under NASA grant 80NSSC17K0241.

Conflicts of Interest: The authors declare no conflict of interest.

Abbreviations

The following abbreviations are used in this manuscript:

AI	artificial intelligence
CDP	Customers Discover Products
DAS	Data as Active Subjects
DDU	Data Discover Users
DFS	Depth-first search
DPO	Data as Passive Objects
EO	Earth observation
FWEN	Food-Water-Energy Nexus
GCI	GEOSS Common Infrastructure
GDPR	General Data Protection Regulation
GEO	Group on Earth Observations
GEOSS	Global Earth Observation System of Systems
IGOS	Integrated Global Observing Strategy
IGOS-P	Integrated Global Observing Strategy Partnership
IAEG-SDGs	Inter-Agency and Expert Group on SDG Indicators
ISDA	Intelligent Semantic Data Agent
LODC	Linked Open Data Cloud
PDC	Products Discover Customers
RDF	Resource Description Framework
SDG	Sustainable Development Goal
SEE-IN KB	Socio-Economic and Environmental Information Needs Knowledge Base
SIDS	Small Island Developing States
UDD	Users Discover Data
UNFCCC	United Nations Framework Convention on Climate Change
UNSC	United Nations Statistical Commission
URR	User Requirements Registry

References

1. Harris, R.; Miller, L. Earth observation and the public good. *Space Policy* **2011**, *27*, 194–201, doi:10.1016/j.spacepol.2011.09.010.
2. Cotton-Barratt, O.; Farquhar, S.; Halstead, J.; Schubert, S.; Snyder-Beattie, A. *Global Catastrophic Risks 2016*; Technical Report; Global Challenge Foundation, Global Priorities Project: Stockholm, Sweden; Oxford, UK, 2016.
3. World Economic Forum. *Global Risks 2019*, 14th ed.; Technical Report; World Economic Forum: Geneva, Switzerland, 2019.
4. United Nations. *Transforming our World: The 2030 Agenda for Sustainable Development*; Technical Report A/RES/70/1; United Nations: New York, NY, USA, 2015.
5. Campbell, J.; Jensen, D.E. *The Promise and Peril of a Digital Ecosystem for the Planet*; Technical Report; United Nations Environment Programme: Nairobi, Kenya, 2019. Available online: https://medium.com/@davidedjensen_99356/building-a-digital-ecosystem-for-the-planet-557c41225dc2 (accessed on 25 September 2019).
6. Ryan, B. Open data for Sustainable Development. *Geospatial World*, 14 August 2016.
7. Jules-Plag, S.; Plag, H.P. Supporting Agenda 2030's Sustainable Development Goals—Agend-Based Models and GeoDesign. *ApoGeoSpatial* **2016**, *31*, 24–30.
8. Taylor, G. *Evolution's Edge—The Coming Collapse and Transformation of our World*; New Society Publishers: Gabriola Island, BC, Canada, 2008.
9. Baum, S.D.; Handoh, I.C. Integrating the planetary boundaries and global catastrophic risk paradigms. *Ecol. Econ.* **2014**, *107*, 13–21, doi:10.1016/j.ecolecon.2014.07.024.
10. Keys, P.W.; Galaz, V.; Dyer, M.; Matthews, N.; Folke, C.; Nyström, M.; Cornell, S.E. Anthropocene risk. *Nat. Sustain.* **2019**, *2*, 667–673, doi:10.1038/s41893-019-0327-x.
11. Wang, M.; Hu, C.; Barnes, B.B.; Mitchum, G.; Lapointe, B.; Montoya, J.P. The great Atlantic Sargassum belt. *Science* **2019**, *365*, 83–87, doi:10.1126/science.aaw7912.
12. Steffen, W.; Rockström, J.; Richardson, K.; Lenton, T.M.; Folke, C.; Liverman, D.; Summerhayes, C.P.; Barnosky, A.D.; Cornell, S.E.; Crucifix, M.; et al. Trajectories of the Earth System in the Anthropocene. *Proc. Natl. Acad. Sci. USA* **2018**, *115*, 8252–8259, doi:10.1073/pnas.1810141115.
13. Rothman, D.H. Thresholds of catastrophe in the Earth system. *Sci. Adv.* **2017**, *3*, e1700906, doi:10.1126/sciadv.1700906.
14. Baum, S.D. The far future argument for confronting catastrophic threats to humanity: Practical significance and alternatives. *Futures* **2015**, *72*, 86–96, doi:10.1016/j.futures.2015.03.001.
15. Barnosky, A.D.; Hadly, E.A.; Bascompte, J.; Berlow, E.L.; Brown, J.H.; Fortelius, M.; Getz, W.M.; Harte, J.; Hastings, A.; Marquet, P.A.; et al. Approaching a state shift in Earth's biosphere. *Nature* **2012**, *486*, 52–58, doi:10.1038/nature11018.
16. Avin, S.; Wintle, B.C.; Weitzdörfer, J.; hÉigeartaigh, S.S.Ó.; Sutherland, W.J.; Rees, M.J. Classifying global catastrophic risks. *Futures* **2018**, *102*, 20–26, doi:10.1016/j.futures.2018.02.001.
17. Dahl, A.L. IGOS from the perspective of the global observing systems and their sponsors. In Proceedings of the 27-th International Symposium on Remote Sensing of Environment: Information for Sustainability, Tromsø, Norway, 8–12 June 1998; Norwegian Space Centre: Oslo, Norway, 1998; pp. 92–94.
18. IGOS-P. *The Integrated Global Observing Strategy (IGOS) Partnership Process*; Technical Report, IGOS Partnership, 2003; IGOS Process Paper, Version of 19 March 2003; World Meteorological Organization: Geneva, Switzerland, 2003
19. IGOS-P Ocean Theme Team. *An Ocean Theme for the IGOS Partnership*; Technical Report, IGOS Integrated Global Observing Strategy; NASA: Washington, DC, USA, 2001.
20. Lawford, R.; The Water Theme Team. *A Global Water Cycle Theme for the IGOS Partnership*; Technical Report, IGOS Integrated Global Observing Strategy, 2004; Report of the Global Water Cycle Theme Team, April 2004; ESA Publications Division: Noordwijk, The Netherlands, 2004.
21. Marsh, S.; The Geohazards Theme Team. *Geohazards Theme Report*; Technical Report, IGOS Integrated Global Observing Strategy; BRGM: Orleans, France, 2004.
22. Townshend, J.R.; The IGOL Writing Team. *Integrated Global Observations of the Land: A Proposed Theme to the IGOS Partnership—Version 2*; Technical Report, IGOS Integrated Global Observing Strategy, 2004; Proposal Prepared by the IGOL Proposal Team, May 2004; FAO: Rome, Italy, 2004.

23. IGOS. *A Coastal Theme for the IGOS Partnership—For the Monitoring of our Environment from Space and from Earth*; IOC Information Document No. 1220; UNESCO: Paris, France 2006; 60p.

24. United Nations Sustainable Development. In Proceedings of the AGENDA 21, United Nations Conference on Environment & Development, Rio de Janerio, Brazil, 3–14 June 1992; Technical Report; United Nations: New York, NY, USA, 1992. Available online: http://sustainabledevelopment.un.org/content/documents/Agenda21.pdf (accessed on 15 August 2019).

25. GEO. *Global Earth Observing System of Systems GEOSS—10-Year Implementation Plan Reference Document*; Technical Report GEO 1000R, Group on Earth Observations; ESA Publications Division: Noordwijk, The Netherlands, 2005. Avaliable online: http://earthobservations.org (accessed on 10 August 2019).

26. LeCozannet, G.; Salichon, J. *Geohazards Earth Observation Requirements*; Technical Report BRGM/RP-55719-FR; BRGM: Orlean, France, 2007.

27. Zell, E.; Huff, A.K.; Carpenter, A.T.; Friedl, L. A user-driven approach to determining critical earth observation priorities for societal benefit. *IEEE J. Sel. Top. Appl. Earth Obs. Remote Sens.* **2012**, *5*, 1594–1602.

28. Plag, H.P.; Rizos, C.; Rothacher, M.; Neilan, R. The global geodetic observing system (GGOS): Detecting the fingerprints of global change in geodetic quantities. In *Advances in Earth Observation of Global Change*; Springer: Berlin, Germany, 2010.

29. Plag, H.P.; Ondich, G.; Kaufman, J.; Foley, G.; Pignatelli, F. The GEOSS User Requirement Registry: A Versatile Tool for the Dialog Between Users and Providers. In Proceedings of the 34th International Symposium on Remote Sensing of the Environment, Sydney, Australia, 10–15 April 2011.

30. Plag, H.P.; Foley, G.; Jules-Plag, S.; Kaufman, J.; Ondich, G. The GEOSS user requirement registry (URR): Linking users of GEOSS across disciplines and societal benefit areas. In Proceedings of the 2012 IEEE International Geoscience and Remote Sensing Symposium IEEE, Munich, Germany, 23–27 July 2012.

31. EAG. *Results-Oriented GEOSS: A Framework for Transforming Earth Observation Data to Knowledge for Decision Making*; Technical Report, Group on Earth Observation, Executive Committee; Report Prepared by the Expert Advisory Group for the 48th Meeting of the Executive Committee; Group on Earth Observation: Geneva, Switzerland 2019.

32. Plag, H.; The Workshop Participants. *Implementing and Monitoring the Sustainable Development Goals in the Caribbean: The Role of the Ocean, 2018, Saint Vincent, Saint Vincent and the Grenadines, 17–19 January 2018*; Technical Report; GEOSS Science and Technology Stakeholder Network (GSTSN): Rossbach, Germany, 2018. Available online: http://www.gstss.org/2018_Ocean_SDGs (accessed on 21 September 2019).

33. Group on Earth Observations. *Task US-09-01a: Critical Earth Observation Priorities*, 2nd ed.; Technical Report; Group on Earth Observations: Geneva, Switzerland, 2012. Available online: http://sbageotask.larc.nasa.gov (accessed on 15 July 15 2019).

34. Valuables. Resources for the Future. Available online: https://www.rff.org/valuables/ (accessed on 15 August 2019).

35. UNISDR. *Sendai Framework for Disaster Risk Reduction 2015–2030*, 1st ed.; Technical Report UNISDR/GE/2015-ICLUX EN5000; UNISDR: Geneva, Switzerland, 2015. Available online: http://www.preventionweb.net/files/43291_sendaiframeworkfordrren.pdf (accessed on 15 July 2019).

36. Rittel, H.W.J.; Webber, M.W. Dilemmas in a general theory of planning. *Policy Sci.* **1973**, *4*, 155–169.

37. UNRISD. *Policy Innovations for Transformative Change—Implementing the 2030 Agenda for Sustainable Development*; Unrisd Flagship Report 2016; United Nations Research Institute for Social Development: Geneva, Switzerland, 2016.

38. Nilsson, M.; Griggs, D.; Visbeck, M. Policy: Map the interactions between Sustainable Development Goals. *Nature* **2016**, *534*, 320–322, doi:10.1038/534320a.

39. Griggs, D.J.; Nilsson, M.; Stevance, A.; McCollum, D. (Eds.) *A Guide to SDG Interactions: From Science to Implementation*; Technical Report; International Council for Science: Paris, France, 2017; doi:10.24948/2017.01.

40. Singh, G.G.; Cisneros-Montemayor, A.M.; Swartz, W.; Cheung, W.; Guy, J.A.; Kenny, T.A.; McOwen, C.J.; Asch, R.; Geffert, J.L.; Wabnitz, C.C.; et al. A rapid assessment of co-benefits and trade-offs among Sustainable Development Goals. *Mar. Policy* **2018**, *93*, 223–231.

41. Alcamo, J.; Chenje, M.; Ghai, A.; Keita-Ouane, F.; Leonard, S.A.; Niamir-Fuller, M.; Nobbe, C. *Embedding the Environment in Sustainable Development Goals*; UNEP Post-2015 Discussion Paper 1, Version 2; UNEP: Nairobi, Kenya, 2013.

42. Leadership Council of the Sustainable Development Solutions Network. *Indicators for Sustainable Development Goals*; Technical Report, Draft Report for Public Hearing; Sustainable Development Solutions Network of the United Nations: New York, NY, USA, 2014.
43. IAEG-SDGs. *Tier Classification for Global SDG Indicators—11 May 2018*; Technical Report; Intern-Agency Expert Group for SDG Inidcators, United Nations: New York, NY, USA, 2018.
44. Jules-Plag, S.; Plag, H.P. Supporting the Implementation of SDGs. *Geospatial World*, 15 August 2016. Available online: http://www.geospatialworld.net/article/supporting--implementation--sdgs/ (accessed on 10 July 2019).
45. Plag, H.P.; Jules-Plag, S.A. A Goal-Based Approach to the Identification of Essential Transformation Variables in Support of the Implementation of the 2030 Agenda for Sustainable Development. *Int. J. Digit. Earth* **2019**. doi:10.1080/17538947.2018.1561761.
46. PANGAEA Team. PANGAEA. Data Publisher for Earth & Environmental Science. Available online: https://pangaea.de (accessed on 28 August 2019).
47. McCrae, J.P.; Abele, A.; Buitelaar, P.; Cyganiak, R.; Jentzsch, A.; Andryushechkin, V.; Debattista, J. The Linked Open Data Cloud. Available online: https://www.lod-cloud.net/ (accessed on 27 August 2019).
48. Christodoulou, P.; Christodoulou, K.; Andreou, A.S. A real-time targeted recommender system for supermarkets. In *Proceedings of the 19th International Conference on Enterprise Information Systems— Volume 2, Porto, Portugal, 26–29 April 2017*; Springer: Berlin/Heidelberg, Germany, 2017; pp. 703–712, doi:10.5220/0006309907030712.
49. The Performance Edge, Inc. Feedback Rewards—Guest Feedback and Rewards Program. Available online: http://www.feedbackrewards.com/ (accessed on 27 August 2019).
50. Russell, S.J.; Norvig, P. *Artificial Intelligence: A Modern Approach*, 2nd ed.; Prentice Hall: Upper Saddle River, NJ, USA, 2003.
51. Weiss, G. *Multiagent Systems*, 2nd ed.; The MIT Press: Cambridge, MA, USA, 2013.
52. Plag, H.P. Implementing and Monitoring the Sustainable Development Goals in the Caribbean: The Role of the Ocean. Presentated at the Meeting of the Steering Committee of the GEO Initiative "Ocean and Society: Blue Planet", Saint Vincent, Saint Vincent and the Grenadines, 15 March 2018.
53. Stevenson, H. Emergence: The Gestalt Approach to Change. Available online: http://www.clevelandconsultinggroup.com/articles/emergence-gestalt-approach-to-change.php (accessed on 15 August 2019).
54. Dietz, J. *Enterprise Ontology - Theory and Methodology*; Springer: Berlin/Heidelberg, Germany, 2006.
55. Pawlak, Z. Rough sets. *Int. J. Parallel Program.* **1982**, *11*, 341–356, doi:10.1007/BF01001956.
56. Bazan, J.; Szczuka, M.; Wojna, A.; Wojnarski, M. On the evolution of rough set exploration system. In *Proceedings of the RSCTC 2004, LNAI 3066, Uppsala, Sweden, 1–5 June 2004*; Tsumoto, S., Ed.; Springer: Berlin/Heidelberg, Germany, 2004; pp. 592–601, doi:10.1007/978-3-540-25929-9_73.
57. Ziarko, W. Rough sets as a methodology for data mining. In *Rough Sets in Knowledge Discovery 1: Methodology and Applications*; Polkowski, L., Skowron, A., Eds.; Physica-Verlag: Heidelberg, Germany, 1998; pp. 554–576.
58. Chen, H.; Li, T.; Luo, C.; Horng, S.J.; Wang, G. A decision-theoretic rough set approach for dynamic data mining. *IEEE Trans. Fuzzy Syst.* **2015**, *23*, 1958–1970.
59. Neukom, R.; Barboza, L.A.; Erb, M.P.; Shi, F.; Emile-Geay, J.; Evans, M.N.; Franke, J.; Kaufman, D.S.; Lücke, L.; Rehfeld, K. Consistent multidecadal variability in global temperature reconstructions and simulations over the Common Era. *Nat. Geosci.* **2019**, doi:10.1038/s41561-019-0400-0.
60. Tarjan, R.E. Depth-first search and linear graph algorithms. *SIAM J. Comput.* **1972**, *1*, 146–160, doi:10.1137/0201010.
61. Cormen, T.H.; Leiserson, C.E.; Rivest, R.L.; Stein, C. *Introduction to Algorithms*, 3rd ed.; The MIT Press: Cambridge, MA, USA, 2009.
62. Soman, J.; Narang, A. Fast community detection algorithm with GPUs and multicore architectures. In Proceedings of the 2011 IEEE International Parallel and Distributed Processing Symposium, Anchorage, AK, USA, 16–20 May 2011; IEEE Computer Society: Washington, DC, USA, 2011; doi:10.1109/IPDPS.2011.61.
63. Adamic, L.A.; Adar, E. Friends and neighbors on the web. *Soc. Netw.* **2003**, *25*, 211–230.
64. Newman, M.E.J. *Networks: An Introduction*; Oxford University Press: Oxford, UK, 2010.
65. Sniedovich, M. Dijkstra's algorithm revisited: The dynamic programming connexion. *J. Control Cybern.* **2006**, *35*, 599–620.
66. Cook, J.; Lewandowsky, S. *The Debunking Handbook*; University of Queensland: St. Lucia, Australia, 2011.

67. Pennycook, G.; Cheyne, J.A.; Barr, N.; Koehler, D.J.; Fugelsang, J.A. On the reception and detection of pseudo-profound bullshit. *Judgm. Decis. Mak.* **2015**, *10*, 549–563.
68. Plag, H.P.; Adegoke, J.; Bruno, M.; Christian, R.; Digiacomo, P.; McManus, L.; Nicholls, R.; van de Wal, R. Observations as decision support for coastal management in response to local sea level changes. In *Proceedings of OceanObs'09: Sustained Ocean Observations and Information for Society (Volume 2), Venice, Italy, 21–25 September 2009*; Hall, J., Harrison, D.E., Stammer, D., Eds.; ESA: Paris, France, 2010; doi:10.5270/OceanObs09.cwp.69.
69. Plag, H.P.; McCallum, I.; Fritz, S.; Jules-Plag, S.; Nyenhuis, M.; Nativi, S. The GEOSS Science and Technology Service Suite: Linking S&T Communities and GEOSS. *E3S Web Conf.* **2013**, *1*, 28003. doi:10.1051/e3sconf/20130128003.
70. Michalski, R.S.; Carbonell, J.G.; Mitchell, T.M., Eds. *Machine Learning: An Artificial Intelligence Approach*; Springer Science & Business Media: Berlin/Heidelberg, Germany, 2013.
71. Hartuv, E.; Schmitt, A.O.; Lange, J.; Meier-Ewert, S.; Lehrach, H.; Shamir, R. An algorithm for clustering cDNA fingerprints. *Genomics* **2000**, *66*, 249–256.
72. Benesty, J.; Chen, J.; Huang, Y.; Cohen, I. Pearson correlation coefficient. In *Noise Reduction in Speech Processing, Springer Topics in Signal Processing 2*; Springer: Berlin/Heidelberg, Germany, 2009; doi:10.1007/978–3–642–00296–0_5.
73. Hameed, M.A.; Al Jadaan, O.; Ramachandram, S. Collaborative filtering based recommendation system: A survey. *Int. J. Comput. Sci. Eng.* **2012**, *4*, 859–876.
74. Linden, G.; Smith, B.; York, J. Amazon. com recommendations: Item-to-item collaborative filtering. *IEEE Internet Comput.* **2003**, *7*, 76–80.
75. Houeland, T.G. An efficient random decision tree algorithm for case-based reasoning systems. In Proceedings of the FLAIRS 24th International Florida Artificial Intelligence Research Society Conference, Palm Beach, FL, USA, 18–20 May 2011; AAAI Press: Menlo Park, CA, USA, 2011.
76. Dalal, S.; Athavale, D.V.; Jindal, K. Case retrieval optimization of case-based reasoning through knowledge-intensive similarity measures. *Int. J. Comput. Appl.* **2011**, *34*, 12–18.
77. Larochelle, H.; Bengio, Y. Classification using discriminative Restricted Boltzmann Machines. In Proceedings of the 25th International Conference on Machine Learning, Helsinki, Finland, 5–9 July 2008; ACM: New York, NY, USA, 2008; pp. 536–543, doi:10.1145/1390156.1390224.
78. Lee, H.; Grosse, R.; Ranganath, R.; Ng, A.Y. Convolutional deep belief networks for scalable unsupervised learning of hierarchical representations. In Proceedings of the 26th International Conference on Machine Learning, Montreal, QC, Canada, 14–18 June 2009; ACM: New York, NY, USA, 2009; pp. 609–616.
79. Wang, H.; Wang, N.; Yeung, D.Y. Collaborative Deep Learning for Recommender Systems. In Proceedings of the 21th ACM SIGKDD International Conference on Knowledge Discovery and Data Mining, Sydney, NSW, Australia, 10–13 August 2015; ACM: New York, NY, USA, 2015; pp. 1235–1244, doi:10.1145/2783258.2783273.
80. Van den Oord, A.; Dieleman, S.; Schrauwen, B. Deep content-based music recommendation. In *Advances in Neural Information Processing Systems 26*; Burges, C.J.C., Bottou, L., Welling, M., Ghahramani, Z., Weinberger, K.Q., Eds.; Curran Associates, Inc.: Red Hook, NY, USA, 2013; pp. 2643–2651.
81. Forum, W.E. *Realizing the Potential of Blockchain—A Multistakeholder Approach to the Stewardship of Blockchain and Cryptocurrencies*; Technical Report; World Economic Forum: Davos, Switzerland, 2017. Available online: http://www3.weforum.org/docs/WEF_Realizing_Potential_Blockchain.pdf (accessed on 13 September 2019).
82. Nakamoto, S. Bitcoin: A Peer-to-Peer Electronic Cash System, 2009. Available online: metzdowd.com (accessed on 10 February 2018).
83. Swan, M. *Blockchain: Blueprint for a New Economy*; O'Reilly Media, Inc.: Sebastopol, CA, USA, 2015.
84. Van Rijmenam, M. The Top 11 Blockchains for Enterprise Organisations, and Why. Available online: https://vanrijmenam.nl/11-blockchains-enterprise-organisations-why/ (accessed on 13 September 2019).
85. Schneier, B. There's No Good Reason to Trust Blockchain Technology, 2019. Wired. Available online: https://www.wired.com/story/theres-no-good-reason-to-trust-blockchain-technology/ (accessed on 13 September 2019).
86. Hijgenaar, S. Not All Blockchains are Created Equal When It Comes to Energy Consumption. Available online: https://www.cgi.com/en/blog/utilities/not-all-blockchains-are-equal-when-it-\protect\discretionary{\char\hyphenchar\font}{}{}comes-to-\protect\discretionary{\char\hyphenchar\font}{}{}energy-\protect\discretionary{\char\hyphenchar\font}{}{}consumption (accessed on 13 September 2019).

87. Matthews, K. 4 Ways to Counter Blockchain's Energy Consumption Pitfall. https://www.greenbiz.com/article/4-ways-counter-blockchains-energy-consumption-pitfall (accessed on 16 August 2019).

88. Boldrini, E.; Craglia, M.; Mazzetti, P.; Nativi, S. The brokering approach for enabling collaborative scientific research. In *Collaborative Knowledge in Scientific Research Networks*; Diviacco, P., Fox, P., Pshenichny, C., Leadbetter, A., Eds.; IGI Global: Hershey, PA, USA, 2015; pp. 283–304, doi:10.4018/978-1-4666-6567-5.ch014.

89. Hsu, L.; Mayorga, E.; Horsburgh, J.; Carter, M.; Lehnert, K.; Brantley, S. Enhancing Interoperability and Capabilities of Earth Science Data using the Observations Data Model 2 (ODM2). *Data Sci. J.* **2017**, *16*, doi:10.5334/dsj–2017–004.

90. Hu, Y.; Janowicz, K.; Prasad, S.; Gao, S. Metadata Topic Harmonization and Semantic Search for Linked-Data-Driven Geoportals: A Case Study Using ArcGIS. *Trans. GIS* **2015**, *19*, 398–416, doi:10.1111/tgis.12151.

91. Khalsa, S.J.S. Data and Metadata Brokering—Theory and Practice from the BCube Project. *Data Sci. J.* **2017**, *16*, doi:10.5334/dsj–2017–001.

92. Campbell, J.; Jensen, D.E. *Could a Digital Ecosystem for the Environment Have the Potential to Save the Planet?*; Technical Report; National Council for Science and the Environment: Washington, DC, USA, 2019. Available online: https://www.ncseglobal.org/ncse-essays/could-digital-ecosystem-environment-have-potential-save-\protect\discretionary{\char\hyphenchar\font}{}{}planet (accessed on 25 September 2019).

93. Barrie, L.A.; The IGACO Writing Team. *An integrated Global Atmospheric Chemistry Observation Theme for the IGOS Partnership*; Technical Report, IGOS Integrated Global Observing Strategy; WMO: Geneva, Switzerland, 2004.

94. Unninayar, S.; Task Team. *GEO Task US-09-01a: Critical Earth Observations Priorities—Water Societal Benefit Area*; Technical Report; Group on Earth Observations—User Interface Committee: Geneva, Switzerland, 2016.

95. Plag, H.P.; Ondich, G.; Kaufman, J.; Foley, G. The GEOSS User Requirement Registry—Supporting a User-Driven Global Earth Observation System of Systems. *Imaging Notes* **2010**, *25*, 28–33.

96. Plag, H.P.; Jules-Plag, S.; Callaghan, C.; McCallum, I. Linking science and technology communities to GEOSS. In *Towards a Sustainable GEOSS (Global Earth Observation System of Systems)—Some Results of the EGIDA Project*; Nativi, S., Mazzetti, P., Plag, H.P., Eds.; Aíon: Florence, Italy, 2013; pp. 13–34, ISBN 978-88-98262-05-2.

97. Yang, X.; Blower, J.D.; Bastin, L.; Lush, V.; Zabala, A.; Masó, J.; Cornford, D.; Díaz, P.; Lumsden, J. An integrated view of data quality in Earth observation. *Philos. Trans. A Math. Phys. Eng. Sci.* **2013**, *371*, 20120072.

98. Lehmann, J.; Isele, R.; Jakob, M.; Jentzsch, A.; Kontokostas, D.; Mendes, P.N.; Hellmann, S.; Morsey, M.; van Kleef, P.; Auer, S.; et al. DBpedia—A Large-scale, Multilingual Knowledge Base Extracted from Wikipedia. *Semant. Web* **2012**, *6*, 167–195.

99. DBpedia Team. DBpedia—Global and Unified Access to Knowledge. Available online: https://wiki.dbpedia.org/ (accessed on 21 September 2019).

100. McCallum, I.; P. Plag, H.; Fritz, S. Data Citation Standard: A Means to Support Data Sharing, Attribution, and Traceability. *E3S Web Conf.* **2013**, *1*, 28002, doi:10.1051/e3sconf/20130128002.

101. W3C. *RDF 1.1 Concepts and Abstract Syntax*; Technical Report; W3C: Keio, Japan, 2014. Available online: https://www.w3.org/TR/rdf11-concepts/ (accessed on 6 June 2019).

102. Oracle. *Oracle Big Data Spatial and Graph—Property Graph: Features and Performance*; Technical Report, ORACLE Technical Whitepaper; Oracle: Redwood City, CA, USA, 2017.

103. Levin, K.; Cashore, B.; Bernstein, S.; Auld, G. Overcoming the tragedy of super wicked problems: Constraining our future selves to ameliorate global climate change. *Policy Sci.* **2012**, *45*, 123–152, doi:10.1007/s11077-012-9151-0.

104. Roberts, N. Wicked Problems and Network Approaches to Resolution. *Int. Public Manag. Rev.* **2000**, *1*, 1–19.

105. Obersteiner, M.; Walsh, B.; Frank, S.; Havlík, P.; Cantele, M.; Liu, J.; Palazzo, A.; Herrero, M.; Lu, Y.; Mosnier, A.; et al. Assessing the land resource–food price nexus of the Sustainable Development Goals. *Sci. Adv.* **2016**, *2*, 10.1126/sciadv.1501499.

106. World Economic Forum. *Global Risks 2016*, 11th ed.; Technical Report; World Economic Forum: Geneva, Switzerland, 2016.

107. García, L.E.; Rodríguez, D.J.; Wijnen, M.; Pakulski, I., Eds. *Earth Observation for Water Resources Management: Current Use and Future Opportunities for the Water Sector*; World Bank Group: Washington, DC, USA, 2016; doi:10.1596/978-1-4648-0475-5.

108. Keskinen, M.; Someth, P.; Salmivaara, A.; Kummu, M. Water-Energy-Food Nexus in a Transboundary River Basin: The Case of Tonle Sap Lake, Mekong River Basin. *Water* **2015**, *7*, 5416–5436, doi:10.3390/w7105416.
109. Lehmann, A.; Giuliani, G.; Ray, N.; Rahman, K.; Abbaspour, K.C.; Nativi, S.; Craglia, M.; Cripe, D.; Quevauviller, P.; Beniston, M. Reviewing innovative Earth observation solutions for filling science-policy gaps in hydrology. *J. Hydrol.* **2014**, *518*, 267–277, doi:10.1016/j.jhydrol.2014.05.059.
110. Brown, V.A.; Harris, J.A.; Russell, J.Y. (Eds.) *Tackling Wicked Problems—Through the Transdisciplinary Imagination*; Earthscan: New York, NY, USA, 2010.

Article

Towards Sentinel-1 SAR Analysis-Ready Data: A Best Practices Assessment on Preparing Backscatter Data for the Cube

John Truckenbrodt [1,2,*], Terri Freemantle [3], Chris Williams [3], Tom Jones [3], David Small [4], Clémence Dubois [1], Christian Thiel [2], Cristian Rossi [3], Asimina Syriou [3] and Gregory Giuliani [5]

1   Department for Earth Observation, Friedrich-Schiller-University Jena, 07743 Jena, Germany
2   Institute for Data Science, German Aerospace Center DLR, 07745 Jena, Germany
3   Satellite Applications Catapult, Harwell Campus, Didcot OX11 0QR, UK
4   Remote Sensing Laboratories, Dept. of Geography, University of Zurich, 8057 Zurich, Switzerland
5   Institute for Environmental Sciences, University of Geneva, 1205 Geneva, Switzerland
*   Correspondence: john.truckenbrodt@uni-jena.de

Received: 15 June 2019; Accepted: 2 July 2019; Published: 5 July 2019

Abstract: This study aims at assessing the feasibility of automatically producing analysis-ready radiometrically terrain-corrected (RTC) Synthetic Aperture Radar (SAR) gamma nought backscatter data for ingestion into a data cube for use in a large spatio-temporal data environment. As such, this study investigates the analysis readiness of different openly available digital elevation models (DEMs) and the capability of the software solutions SNAP and GAMMA in terms of overall usability as well as backscatter data quality. To achieve this, the study builds on the Python library pyroSAR for providing the workflow implementation test bed and provides a Jupyter notebook for transparency and future reproducibility of performed analyses. Two test sites were selected, over the Alps and Fiji, to be able to assess regional differences and support the establishment of the Swiss and Common Sensing Open Data cubes respectively.

Keywords: Sentinel-1; SAR; analysis ready data; ARD; interoperability; data cube; Earth observation; pyroSAR

---

1. Introduction

Global Earth systems are facing increased pressure—over-exploitation of resources, climate change, environmental and ecological degradation, and overpopulation—meaning that the ability to measure and monitor Earth surface change is of ever-increasing value [1]. Advances in technology, the democratization of space and recognition of the value of Earth Observation (EO) in providing insights—e.g., for the Sustainable Development Agenda—have led to an increase in the availability of EO data worldwide, and with this, a growing interest globally in efficient exploitation of EO data at scale [2]. Global monitoring programs coupled with an extensive archive of historical remotely sensed imagery have paved the way for both historic time-series analysis and operational routine monitoring [3].

The launch of the set of Sentinel satellites by the European Space Agency (ESA) as part of the European Commission's Copernicus Program has been a catalyst for this change and is generating ever-increasing interest across governments and different market sectors, each with different user requirements [4]. It becomes quickly apparent that it is simply not technically feasible or financially affordable to consider traditional methods of storing, handling and manipulating EO data. Local processing and data distribution methods currently exploited by industry and government are not suitable to address the challenge of scalability, increases in the size of data volumes, and the growing complexities in the preparation, handling, storage, and analysis required to meet user requirements. To allow immediate analysis of the data without additional significant user effort, these barriers need to be addressed. In response, there has been a drive within the EO community to find faster, cost-effective ways to process EO data at scale, whilst facilitating access to EO-derived insights. Two concepts addressing this challenge are Earth Observation Data Cubes (EODC), coupled with the concept of Analysis-Ready Data (ARD) [5,6].

As a geodata infrastructure technology for convenient storage and analysis of large amounts of raster data, the concept of data cubes has been gaining ground. In particular, the Open Data Cube (ODC), originally developed by Geoscience Australia (GA) and having evolved as an international initiative supported by the Committee of Earth Observation Satellites (CEOS), has found wide application in part due to its user-friendly Python application programming interface (API). Several efforts from CEOS and national ODC initiatives, such as the Swiss Data Cube (SDC) [7], the Common Sensing Data Cube (CSDC), the Ghana Data Cube [8], and Digital Earth Australia (DEA) [9] are working towards making EO data accessible and are discussing which data specifications need to be met to optimally provide data over this new infrastructure.

Synthetic Aperture Radar (SAR) is an EO system that has the advantage of being almost weather and solar illumination independent. Therefore, SAR data, and more specifically, Sentinel-1 is becoming popular, as many regions have issues with cloud cover, and would benefit from denser temporal sampling, e.g. for applications such as forest change detection and coastal monitoring. Through the Copernicus program, Sentinel-1 data is routinely and freely available. With the availability of ESA's Sentinels Application Platform (SNAP) open-source software, the access barrier to SAR data has been significantly lowered. Large amounts of SAR data acquired with a repeat interval of twelve days (reduced to six using both Sentinel-1A and Sentinel-1B) can be freely downloaded and conveniently processed [10,11].

However, to access the valuable information contained within EO data, users are required to undertake a series of complex pre-processing steps to turn the data from a 'raw' unprocessed format into a state that can be analyzed. Unless the user has the expertise, software and infrastructure to handle and process this information, efficient exploitation of the data is not realized.

A term that is now frequently used in this context is Analysis-Ready Data (ARD). According to the Committee on Earth Observation Satellites (CEOS), this is defined as 'satellite data that have been processed to a minimum set of requirements and organized into a form that allows immediate analysis without additional user effort and interoperability with other datasets both through time and space' (http://ceos.org/ard/).

Originally defined for optical satellite imagery, this generally describes data that is corrected for atmospheric effects and thus contains measurements of surface reflectance. Currently, the majority of known Data Cube implementations rely on optical imagery [9,12,13] and only a few of them offer access to SAR ARD products. One example of the use of SAR data in a Data Cube framework is the Water Across Synthetic Aperture Radar Data (WASARD) for water body classification [14]. Having SAR data in an Earth Observation Data Cube (EODC) can be an excellent complement to optical imagery and can overcome limitations such as cloud coverage. The main reason that there is currently little SAR data available in EODCs comes from the fact that there was, until recently, no common definition of the ARD level. CEOS is leading an effort to define the minimum set of requirements to allow immediate analysis with minimum additional user effort. The CEOS Analysis-Ready Data for Land (CARD4L—http://ceos.org/ard/) provides specifications for Optical, Thermal, and SAR imagery. Regarding SAR, the ARD level was recently defined only for terrain-corrected radar backscatter. Polarimetric and interferometric specifications are under development and are expected for 2019. CARD4L SAR products will be: (1) Normalized Radar Backscatter; (2) Geocoded Single-Look Complex; (3) Polarimetric Radar Decomposition; (4) Normalized Radar Covariance Matrix, and (5) Differential Interferometry Products [15]. To be considered as ARD, the Normalized Radar Backscatter product should be Radiometric Terrain Correction (RTC) and provided as gamma0 (γ0) backscatter, which mitigates systematic contamination that would otherwise still be present in sets of data acquired with multiple geometries [16].

While there is little dispute over the individual processing steps necessary to convert the original level 1 backscatter products provided by, e.g., Copernicus to RTC [16], different software implementations may lead to significant differences in final product quality. The Committee on Earth Observation Satellites (CEOS) recently published a comprehensive guide on how to produce Analysis-Ready SAR RTC data for land mapping applications, listing a large number of requirements for metadata specifications and necessary corrections to obtain a well-documented data set of high quality [15]. However, what remains missing, is a straightforward implementation in commercial and open-source software such that a user can directly select a certain level of quality and "analysis readiness". Once the data is prepared to the highest standard possible, the influence of SAR-specific imaging effects is to be assessed. Effects such as geometric decorrelation are inherent to SAR data and while they are relevant to some applications, they can be considered as disturbances for others. Therefore, the user needs to be aware of them when analyzing any specific backscatter ARD product.

The overall aim of this study is to evaluate how far Sentinel-1 data is interoperable in terms of geometry and software. It picks up on findings reported in [17], however, investigating two new test areas; in the Alps and in Fiji. These test areas were selected as there is currently work underway on developing operational data cubes for these two areas (SDC—http://www.swissdatacube.ch, CSDC—http://commonsensing.org.gridhosted.co.uk/).

The SDC is an innovative analytical cloud-computing platform allowing users to access, analysis and visualization of 35 years of optical (e.g., Sentinel-2; Landsat 5, 7, 8) and radar (e.g., Sentinel-1) satellite EO ARD over the entire country [5,18]. Importantly, the SDC minimizes the time and scientific knowledge required for national-scale analyses of large volumes of consistently calibrated and spatially aligned satellite observations. The SDC is based on the Open Data Cube software stack [19,20] and is updated continually. It contains approximately 10,000 scenes for a total volume of 6TB and more than 200 billion observations over the Alps. The objective of the SDC is to support the Swiss government for environmental monitoring and reporting, as well as enabling Swiss scientific institutions to benefit from EO data for research and innovation. Additionally, the SDC allows for medium/high spatial and temporal resolution environmental monitoring, thereby providing synoptic, consistent and spatially explicit information sufficiently detailed to capture anthropogenic impacts at the national scale. The SDC is supported by the Swiss Federal Office for the Environment (FOEN) and currently is being developed, implemented and operated by the United Environment Program (UNEP)/GRID-Geneva in partnership with the University of Geneva (UNIGE), the University of Zurich (UZH) and the Swiss Federal Institute for Forest, Snow and Landscape Research (WSL). Ultimately, the SDC will deliver a unique capability to track changes in unprecedented detail using EO satellite data and enable more effective responses to problems of national significance [18]. To our knowledge, the Swiss Data Cube is the first Data Cube to contain almost the entire Sentinel-1 ARD archive. It contains five years of 12-day terrain-flattened backscatter Sentinel-1 backscatter composites generated using the methodology described in [16] in the initial stage.

For the SDC, Sentinel-1 data will be useful to enhance the Snow Observations from Space (SOfS) algorithm that currently only uses optical imagery (e.g., Landsat, Sentinel-2) [20]. Preliminary results have shown a clear decrease in snow cover over the Alps in the last 30 years. However, to provide an integrated and effective mechanism to monitor snow cover and its variability, SAR data will help identify snowmelt processes [21]. In addition to snow mapping, Sentinel-1 analysis-ready data has been shown to be useful for vegetation mapping and dynamics [22], rapid assessment after a storm event [23], and melt-onset mapping using multiple SAR sensors [24].

The Common Sensing project is an international development project that aims to improve climate change and disaster risk resilience in the Small-Island Developing States (SIDS) of Fiji, Vanuatu and the Solomon Islands (http://commonsensing.org.gridhosted.co.uk/) with the support of EO data and tools. As part of this multi-year project, an EODC for Fiji is being developed on the Open Data Cube software stack [19,20] and will contain Analysis-Ready Data for Sentinel-2, Landsat-5-8, SPOT 1-5 (surface reflectance) and Sentinel-1 (normalized radar backscatter). Much like the SDC, the Common Sensing Data Cube (CSDC) will be built with an aim to break down barriers to the use of EO data by policymakers through the provision of data and tools to facilitate rapid generation of EO-derived products and insights through both time and space. The objective is to support government and non-government stakeholders to undertake routine monitoring and reporting on Earth surface dynamics in rapidly changing environments in the South Pacific SIDS. For example, the CSDC will focus on exploitation of S1 data for vegetation mapping, coastal erosion, water resource management and disaster response, e.g., flooding.

As part of this study, different existing pre-processing workflows from different software solutions were evaluated for their interoperability. For instance, changes in backscatter from these differences might be too severe for certain mapping applications. Therefore, this study aimed at investigating the signal stability over different land cover classes to observe whether temporal backscatter variability originates from actual changes over land or are in fact the result of, e.g., different viewing angles or acquisition times. Of further interest in this context is a thorough analysis of how far the quality of the Digital Elevation Model (DEM) used in the processing affects the quality of the resulting products. Finally, this study provides an open-source assessment framework via a Python package including a Jupyter notebook (see Supplementary Materials).

Although compatibility with the CEOS ARD backscatter standard is, ultimately, to be reached, this study does not perform a formal assessment of the extent to which the specific requirements are met.

2. Study Outline and Description of Test Sites

This study investigated the use of two SAR processing software solutions, SNAP and GAMMA, for producing radiometrically terrain-corrected Sentinel-1 SAR backscatter. In particular, the influence of the resampling method and the DEM on the resulting topographic normalization was assessed. This section guides the reader through the paper's structure. First, Chapter 3 describes the technical methodology of the study. In Chapter 4, the results from the analyses performed are presented. Chapters 5 and 6 discuss the findings and conclude. Motivated by the activities around the Swiss and Common Sensing Data Cubes, two test sites were selected, in the Alps and in Fiji, respectively.

In a first major component, two single S1A ground-range detected (GRD) scenes, acquired over the two test sites, were processed using two software solutions with different parametrizations and DEMs to assess the quality of the resulting RTC backscatter. This is described in Sections 4.1–4.3. The identifiers of the two scenes are presented in Table 1; their footprints are shown in Figures 1 and 2. The footprints shown in these plots were used throughout the three mentioned sections to create DEMs of the same size and used the same inputs for the SAR processing.

Table 1. Identifiers of the scenes used for single image analysis.

Test Site	Scene Identifier
Alps	S1A_IW_GRDH_1SDV_20180829T170656_20180829T170721_023464_028DE0_F7BD
Fiji	S1A_IW_GRDH_1SDV_20181229T064000_20181229T064036_025236_02CA47_4B57

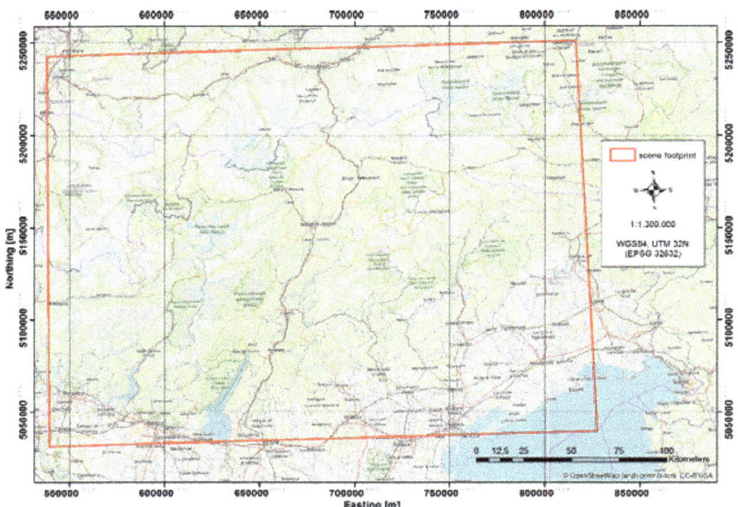

Figure 1. Footprint of the S1 ground-range detected (GRD) scene over the Alps used throughout this study.

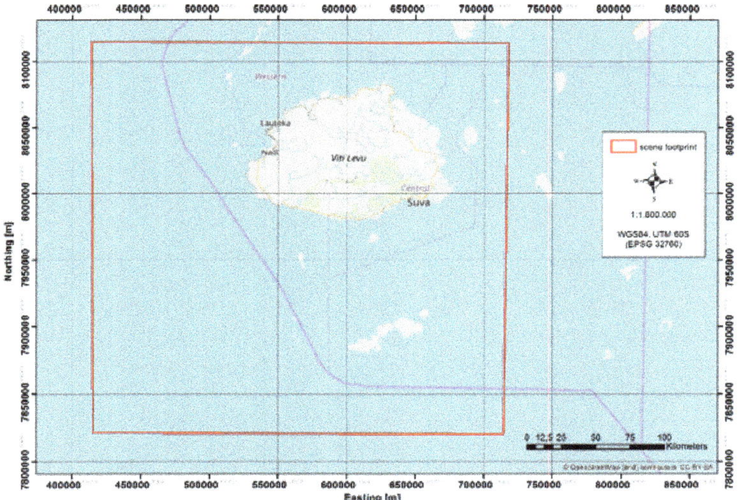

Figure 2. Footprint of the S1 ground-range detected (GRD) scene over Fiji used throughout this study.

These scenes were processed with the steps described in Section 3.3 to UTM Zone 32N (EPSG 32632) and Zone 60S (EPSG 32760), respectively, with a spatial resolution of 90 m. This resolution was chosen as a common denominator of all DEMs used in order to more objectively compare their quality independent of the differences in spatial resolution.

After thorough analysis of single image processing results, focus shifted to time series analysis in Section 4.4. For this, the study area was scaled down slightly to the island of Viti Levu, which can also be seen in Figure 2. *Viti Levu* is the largest island in the Fijian archipelago and the most populated. Sentinel-1 data is being routinely collected over *Viti Levu* from ascending and descending tracks every ~6 days; however, data is currently only being collected by Sentinel-1A. For this study, 62 Sentinel-1 GRD scenes acquired between April 2018 and April 2019 were processed in accordance with steps described in Section 3.3 and re-projected onto a UTM Zone 60S (EPSG 32760) grid with a sample interval of 20 m. An overview of Fiji, the study area for time series analysis, and the relevant S1 acquisition frames are displayed in Figure 3.

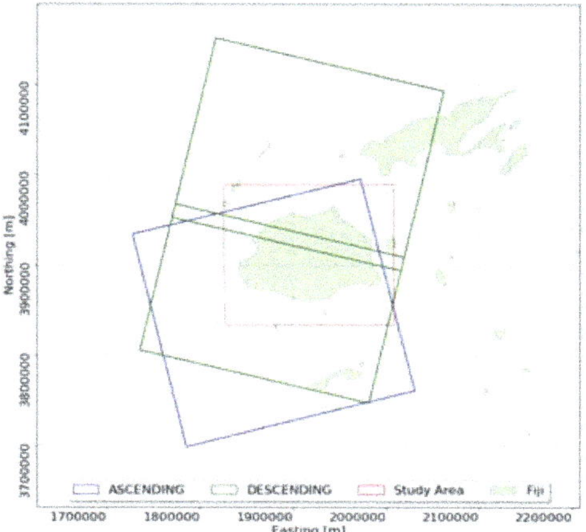

Figure 3. Sentinel-1 acquisition frames covering Fiji study area. Spatial reference system: Fiji 1986 Map Grid (EPSG 3460).

3. Methods

3.1. Software

This study and the accompanying Jupyter notebook (see Supplementary Materials) build on two Python packages pyroSAR [25,26] and spatialist [27]. PyroSAR is a framework for organizing and processing SAR data with APIs to the ESA Sentinel Application Platform (SNAP) [28] and GAMMA [29]. It serves the purpose of wrapping the image processing into convenient Python functions so that processing in SNAP and GAMMA can be operated in a similar way. Spatialist offers general spatial data handling functionality for pyroSAR by providing a convenient wrapper for the Python bindings of the Geodata Abstraction Library (GDAL) [30], offering a collection of general spatial data handling tools.

Several additions have been made to pyroSAR during this study, which are reflected in versions 0.7 to 0.9.1. A changelog summarizing these changes is available in pyroSAR's online documentation [26].

While the workflows used during this study for processing with GAMMA and SNAP already existed, several additions and modifications were made to further improve the accuracy of the processing result and the usability of the routines within a Jupyter notebook.

Throughout this study, images were processed using SNAP version 6.0.9 and a GAMMA version released in November 2018. SNAP7 is due to be released in Summer 2019 with announced improvements to the terrain flattening procedure [31]. This processing step is particularly important for creating RTC products and processing results will be considered as soon as this new version is released to update the findings accordingly.

During this study, it was observed that SNAP processing of large workflows, i.e., with many processing steps in sequence, takes disproportionately longer, the more processing steps are added to it. For this reason, a mechanism was added to pyroSAR which splits a workflow into several groups, writes each group to a new temporary workflow XML file and executes these new workflows in sequence. Temporary products are written by the intermediate workflows, which are then passed to the succeeding workflow. Once finished, the directory containing the temporary workflows and products is deleted. This was observed to drastically increase processing speed, but no dedicated benchmarking was performed.

In addition to the Jupyter notebook technology [32], this study further builds on several open-source Python packages, in particular, Numpy [33] for general array handling, Matplotlib [34] for visualization, Scipy [35] and Astropy [36] for specific array computations and Scikit-Learn [37] for computation of performance statistics.

3.2. DEM Preparation

The choice of a high-quality DEM is crucial for accurate SAR processing, in particular for the correction of topographic effects such as foreshortening. According to the CEOS recommendations on producing analysis ready normalized backscatter for land [15], the selected DEM optimally has a spatial resolution as good or better than the resolution of the SAR image. Furthermore, it is recommended to assess whether the topography had changed between the acquisition of the DEM and that of the SAR scene to ensure that changes in backscatter are not related to changes in topography.

Thus, in this study, different DEMs were compared in order to assess the extent to which newer options are better suited for processing SAR imagery than older ones. Although newer DEMs might be closer in acquisition date to the SAR scene, older options have likely undergone more processor updates and manual edits to correct processor shortcomings over the years and might still be the better choice. See, e.g., [38] and [39] for details on SRTM quality enhancement.

The SAR processing was performed with four different DEMs, SRTM in 1 arcsec and 3 arcsec resolution [40], the 30 m ALOS World DEM (AW3D30) [41] and the TanDEM-X DEM in 90 m resolution [42]. An overview of DEM download URLs is given in Table 2. For this study, routines were developed in pyroSAR to automatically prepare these different DEM types for processing in both SNAP and GAMMA software. This includes downloads of respective DEM tiles for a defined geometry, e.g., the footprint of a SAR scene, mosaicking and cropping, as well as re-projection and conversion from EGM96 geoid heights to WGS84 ellipsoid heights if necessary. Adopting this methodology, it was guaranteed that all DEM mosaics were created in an identical way—thereby mitigating inconsistencies introduced by the DEM preparation itself. The DEM files used for SNAP and GAMMA were thus identical aside from the file format, which was GeoTIFF for SNAP and the GAMMA file format in the latter.

Table 2. Digital Elevation Models (DEMs) used in this study and their sources.

DEM	Source
ALOS World 3D 30 m	ftp://ftp.eorc.jaxa.jp/pub/ALOS/ext1/AW3D30/release_v1804
SRTM 1 arcsec	https://step.esa.int/auxdata/dem/SRTMGL1
SRTM 3 arcsec	http://srtm.csi.cgiar.org/wp-content/uploads/files/srtm_5x5/TIFF
TanDEM-X 90 m	ftpes://tandemx-90m.dlr.de

To fully evaluate the quality of the respective DEMs, a high-resolution reference DEM, e.g., from a LiDAR flight campaign, is required for error analysis. Such products were not available in this study for either test site. By relying on openly available DEM products, however, the overall reproducibility of the study is increased. Hence, the analysis focused on identifying which DEM deviated most from the others to provide an indication of relative DEM consistency within the area under investigation. For this, the median of all DEMs was computed and an index map created identifying which DEM deviated most from the median at respective pixels and to what magnitude. This analysis served the purpose of quantifying outliers in the four DEMs. In a second analysis, the impact of the DEM choice on SAR processing was investigated, whose methodology is described in Section 3.4. The optimal DEM over a specific test site has only a few outliers of small magnitude, resulting in high-quality SAR products. The results of both analyses are presented in Section 4.3.

Generally, this analysis aims to be reproducible for every area worldwide in order to assess which DEM is best suited for a specific area of interest—we do not intend here to make a general global DEM recommendation.

3.3. SAR Processing

For this study, processing of Sentinel-1 Ground-Range Detected (GRD) imagery to radiometrically terrain corrected gamma0 backscatter (γ0 RTC)—in line with the CARD4L SAR backscatter specification—was investigated. The correction of topographic effects is seen as essential for storing SAR datasets from multiple acquisition geometries in a data cube in order to create a consistent interoperable product. The superior interoperability of gamma0 in comparison to sigma0 was previously investigated in a previous study [17].

The workflows in SNAP and GAMMA were designed to match each other as closely as possible. This includes removal of border noise and thermal noise, calibration, multilooking, update of orbit state vectors, terrain flattening according to [16], geocoding and conversion to logarithmic (dB) scaling. See Figure 4 for a visualization of the GAMMA and SNAP workflows used.

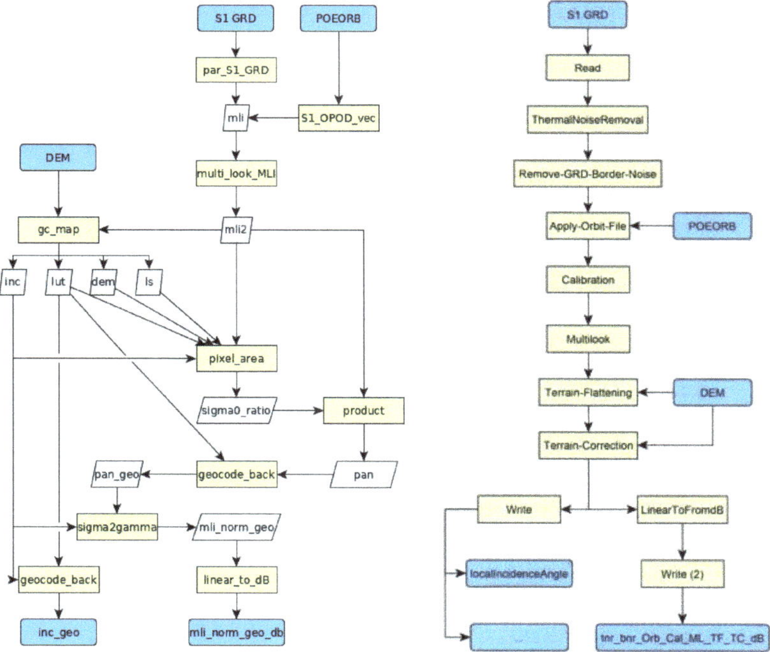

Figure 4. pyroSAR's GAMMA (**left**) and SNAP (**right**) workflows for producing Sentinel-1 radiometrically terrain corrected (RTC) backscatter (from [26]). Dark blue: input and output products; light blue: intermediate products; green: processing steps. *POEORB*: Precise Orbit Ephemerides orbit state vector files; intermediate products created by GAMMA: *mli*: multi-looked image in slant range; *inc*: local incident angle map; *lut*: geocoding lookup table; *ls*: layover-shadow mask; *pan*: pixel area normalized backscatter; *sigma0_ratio*: ratio between ellipsoid and DEM-based sigma0 normalization areas. The suffix *_geo* depicts products in map geometry. No intermediate products are created by SNAP.

One major difference in the overall approach between the two software solutions is the handling of the input ground range detected (GRD) imagery. While in SNAP, all processing steps are directly performed on the original GRD images, in GAMMA, the images are per default first converted back to slant range before further processing steps such as multi-looking and terrain flattening are performed. According to [16], the topographic normalization can be performed in either ground or slant range geometries after appropriate conversions. Differences between the backscatter estimates from alternatively processing the images in ground range or slant range were not investigated in this study.

Further differences can occur due to the different resampling methods used by SNAP and GAMMA. While in GAMMA, an input DEM is either left as it is or oversampled by user-defined factors, a SNAP user has several options of standard resampling methods of which one has to be selected. Therefore, the input DEM is modified by SNAP at all times while in GAMMA, the DEM can be left unaltered, possibly reducing additional inaccuracies caused by resampling. The latter approach is preferred, since all images processed with a certain DEM will be in exactly the same pixel grid while they can be shifted relative to each other if the DEM is resampled to the exact extent of the SAR scene during processing.

The same methods that are available in SNAP for resampling of the input DEM are also available for geocoding the final SAR image. For this study, bilinear resampling was chosen based on the overall quality of the result and processing time. For resampling multi-looked images with the GAMMA software, B-spline interpolation on the square root of the SAR data (sqrt(data)) was the recommended method according to the GAMMA documentation. By first transforming the data to the square root, interpolation errors are reduced due to reduced dynamic range and effective spectral bandwidth. After interpolation the data is transformed back to its original linear scale [43].

In addition to the processing capabilities of both software packages, one critical step was executed directly in pyroSAR. A feature inherent to S1 GRD images acquired before IPF version 2.9 released in early 2018 was the border noise, which needs to be masked prior to processing. A processing step is available in SNAP, which is a direct implementation of the recommendations for removal by ESA [44]. No such implementation is offered in GAMMA. While the SNAP implementation certainly reduces the noise, it is not sufficient to completely remove it. Hence, a custom implementation is used by pyroSAR, which also follows the official ESA recommendations but applies additional corrections to the results, thus creating a cleaner image border and reducing the noise to a minimum. The correction consists of three major steps. First, a line is generated marking the border between valid and invalid pixels from the ESA masking. Second, this line is simplified using the *Visvalingam-Whyatt* (VW) method of polyline vertex reduction [45]. Third, the VW-simplified line is shifted so that all areas masked by the original line are again covered. This process is exemplarily shown in Figure 5. While earlier versions of pyroSAR already featured this removal for GAMMA processing, it was added to the default SNAP workflow during this study to further match the workflows across both software packages and increase the quality of SNAP products.

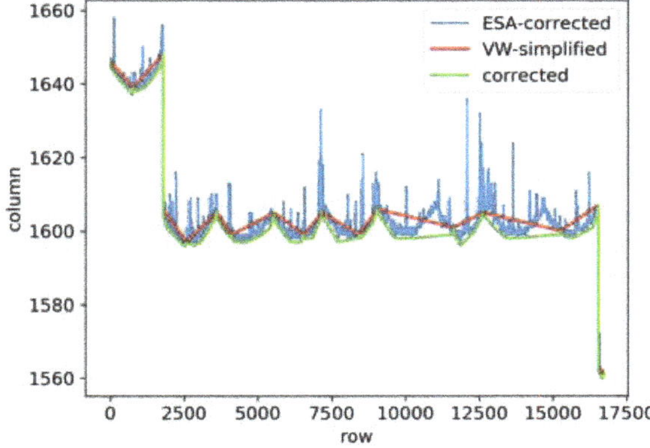

Figure 5. Demonstration of the border noise removal for a left image border. The area under the respective lines covers pixels considered valid, everything above will be masked out. From [26].

3.4. Assessment of Topographic Normalization Quality

In order to compare different processing workflows for their ability to correct for backscatter differences originating from the orientation of the terrain towards the sensor, the RTC backscatter products were compared with the local incident angle (INC). An image which has not been corrected for terrain effects will show a negative correlation with the INC product such that areas tilted towards the sensor are brighter than shadowed areas tilted away from the sensor [16]. In order to compare all processed images to a common INC product, an image was created according to the descriptions by Small 2011 and Meier et al. 1993 [46] in 30 m resolution and UTM projection for both study sites, respectively. In the following sections, this product is referred to as a UZH (University of Zurich) incident angle product. During processing, the products created by SNAP were found to be aligned to a different pixel grid than that defined by the input DEM, while in GAMMA, this exact grid was preserved. This is explained by the above-described additional resampling, which is always applied in SNAP. For this reason, two different INC products were resampled from the original UZH product to match the respective grids and the spatial resolution of 90 m used throughout this study for comparison with single image results. By up-sampling the product from 30 m to 90 m, nearly identical products were used for the SNAP and GAMMA grids. Otherwise, an additional resampling step would have had to be applied directly to the backscatter results of either software to match the grid of the other, potentially introducing additional errors and impairing comparability of the results.

By default, the INC products created by SNAP and GAMMA internally are also created as GeoTIFF files together with the SAR backscatter files in the accompanying Jupyter notebook using the processors described in Section 3.3. While the large UZH product could not be integrated with this otherwise open-source approach, similar products can thus be created in the notebook and alternatively be used for comparisons.

While it was expected that these three INC products, UZH, GAMMA and SNAP, would be nearly identical due to their simple computation, large differences were observed. Figure 6 shows the general differences between the three. All products were resampled to a common grid at 90 m resolution.

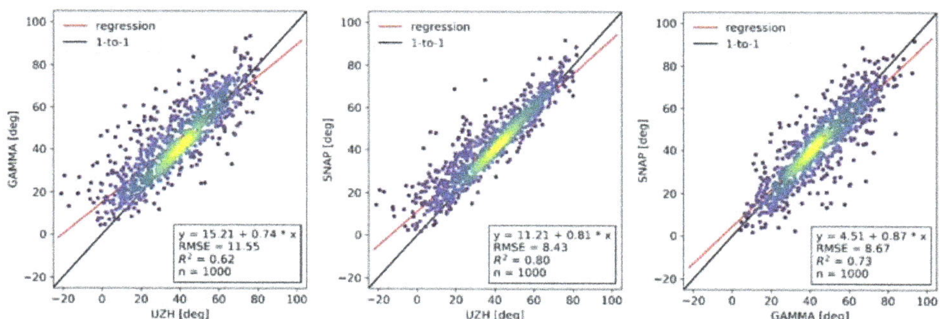

Figure 6. Comparison of local incident angle maps produced by University of Zurich (UZH), GAMMA and SNAP.

One large difference is the value range of the angles found in the maps. While the UZH product contained negative values lower than −20°, the other two products contained only positive values. An angle of 0° would be found on slopes oriented vertically to the sensor's line of sight with a value identical to that of the sensor's incident angle. Any slope tilted even further would thus be negative. The reason that these values do not occur in either SNAP or GAMMA may be that they employ a different solution compared to the angle between two 3D vectors, as done by UZH.

A much larger divergence between the GAMMA and UZH products was observed compared to the other two juxtapositions. The highest similarity, both in Root Mean Square Error (RMSE) and coefficient of determination (R^2), was found between the UZH and SNAP product. On closer inspection, a shift of the GAMMA product relative to the other two products of about $\frac{1}{2}$ to one pixel to the east was observed. The SNAP product appeared much smoother than the other two, suggesting some additional spatial filtering is internally applied by the software. Overall, the UZH product visually contained the most detail and spatial variation, particularly, on slopes tilted towards the sensor.

Furthermore, the linear regression slope significantly deviated from one with values between 0.74 and 0.87. The value closest to one was observed in the SNAP vs. GAMMA comparison, both being similar in value range between 0 and 90. Larger deviations were observed in the UZH comparisons, which can be explained by the different value ranges.

3.5. Masking of Land Cover Classes

In order to compare the quality of topographic normalization of the different DEMs, backscatter was masked to forested areas which generally return more stable backscatter than most other land cover classes [16]. In the Alps, the European CORINE 2018 product (CLC) was used [47]. Being available with 100 m resolution, it was up-sampled to 90 m resolution and the grid of the respective SAR images to be masked. Binary forest masks were created combining broad-leaved, coniferous and mixed forest.

Over the Fiji study area, a land cover dataset made available by Fiji's Ministry of Lands, sourced from PacGeo (http://www.pacgeo.org/layers/geonode:fiji_vector), was used. The Land Use/Land Cover dataset was created by AIR Worldwide for the Pacific Catastrophe Risk Assessment & Financing Initiative (PCRAFI—http://pcrafi.spc.int/).

To reliably assess ARD interoperability, it is essential that backscatter properties of corresponding land cover surfaces remain relatively consistent over the period of analysis. To discriminate regions of moist (evergreen) and dry (deciduous) tropical forest, forested-area polygons were fused with average annual precipitation derived from 1 km resolution WorldClim version 2 climate datasets—http://www.worldclim.org/. The fused dataset is shown in Figure 7. Interoperability analysis subsequently focused on areas provisionally identified as evergreen forest and regions of grassland located to the north west of *Viti Levu*—the main island of Fiji.

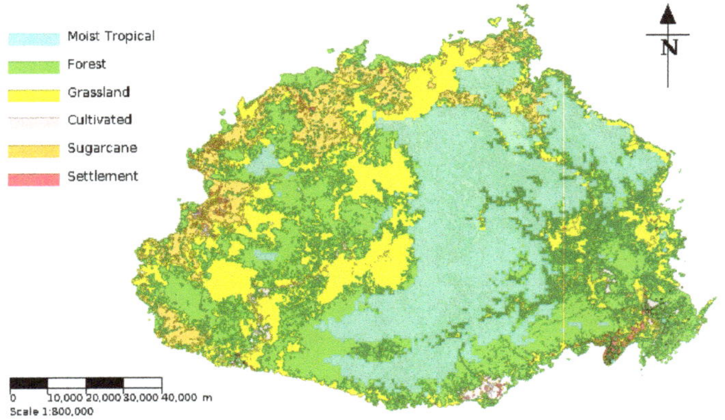

Figure 7. *Viti Levu* land cover. Spatial reference system: Fiji 1986 Map Grid (EPSG 3460).

3.6. Time Series Analysis

A 12-month time series (April 2018–April 2019) of Sentinel-1 GRD gamma0 at 20 m resolution was generated using the GAMMA- and SNAP-derived workflows outlined in Section 3.3. For the latter software, the SNAP6 SRTM 1 arcsec *auto-download* setup was used. In total, 62 Sentinel-1A raw scenes from orbital tracks 44 (descending) and 139 (ascending) were processed, providing a regular six-day sampling of spatiotemporal variation in C-band backscatter across the island of *Viti Levu*. Sentinel-1B is currently not being tasked to acquire imagery of this part of the South Pacific.

To manage large raster time series and retrieve image statistics coincident with shapefile geometries, the study leveraged the functional power of open-source spatiotemporal database technologies (PostgreSQL, PostGIS, TimescaleDB). Land cover datasets, outlined in Section 3.5, were re-projected into a local UTM coordinate reference system and loaded into ancillary PostgreSQL/PostGIS data tables. On a scene by scene basis, gamma0 image files were compiled into a multi-band, XML-based virtualized raster (VRT) file and loaded as out-of-database raster objects into TimescaleDB hypertables partitioned according to acquisition datetime. Analysis was, therefore, supported by a highly optimized, data abstraction platform, where complex multi-dimensional queries were rapidly executed against raster and vector datasets using a single SQL command. This approach allows for convenient repeatability and scalability of queries for statistical analysis.

4. Results

4.1. Software Parameterization

Several tests were performed to ensure optimal parameterization of the processing routines. A large number of options exist for preparing the DEM, adjusting its resolution, the choice of DEM resampling during processing and choice of interpolation of the SAR scene during geocoding. While optimizing all these processing parameters is outside the scope of this study and the results are likely different for other SAR scenes, a quick comparison was judged necessary in order to approximate optimal processing parameters. This analysis was performed for the Alps test site only.

4.1.1. GAMMA

Of particular interest was the effect of DEM resolution choice on the normalization. For example, if a SAR scene is to be processed to 90 m resolution, users have the option to resample the DEM to this resolution prior to processing, or alternatively, during processing. Furthermore, users have the option to convert DEM heights from EGM96 geoid to WGS84 ellipsoid in GDAL or, alternatively, in GAMMA. While for the sake of optimal comparison with SNAP it is seen preferential to do the conversion in GDAL, it was judged necessary to assess whether the SAR image quality is similar to the result using GAMMA's internal conversion.

A second comparison was made between UTM and WGS84 LatLon to assess the software's sensitivity to different coordinate reference systems. The UTM DEM was used in 30 m resolution, the WGS84 DEM was left at its original resolution of approximately 30 m north–south and 21 m east–west. Both were internally resampled in GAMMA so that a target resolution of 90 m in both directions was approximated.

As a third assessment, several geocoding interpolation modes were compared, which are listed in Table 3.

Table 3. GAMMA geocoding interpolation modes compared in this study.

Identifier	Description
1	Bicubic Spline (GAMMA Default)
2	bicubic-log spline, interpolates log(data)
3	bicubic-sqrt spline, interpolates sqrt(data)
4	B-spline interpolation (degree: 5)
5	B-spline interpolation sqrt(x) (degree: 5)

The degree of option 4 and 5 were left at their defaults, as further optimization was considered to be outside the scope of this study.

It should be mentioned that in order to compare the backscatter to the UZH local incident angle product, three different subsets had to be created for the latter each at a 90 m resolution but in three different pixel grids, for the utm_30, utm_90 and wgs84_30 DEMs, respectively. Two different UZH base products were used, one in WGS84 LatLon, the other in UTM but both with a resolution of 30 m in order to keep the differences as small as possible. Similarly, the CLC product was resampled to these three different grids for masking forest.

The results computed for the SRTM 1 arcsec DEM are shown in Figure 8. In the first row, several DEM setups were tested with different coordinate reference systems, conversions from geoid to ellipsoid and resampling of the DEM to 90 m target resolution directly in GDAL, or alternatively, internally in GAMMA. The nomenclature used to describe the different setups is listed in Table 4.

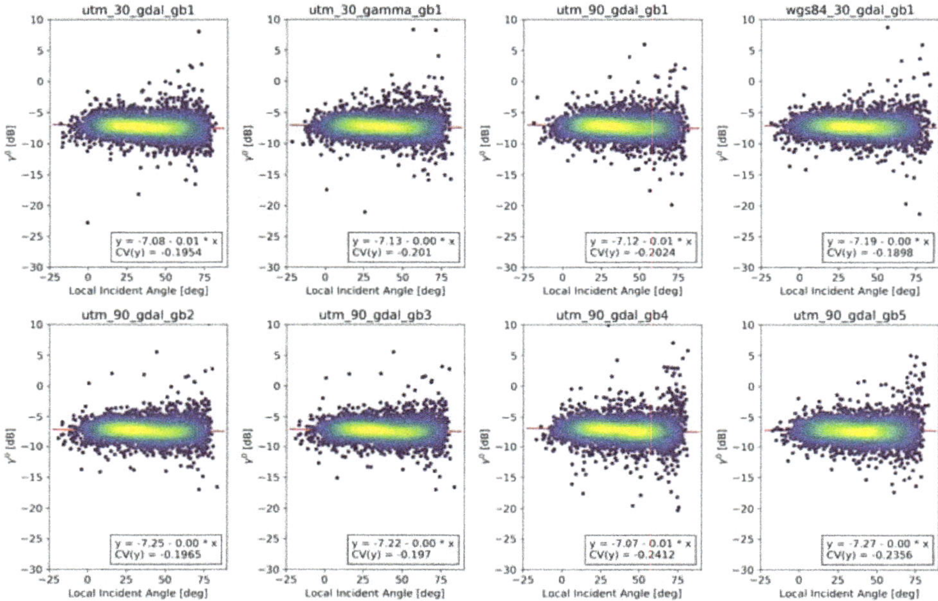

Figure 8. Comparison of topographic normalization quality for several GAMMA parameter settings. In all cases, the SRTM 1 arcsec DEM was used. Only backscatter acquired within the forest mask is shown.

Table 4. Nomenclature used for describing the different GAMMA image processing setups.

Identifier	Description
utm|wgs84	The Coordinate Reference System
30|90	the resolution of the DEM input into the processor in meters
gdal|gamma	the geoid/ellipsoid height conversion process
Gb [1–5]	the GAMMA interpolation method

While differences in the top row were marginal, the optimal configuration was selected to be a DEM in UTM and 90 m resolution with geoid heights converted using GDAL, which was option utm_90_gdal_gb1 in Figure 8. While the other configurations showed slightly better values for the coefficient of variation, the values were only marginally different from the selected choice. The configuration of first resampling the DEM to the target resolution was seen as preferential, as any further resampling during processing is specific to the scene extent and thus introduces shifts in pixel grids between different images. Since several images processed with the 30 m UTM DEM would be in different grids relative to each other, further resampling would be necessary to align the grids after processing, introducing additional inaccuracies. UTM was selected since working with a resolution in meters with same values for x and y resolution was seen to be more convenient for interpreting and visualizing results. As expected, only small differences could be detected between UTM and WGS84 LatLon with otherwise identical parametrization. Differences between using the GAMMA and GDAL geoid conversion were also negligible; the latter was preferred in order to enable more consistent treatment with SNAP.

Having selected an optimal DEM set up in the first row, the second row only compares differences in geocoding resampling algorithms. Contrary to the recommendations given in [43], the optimal result was achieved using the bicubic-log spline method, which can be identified by utm_90_gdal_gb2 in Figure 8. This setup, using a DEM in 90 m resolution projected to DEM and geocoding the SAR images using bicubic-log spline interpolation was used for further GAMMA processing of SAR scenes throughout this study.

4.1.2. SNAP

During this study, the utilization of external DEMs for SNAP6 terrain flattening was not possible. An error message indicating a bug in reading the DEM files was identified, which occurred in the default parametrization of the workflow in the SNAP GUI, as well as all possible parameter combinations available in the workflow XML files. The cause of this was not further investigated since several bug reports were found in SNAP's online ticketing system describing similar problems and a fix of the problem is thus soon to be expected.

To still be able to compare different DEMs as intended in this study, a mechanism was developed in pyroSAR to execute individual processing nodes in different versions of SNAP. The workflow was set up such that all processing nodes are executed in SNAP6 except Terrain-Flattening, which was executed in SNAP5. The authors are well aware that the terrain flattening has been fundamentally improved in SNAP6 and will further be improved in the soon-to-be-released SNAP7. However, in order to compare different DEMs and their processing results, the older version with a less accurate result had to be used. The SNAP6 terrain flattening could still be run with, e.g., the SRTM 1-Sec *auto-download* option and thus, a visualization of the improvement from one version to the other was still possible. This is shown in Figure 9. Once the bugs in SNAP6 are fixed or SNAP7 is released, the analyses can be run again and the conclusions updated accordingly.

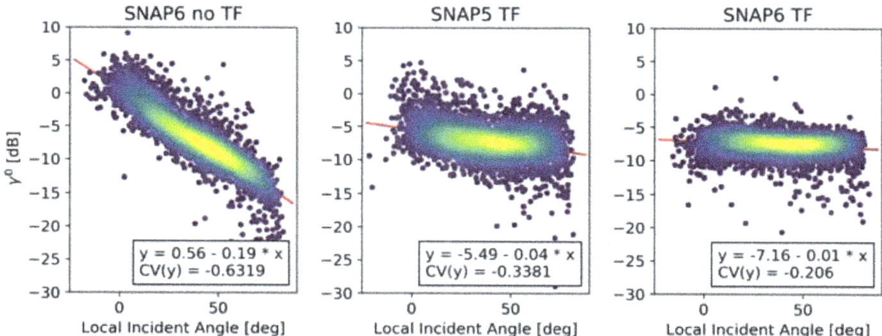

Figure 9. Comparison of SNAP processing results without terrain flattening (TF) and with the implementations of SNAP5 and SNAP6 over forested areas; all processing steps aside from Terrain-Flattening were performed in SNAP6.

While this feature to replace individual processing nodes with other versions was developed as a work-around for this study, it could also be beneficial in future studies, as it enables users to selectively assess the impact of single processing steps within future releases on the processing result. Once SNAP7 is released, a user could continue processing with SNAP6 and replace individual nodes with the SNAP7 version to selectively compare the impact of each on the processing result.

In order to assess which resampling method of the seven different options available in SNAP is best suited to the task at hand, the GRD product was processed with all combinations of DEM resampling and SAR image resampling. The RTC backscatter of all 49 images over forest was then compared to the UZH incident angle product. The resulting coefficient of variation is shown in Table 5. The slope values are not shown here as only very small differences were found, with nearly all values being 0.01 and 0.0 or 0.02 in few cases. DEM resampling can be defined for terrain flattening and terrain correction, while image resampling is only relevant for the terrain correction in the geocoding step. The same option for DEM resampling was applied in both flattening and orthorectification. Only options available for both DEM and image resampling in both processing steps were selected. For terrain correction, Delaunay interpolation is available for DEM resampling—it was excluded from this analysis as it was not available for DEM resampling in the flattening step and also not for the image resampling. For this experiment, the images were processed using the SRTM 1 arcsec HGT *auto-download* option in SNAP6. All produced images were of the exact same size and pixel grid, thus only a single subset was necessary for the UZH and CLC products, respectively.

Table 5. Coefficient of variation (CV) of SAR backscatter compared to UZH local incident angle for different combinations of DEM resampling (rows) and SAR image resampling (columns). Less variation with a CV closer to zero describes a better normalization, which is highlighted in color accordingly. Range depicts the CV value range for respective rows and columns.

	NEAREST	BILINEAR	CUBIC	BISINC_5	BISINC_11	BISINC_21	BICUBIC	Range
NEAREST	−0.26	−0.21	−0.26	−0.24	−0.3	−0.38	−0.26	0.17
BILINEAR	−0.26	−0.19	−0.23	−0.22	−0.3	−0.34	−0.23	0.15
CUBIC	−0.25	−0.2	−0.23	−0.23	−0.32	−0.36	−0.23	0.16
BISINC_5	−0.25	−0.19	−0.23	−0.21	−0.28	−0.34	−0.23	0.15
BISINC_11	−0.24	−0.19	−0.22	−0.22	−0.3	−0.36	−0.23	0.17
BISINC_21	−0.25	−0.19	−0.22	−0.21	−0.32	−0.35	−0.22	0.16
BICUBIC	−0.25	−0.2	−0.23	−0.22	−0.32	−0.36	−0.23	0.16
range	0.02	0.02	0.04	0.03	0.04	0.04	0.04	

Only small differences in coefficient of variation (CV) were visible when changing the DEM resampling method with ranges of 0.02 to 0.04. Changing the image resampling method, on the other hand, resulted in much higher differences in CV ranging from 0.15 to 0.17. The best results were achieved for bilinear image resampling and bilinear or BSINC DEM resampling with a CV of −0.19.

Table 6 shows the processing times needed to achieve the results using the different methods. For this test, a laptop with 16 GB of RAM and a 1.9 GHz × 8 intel i7 CPU was used. This test was not intended as a formal benchmarking but rather a quick comparison and hence the numbers only show an approximation. Interestingly, several methods needed slightly less processing time than the simplest and presumably fastest nearest-neighbor method during DEM resampling while they required significantly more time during image resampling. The reason for this was not further investigated. Likely, the small relative time differences for DEM resampling will change with repeated runs. The differences in time between the DEM resampling methods varied significantly, with a range of 1324 to 1435 s, while the choice of image resampling had a much smaller impact, with ranges of 29 to 112 s.

In summary, changing the DEM resampling method did not result in large differences in the quality of the topographic normalization but had an impact on the processing time needed. In contrast, the image resampling method did not impact the CV quite as strongly but differences in processing time were much smaller. As a best compromise of processing time and lowest coefficient of variation, bilinear resampling was chosen for both DEM and image resampling and was used for further processing throughout this study.

Table 6. Processing time in seconds for different combinations of DEM resampling (rows) and SAR image resampling (columns). The color coding highlights the overall processing time differences between all 49 runs. Range depicts the time value range for respective rows and columns.

	NEAREST	BILINEAR	CUBIC	BISINC_5	BISINC_11	BISINC_21	BICUBIC	Range
NEAREST	277.59	262.54	258.86	266.74	274.56	287.68	265.8	28.82
BILINEAR	272.3	271.82	320.97	297.16	309.95	303.75	285.69	49.15
CUBIC	291.67	286.74	289.61	296.76	304.83	319.24	292.17	32.5
BISINC_5	532.42	549.04	511.16	509.3	517.1	528.16	507.99	41.05
BISINC_11	866.05	839.12	859.91	865.12	885.41	915.87	875.77	76.75
BISINC_21	1707.67	1595.32	1604.46	1608.85	1614.23	1611.57	1601.53	112.35
BICUBIC	387.02	388.06	395.55	397.98	400.79	417.86	400.02	30.84
range	1435.37	1332.78	1345.6	1342.11	1339.67	1323.89	1335.73	

4.2. DEM Assessment

4.2.1. Alps

Quantification of Outliers

The maximum deviation from the median for the area of the scene under investigation is shown in Figure 10. While a higher level of deviation can generally be seen in mountainous regions compared to the flatland to the Southeast of the image, a particularly high deviation was observed around Lake Garda to the southwest and a mountain range to the east. These two regions are highlighted in Figure 10 and inspected more closely in Figure 11.

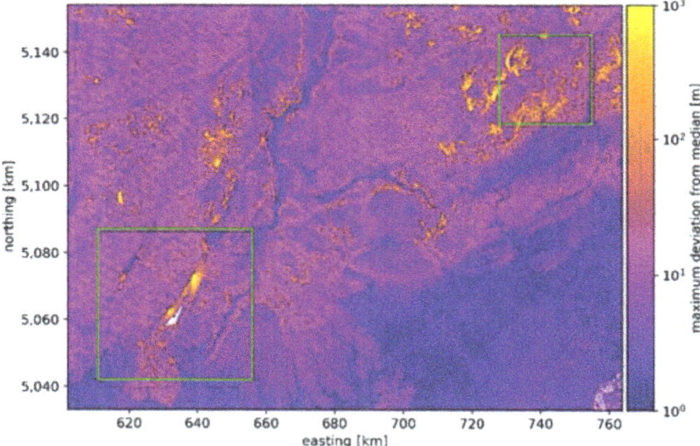

Figure 10. Alps: magnitude of maximum deviation from the median of all four DEMs. The area shown is a subset of the scene footprint in Figure 1. The highlighted squares are magnified in Figure 11. WGS84, UTM zone 32N (EPSG 32632).

The mentioned index map identifying which of the DEMs contained the height value that deviated most from the median is not shown here, as no areal patterns could be visually identified due to frequent near-random deviations of lower magnitude. However, a difference becomes visible above 100 m deviation, which is shown in Figure 12. Unexpectedly, deviations on the order of several hundreds of meters were seen, with the ALOS World DEM deviating more than 2500 m in several cases. On closer inspection, several artifacts could be identified in this particular DEM over Lake Garda, which would need to be masked out prior to SAR processing. In all other DEMs, the lake was either masked out or contained the actual height of the lake. While these deviations were very high in magnitude, the ALOS World DEM showed only a few maximum deviations in comparison to the TanDEM-X DEM. In particular, the mountain range with high deviations visible to the east shown in Figure 10 and the bottom row of Figure 11 can be attributed to this particular DEM. While this DEM showed the lowest magnitude of deviations in the boxplot of Figure 12, it was by far the option with the highest number of maximum deviations, contributing 86% to the overall random sample. The ALOS DEM, on the other hand did contain only a few, but extreme, outliers centered around Lake Garda.

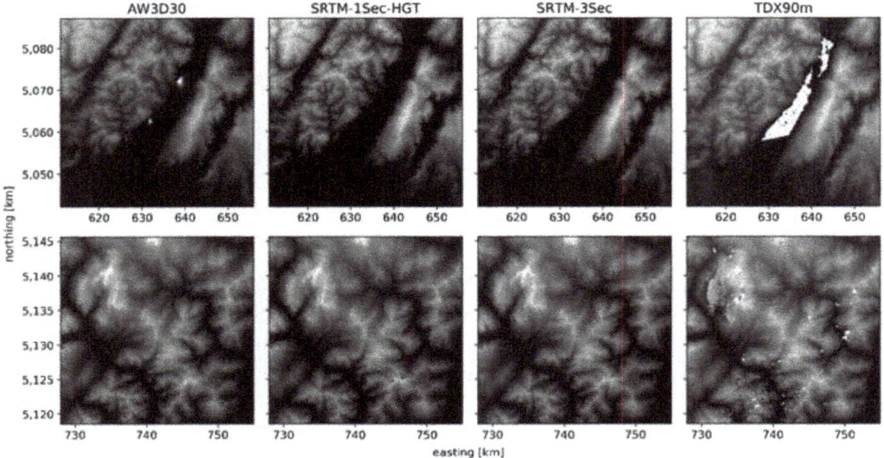

Figure 11. Exemplary image chips showing DEM inconsistencies over Lake Garda (**top**) and in a mountain range northeast of the Italian town of *Belluno* (**bottom**). The position of the image chips is highlighted in Figure 10.

It is not clear what caused these artifacts, and it is expected that the processing result will significantly improve in future versions of this product initially released in October 2018. We stress that the SRTM DEM was manually edited to correct processor deficits, likely more than the current version of the TanDEM-X DEM, as previously mentioned in Section 3.3.

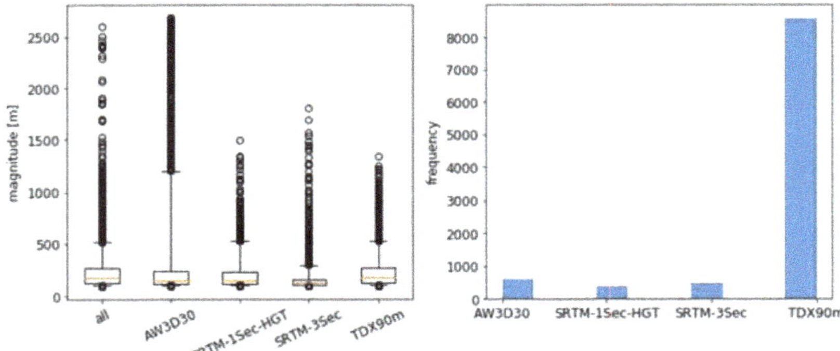

Figure 12. Alps: value distribution of DEM height deviations greater 100 m. For each of the box plots, 10,000 samples were selected randomly wherever available. The whiskers represent the 5th and 95th percentile. The histogram shows the distribution of DEM IDs for the 'all' sample shown in the box plot to the left.

Table 7 summarizes statistics of the samples selected for Figure 12. The lowest overall deviation magnitude and frequency can be observed for the two SRTM DEMs, with both showing the two lowest values for the 95th percentile of maximum deviations and containing the least maximum deviations across the image. It is therefore concluded that, in particular, the higher resolution SRTM 1 arcsec DEM is a viable choice for any further processing. However, this was only a quick test and not an in-depth investigation and is thus not made as a general recommendation. In many regions, the SRTM 1 arcsec is likely a suboptimal choice due to ground movements that occurred between its acquisition and that of the SAR scene. One aim of the accompanying Jupyter notebook is to enable quick and convenient assessments of DEM quality for any study site.

Table 7. Alps: size (n) and 95th (p95) percentile of the samples drawn for the boxplot in Figure 12, as well as the histogram values of the 'all' sample in percent (%). If n is smaller than 10,000, all pixels where the individual DEM showed the highest deviation were selected.

Identifier	n	p95	%
all	10,000	522.92	
AW3D30	5629	1197.80	5.90
SRTM-1Sec-HGT	3423	526.48	3.69
SRTM-3Sec	4826	298.47	4.74
TDX90m	10,000	530.46	85.67

Comparison of Single Image Processing Results

In order to assess the quality of the topographic normalization between SNAP and GAMMA, as well as between the four different DEMs, backscatter was compared to the local incident angle at each pixel location in forest areas which were masked as described in Section 3.5.

As measures of quality of topographic normalization, the slope of the linear regression function and the coefficient of variation were used. The former is optimal at zero, as no significant dependence on the local angle of incidence was found in the data. An uncorrected backscatter image correlates negatively with the incident angle, containing lower values with increasing angle of incidence.

The coefficient of variation was used to quantify the scattering around the mean backscatter as an indicator of insufficiently corrected pixels. The mean is also displayed so that the overall level of backscatter can be compared between images. The result is displayed in Figure 13.

It is understood, that this presents only a basic assessment of image quality, which does by far not cover all aspects of SAR processing and the resulting differences in RTC backscatter. A more formal assessment was made by [48].

In images processed with GAMMA using the AW3D30 and SRTM 1 arcsec DEMs, the overall dependency on the local incident angle was completely removed, reflected in an overall slope of 0. The two other GAMMA cases showed a slight under-correction of the incident angle dependency with slopes of 0.01. In terms of variation, the best GAMMA result was achieved using the SRTM 1 arcsec DEM showing the lowest CV of −0.1965. This DEM thus presented the optimal choice for this test site as it had the lowest values for both slope and variation. Of similarly high quality was the AW3D30 DEM result, with the same slope and only marginally higher variation of −0.208. The TDX DEM performed worst, with a slope of 0.01 and a CV of −0.27.

In all SNAP images, a larger dependency on the incident angle was still present with higher slopes of 0.04 and 0.05. Contradicting the GAMMA results, the SRTM 3 arcsec and TDX DEMs yielded the best SNAP results, with the former showing the overall best values for slope (0.04) and CV (−0.315).

Due to the use of the deprecated SNAP5 processor, a direct comparison between SNAP and GAMMA was refrained from at this point. However, relative results for the different DEMs were expected to be similar between SNAP and GAMMA.

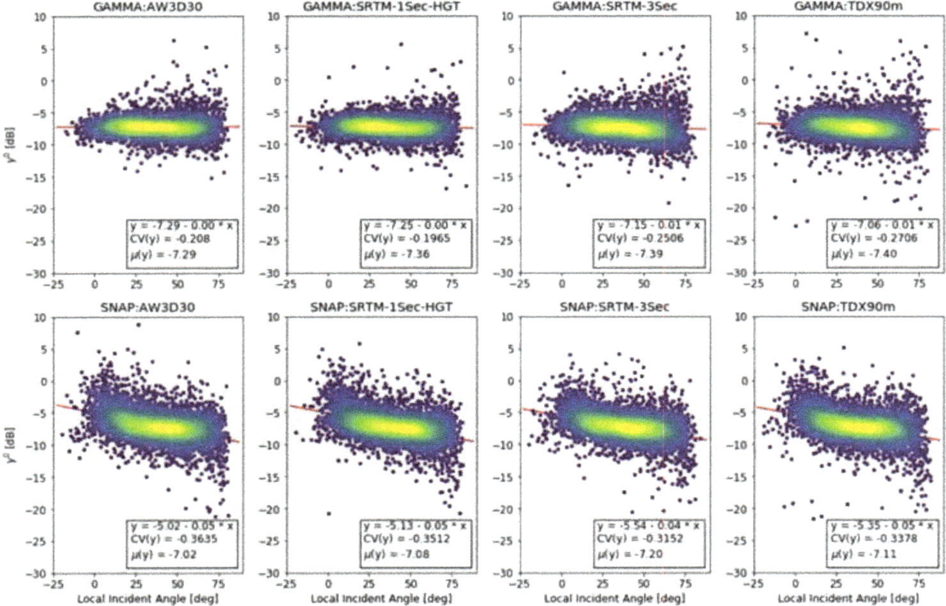

Figure 13. Alps: backscatter processed with GAMMA and SNAP using four different DEMs compared to the UZH local incident angle. Only pixels acquired over forests are shown.

4.2.2. Fiji

Quantification of Outliers

The maximum deviation from the DEM median for Fiji is shown in Figure 14. Multiple noise features across the water body around the islands were observed. Several rectangular features are visible along the coast of Fiji which likely present artifacts of the automated DEM processing. A closer inspection of the southwest coast of *Viti Levu* and several islands to the east are shown in Figure 15.

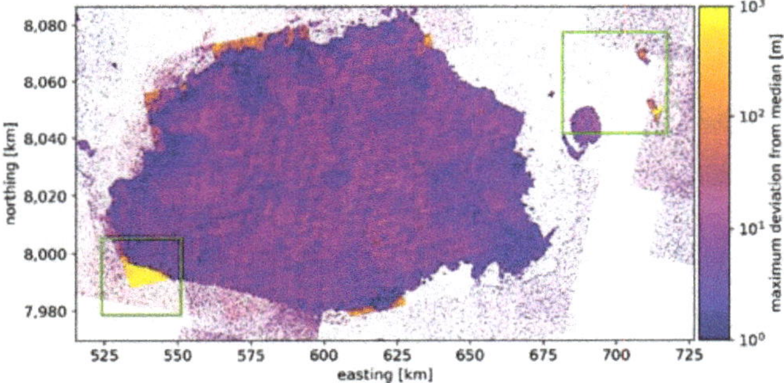

Figure 14. DEM differences in Fiji: magnitude of maximum deviation from the median with areas shown in Figure 15 highlighted in green. The area shown is a subset of the scene footprint in Figure 2. WGS84, UTM zone 60S (EPSG 32760).

It was observed that the mentioned noise features, as well as the rectangular artifacts along the Fiji coast, are contained in the TDX data. In the bottom row of Figure 15, several smaller features of high deviation are shown next to the mentioned noise. These features, although only rarely occurring, highlight very large deviations of up to more than 15,000 m in several cases. While these features are mostly contained in the AW3D30 product, they can also be observed in the SRTM 1 arcsec product, which was found to be the most viable choice for SAR processing in the Alps.

Furthermore, it needs to be noted that the DEMs differ in their representation of water bodies. While in the AW3D30 and SRTM 1-Sec DEMs water is set to 0 in the original products, they are represented by no data in the other two. Due to the conversion of DEM heights from geoid to ellipsoid, the former two DEMs will contain a mean value of 56 m across the image, varying slightly with the local geoid height.

Since DEM no-data areas will also be set to no data in resulting SAR products, the latter two DEMs are not suited for processing without further modifications if water bodies are of interest. The water mask of the TDX DEM contains a high omission error not only across the water body with the aforementioned noise and the rectangular features along the coast but also, with a general overestimation of the island size wherein the water mask shows an average distance of approximately 500 m on to the actual coast. The TanDEM-X DEM was delivered with several ancillary products, including a water indication layer (referred to as WAM in the product guide [42]). This layer was extracted for the two sites to investigate whether water bodies could easily be masked in the actual DEM. A binary water mask was extracted by thresholding the WAM product, which contains several water indication metrics, and setting all values between 3 and 127 to water. This was a quick method for decoding the bitmask values contained in the product to a binary water mask, as recommended by the TDX90m product guide [42]. Unfortunately, it was observed that this resulted in a high commission error of detected water bodies across mountainous areas in both study sites. Instead, the SRTM 3 arcsec DEM was used for masking, which directly contains a reliable water mask and thus presented a quick and accurate solution.

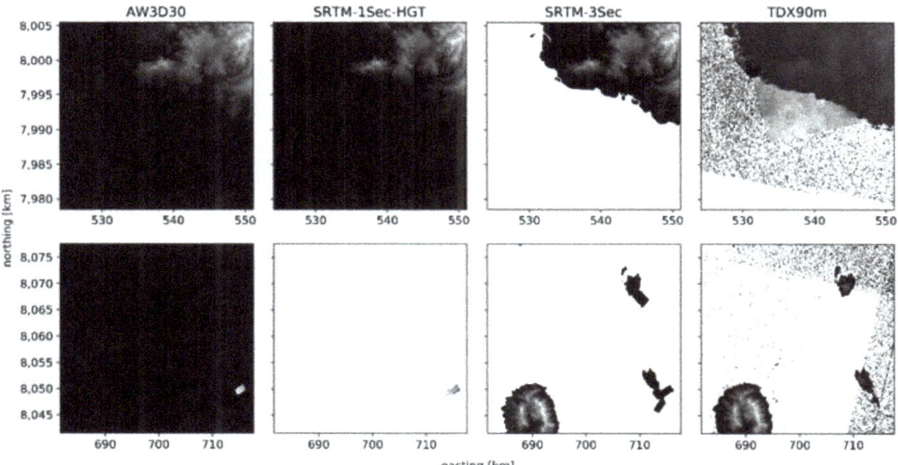

Figure 15. Exemplary image chips as highlighted in Figure 14 showing DEM inconsistencies on the southwest shore of *Viti Levu* (**top**) and around the small islands of *Makogai* and *Wakaya* to the east (**bottom**).

The general quantitative overview of median deviation statistics is shown in Figure 16; the corresponding statistics are shown in Table 8. Only areas not masked as water in the SRTM 3 arcsec product were considered. Both the SRTM 3 arcsec and the TDX DEMs contained only a few maximum deviations, which were also of low magnitude. The AW3D30 and SRTM 1 arcsec DEMs contained deviations of more than 15,000 m with the former showing a higher frequency of deviations of a particularly high magnitude.

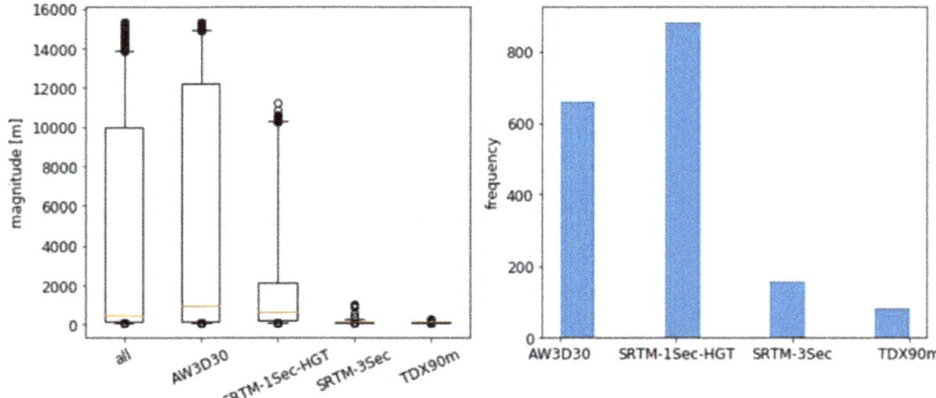

Figure 16. Fiji: distribution of DEM height deviations greater than 100 m. Due to the small size of the land areas in the scene, only 1783 samples could be selected. The number of samples in the box plots are thus the same as shown in the histogram. The whiskers represent the 5th and 95th percentile. The histogram shows the distribution of DEM IDs for the 'all' sample shown in the box plot to the left. Only samples not masked as water in the SRTM 3 arcsec product were used for all DEM options.

Table 8. Size (n) and 95th percentile (p95) of the samples drawn for Figure 16, as well as the histogram values of the 'all' sample in percent (%). All sample sizes are the maximum number of pixels available for the respective selection.

Identifier	n	p95	%
all	1783	13890.79	
AW3D30	661	14913.00	37.07
SRTM-1Sec-HGT	881	10275.38	49.41
SRTM-3Sec	157	296.20	8.81
TDX90m	84	170.16	4.71

Based on the low frequency and magnitude of maximum deviations, the TDX DEM presented the best option for processing SAR data over Fiji. However, based on the need to include an ancillary water mask, which requires an additional pre-processing step, the SRTM 3 arcsec DEM was selected as the best choice. If the higher resolution of the AW3D30 and SRTM 1 arcsec DEM are required, further masking is recommended to eliminate the mentioned artifacts in order to avoid propagation of errors into the SAR backscatter products.

Comparison of Single Image Processing Results

The SAR images processed over Fiji using different DEMs were compared to the UZH local incident angle in Figure 17 to assess the quality of topographic normalization. The value range of the incident angle was smaller than in the Alps, reflecting the overall flatter slopes in this study area. As compared to the Alps, higher slopes and variation were observed for both SNAP and GAMMA, yet the same trends were present with GAMMA being very slightly under-corrected and SNAP heavily under-corrected.

For GAMMA, the best performing DEM was the SRTM 1 arcsec DEM with the lowest slope and variation. The SRTM 3 arcsec and TDX DEMs are, equally, the worst performers, with the former showing the higher slope and the latter the highest variation of the two.

In line with the findings in the Alps, the two best-performing DEMs using SNAP were the SRTM 3 arcsec and the TDX. However, the latter performed slightly better with a marginally lower variation.

4.3. Comparison of SNAP and GAMMA by Terrain

Although differences in the quality of the topographic normalization were observed between SNAP and GAMMA, this could be of little concern for many users who are interested in areas with flatter terrain only. It was thus of great interest to investigate the similarity of images from the two processors depending on the orientation of the terrain towards the sensor to quantify differences originating from the terrain flattening procedure. For this analysis, the SNAP6 SRTM 1 arcsec *auto-download* result was used and compared to the GAMMA result processed with the same DEM.

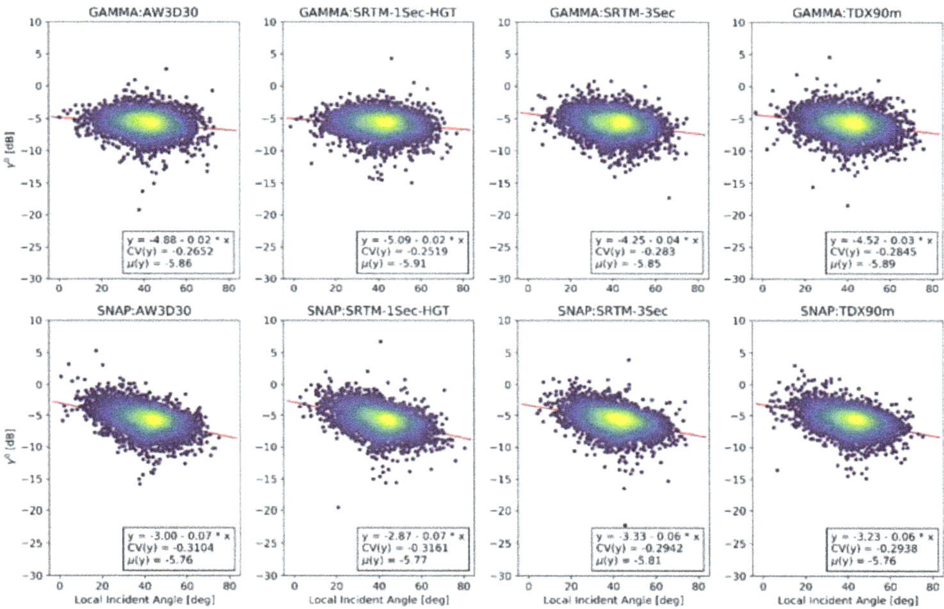

Figure 17. Fiji: backscatter processed with GAMMA and SNAP using four different DEMs compared to UZH local incident angle. Only pixels acquired over forests were selected.

Figure 18 displays the similarity of the processing results from SNAP6 and GAMMA for the whole scene without stratification to the left and dependent on the local angle of incidence to the right. A differentiation was made between samples collected across the whole image and all present CLC classes (black dashed and solid lines for SNAP5 and SNAP6, respectively) and samples of forested areas only (green line). The color bars represent the composition of CLC classes for the specific incidence angle ranges.

The overall RMSE of 2.29 dB in the image to the left was also reflected in the SNAP6 RMSE values for individual terrain classes up to 60° in the image to the right, ranging from 2.11 (>20–30) to 2.56 dB (>−10–0) for samples from all classes combined. Hence, in this terrain, the differences between the two could not be explained by the terrain, since this range of incident angles includes regions of layover and foreshortening over flat terrain to moderate shadows, but showed only little variation in the RMSE. The mean incident angle of the acquired SAR scene over the island was 39°, thus this angle represents approximately flat terrain in the plot.

Only a slight increase in RMSE was observed with incident angles lower than 0°, showing different yet very similar qualities of normalization of layover and foreshortening for both SNAP and GAMMA. In contrast, the SNAP5 normalization exhibited a strong increase of RMSE from approximately

2.6–3.1 dB in flat areas to nearly 5 dB at angles lower than 0°, again confirming the large improvement in SNAP6.

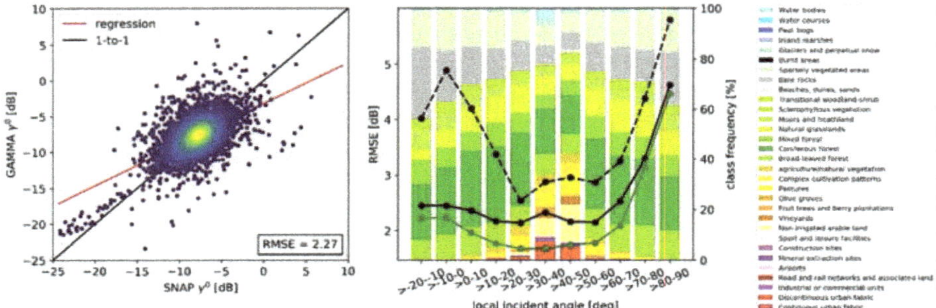

Figure 18. Comparison of SNAP vs. GAMMA 90 m processing results for the scene acquired over the Alps using the SRTM 1 arcsec DEM. Left: scatter plot of 10,000 samples drawn from across the whole scene; the red and black lines represent the linear regression and one-to-one lines, respectively. Right: RMSE of backscatter comparison for different local incident angle classes and the class distribution of the drawn samples. The solid black and green lines show RMSE values comparing SNAP6 and GAMMA for all classes and for forest, respectively. The dashed black line shows the result for all classes using the SNAP5 product. For the right plot, 2500 samples were used for each incident angle class. Although incident angles higher than 90° are present in the scene, these classes were excluded due to an insufficient number of samples.

From 50° onwards, a strong increase in RMSE up to 4.5 at >80–90° for SNAP6 was observed, demonstrating larger differences in normalization in areas close to radar shadow. A similar pattern was observed with the SNAP5 results, yet the maximum RMSE value was much higher at 5.8 dB, showing an improvement of SNAP6 in this region as well. As was expected, the SNAP6 forest RMSE progression line followed the same pattern of the overall RMSE, however, with generally smaller values, as low as 1.70 in flat terrain. The corresponding SNAP5 line, showing a similar trend, is omitted here for clarity.

4.4. Time-Series Analysis

The scope of this study was expanded to assess the interoperability of pyroSAR-derived GAMMA and SNAP Analysis-Ready Datasets over space and time—including consideration of internal (software tool set, DEM selection, pre-processing) and external (local topography, orbit direction) factors. To align with the objectives and parallel activities of the UK International Partnership Program (IPP) Common Sensing initiative, a test site was selected, encompassing the island of *Viti Levu* in Fiji. Preliminary analysis focused on benchmarking monthly and seasonal variability in gamma0 VV- and VH-polarized backscatter across the 12-month time series for selected land cover classes. To identify and quantify systematic inconsistencies caused by acquisition parameters, it was vital to conduct statistical analysis against land surfaces exhibiting a low frequency variability in their radar backscatter properties.

Mean and standard deviation of VV and VH gamma0 backscatter were computed for all point geometries (~5000 samples) across every scene in the GAMMA and SNAP time series, which is shown in Figure 19. Due to the stable canopy structure and climatic conditions, moist tropical forests in *Viti Levu* demonstrated minimal variability in mean backscatter for the duration of the gamma0 time series (−6 dB VV, −12 dB VH). Conversely, the temporal backscatter signature of grassland regions demonstrated a clear seasonal variation caused by the transition from cooler dry conditions (May to September) to the wet, warmer season (November to March). Variations in biomass and surface moisture content have strong influences on microwave backscatter properties of vegetated land surfaces [49].

In compliance with CEOS ARD guidelines, no speckle reduction filtering was incorporated into the SNAP and GAMMA workflows (selection and configuration of speckle filter dependent on application). The large sample population (~5000 points per land cover class per scene), therefore, exhibited high levels of deviation (±2 dB) around mean backscatter. Visual inspection revealed that mean VV and VH backscatter signatures computed for GAMMA and SNAP gamma0 time series demonstrated near equivalence as a function of time.

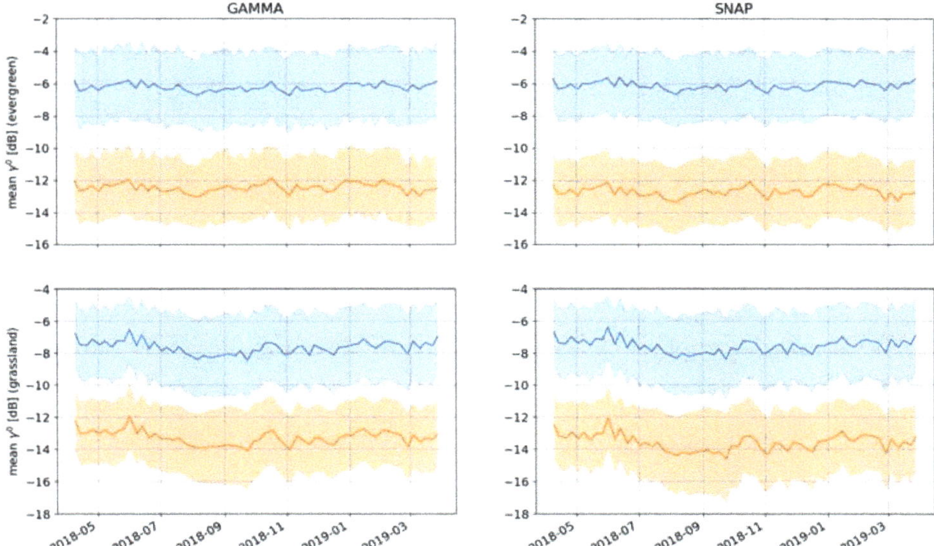

Figure 19. Mean (**line**) and standard deviation (**shaded area**) VV (**blue**) and VH (**orange**) gamma0 backscatter computed for the time series using 5000 randomly selected point geometries coincident with moist tropical forest (**top row**) and grassland (**bottom row**) areas across in *Viti Levu*.

Workflow interoperability was further examined by computing error statistics between spatially coincident gamma0 backscatter values retrieved from GAMMA and SNAP raster time series. As indicated in Table 9, GAMMA and SNAP gamma0 demonstrated a high level of consistency—minimal deviation was evident in temporally averaged co-polarized backscatter computed for evergreen forest and grassland classes (<0.1 dB). Statistical analysis revealed a greater level of inconsistency between GAMMA and SNAP cross-polarized products—~0.25dB differences in mean gamma0.

Table 9. Temporally averaged gamma0 backscatter and RMSE statistics derived from GAMMA and SNAP time series for different land categories over *Viti Levu*.

	VV gamma0			VH gamma0		
	GAMMA μ	SNAP μ	RMSE	GAMMA μ	SNAP μ	RMSE
Evergreen	−6.11	−6.20	0.67	−12.65	−12.42	0.64
Grassland	−7.54	−7.63	0.58	−13.50	−13.22	0.63

As shown in Figure 20 and summarized in Table 10, variance between GAMMA and SNAP backscatter was evaluated by fitting a Gaussian probability function (PDF) to the frequency distribution of mean signed differences. From Table 10, it can be seen that the deviation between

GAMMA and SNAP-derived backscatter closely approximated a random variable with a normal distribution, exhibiting minimal bias around zero mean.

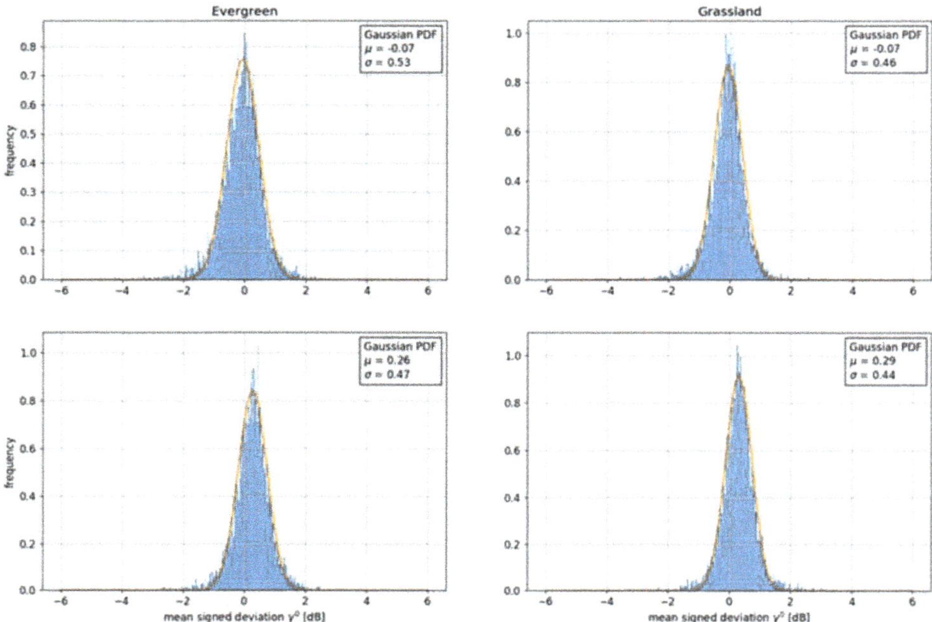

Figure 20. Frequency distribution of mean signed difference between temporally averaged GAMMA and SNAP backscatter plotted against best fit Gaussian probability density function.

Table 10. Best-fit Gaussian PDF fitted to frequency distribution of mean signed difference between gamma0 values extracted from *Viti Levu* GAMMA and SNAP time series.

	VV gamma0		VH gamma0	
	PDF μ	**PDF σ**	**PDF μ**	**PDF σ**
Evergreen	−0.07	0.53	0.26	0.47
Grassland	−0.07	0.46	0.29	0.44

Follow-on analysis evaluated the interoperability between SNAP and GAMMA gamma0 products as a function of the underlying topography, represented by the SRTM 1 arcsec slope. Randomized point geometries generated for evergreen and grassland regions were stratified into two categories—one group coincident with areas of relatively flat terrain (0 to 12 percent slope); the other set aligned with locations of rapidly varying elevation (20 percent slope and higher).

The RMSE between GAMMA and SNAP VV and VH gamma0 backscatter values coincident with flat and steep moist forest and grassland areas was subsequently computed on a scene by scene basis and rendered as a time series plot. Figure 21 visualizes the variation in GAMMA vs SNAP RMSE as a function of time (x-axis) where inter-comparison between coincident VV (top row) and VH (bottom row) gamma0 backscatter values was stratified into steep (blue) and flat (orange) locations. A summary is given in Table 11. Analysis revealed that the underlying slope significantly affected the degree of consistency between GAMMA and SNAP gamma0 products.

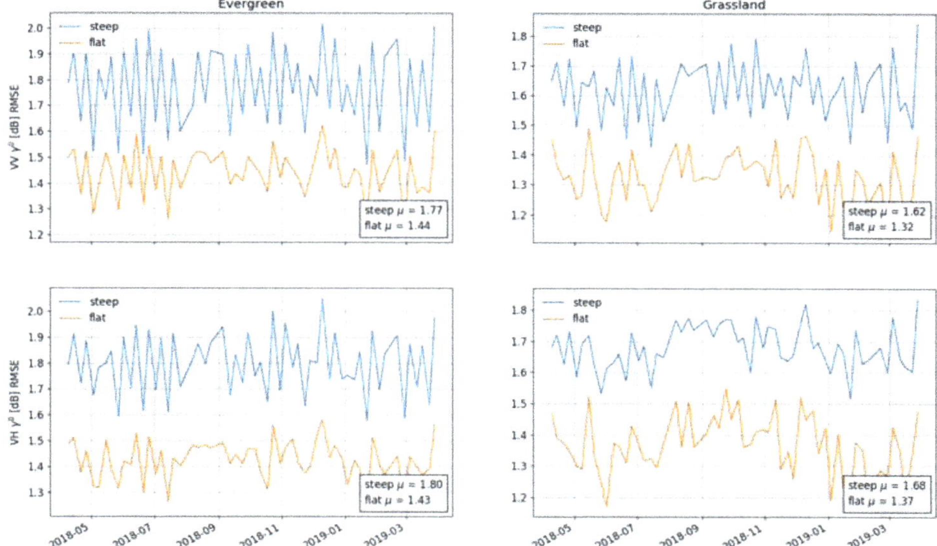

Figure 21. Temporal variation in RMSE between GAMMA and SNAP VV (**top row**) and VH (**bottom row**) gamma0 values computed for 5000 randomly selected point geometries coincident with steep (**blue**) and flat (**orange**) evergreen (**moist forest**) and grassland areas of *Viti Levu*.

Table 11. Aggregated RMSE statistics quantifying level of consistency between gamma0 backscatter generated from GAMMA and SNAP workflows for *Viti Levu* raster time series.

	VV gamma0 RMSE (dB)		VH gamma0 RMSE (dB)	
	Steep	**Flat**	**Steep**	**Flat**
Evergreen	1.77	1.44	1.80	1.43
Grassland	1.62	1.32	1.68	1.37

Additionally, the time series was subdivided into ascending and descending scenes and the analysis was run again. Tables 12 and 13 indicate increased variability between GAMMA and SNAP gamma0 backscatter when comparing locations with a high slope in descending scenes.

Table 12. RMSE statistics quantifying consistency between GAMMA and SNAP gamma0 for ascending scenes and locations of flat and steep terrain.

	VV gamma0 RMSE (dB)		VH gamma0 RMSE (dB)	
	Steep	**Flat**	**Steep**	**Flat**
Evergreen	1.63	1.39	1.70	1.40
Grassland	1.54	1.34	1.65	1.40

Table 13. RMSE statistics quantifying consistency between GAMMA and SNAP gamma0 for descending scenes and locations of flat and steep terrain.

	VV gamma0 RMSE (dB)		VH gamma0 RMSE (dB)	
	Steep	**Flat**	**Steep**	**Flat**
Evergreen	1.91	1.49	1.89	1.45
Grassland	1.70	1.31	1.71	1.34

The final phase of the *Viti Levu* study evaluated capabilities of GAMMA and SNAP processing tools to normalize gamma0 backscatter for changes in viewing geometry and effects of the underlying topography. With even distribution of ascending and descending scenes across the time series period with a repeat cycle of ~six days, the analysis evaluated the variability of temporally averaged gamma0 backscatter values as a function of orbit direction and slope.

As indicated in Figure 19, radar backscatter properties of the moist tropical forest canopy remained relatively consistent throughout the year. To quantify variability introduced by the direction of the satellite platform, temporally averaged gamma0 values were derived for a collection of randomly selected geometries from sub-divided ascending and descending time series. With over-sampling dampening effects of natural variability, it was hypothesized that mean backscatter measured at locations across the *Viti Levu* evergreen forest should eventually approach a one-to-one relationship when comparing ascending and descending time series.

For evergreen land cover, Figure 22 indicates the level of interoperability between temporally averaged gamma0 backscatter derived from ascending and descending scenes of GAMMA and SNAP time series. Low correlation was recorded when comparing mean gamma0 for ascending and descending scenes. This can be explained by the overall low variation of backscatter over the homogeneous evergreen forest around a mean of −6 dB. Differences in the local incident angle between the orbits and differences in dominant scattering processes (double bounce vs. volume scatter) in an inhomogeneous forest structure caused frequent outliers, whose origin was not further investigated. A much stronger linear relationship would be expected for L-Band SAR penetrating deeper into the forest canopy, thus causing a higher variability in backscatter as compared to C-Band. Products generated by GAMMA demonstrated a slightly greater error variance compared to SNAP gamma0 products.

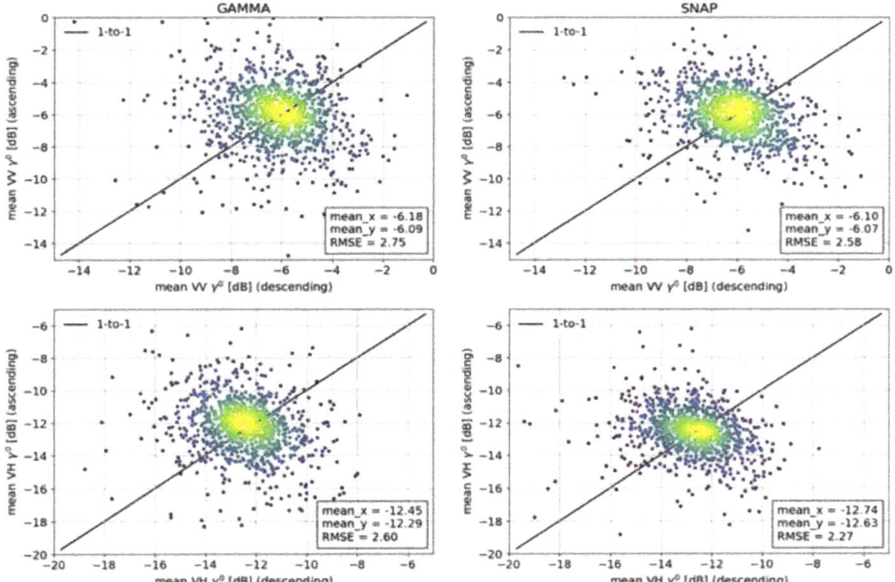

Figure 22. Relationship between temporally averaged VV (**top row**) and VH (**bottom row**) gamma0 backscatter values derived from ascending and descending 20 m scenes for 5000 randomly selected point geometries across evergreen moist forest of *Viti Levu*.

The analysis was subsequently repeated for collections of randomly selected point geometries coincident with locations of flat and steep terrain. Tables 14 and 15 indicate an increased level of inconsistency between temporally averaged gamma0 across steep terrain. Tables 16 and 17 summarize the results of re-executing the analysis with randomly selected point geometries coincident with flat and steep grassland areas.

Table 14. Inter-comparison of temporally averaged gamma0 backscatter derived from ascending and descending 20 m scenes for 5000 randomly selected point geometries coincident with flat terrain in moist tropical forests of *Viti Levu*.

	VV gamma0			VH gamma0		
	Ascend μ	Descend μ	RMSE	Ascend μ	Descend μ	RMSE
GAMMA	−5.99	−6.43	2.73	−12.27	−12.63	2.32
SNAP	−5.95	−6.33	2.33	−12.51	−12.84	2.18

Table 15. Inter-comparison of temporally averaged gamma0 derived from ascending and descending 20 m scenes for 5000 randomly selected point geometries coincident with steep terrain across moist tropical forests of *Viti Levu*.

	VV gamma0			VH gamma0		
	Ascend μ	Descend μ	RMSE	Ascend μ	Descend μ	RMSE
GAMMA	−6.30	−6.24	3.10	−12.57	−12.49	2.81
SNAP	−6.20	−6.11	2.71	−12.67	−12.81	2.48

Table 16. Inter-comparison of temporally averaged gamma0 derived from ascending and descending 20 m scenes for 5000 randomly selected point geometries coincident with steep terrain across grassland areas of *Viti Levu*.

	VV gamma0			VH gamma0		
	Ascend μ	Descend μ	RMSE	Ascend μ	Descend μ	RMSE
GAMMA	−7.84	−7.51	2.12	−13.14	−13.41	1.63
SNAP	−7.56	−7.88	1.81	−13.43	−13.61	1.59

Table 17. Inter-comparison of temporally averaged gamma0 derived from ascending and descending 20 m scenes for 5000 randomly selected point geometries coincident with flat terrain in grassland areas of *Viti Levu*.

	VV gamma0			VH gamma0		
	Ascend μ	Descend μ	RMSE	Ascend μ	Descend μ	RMSE
GAMMA	−7.54	−7.57	2.59	−13.30	−13.30	2.43
SNAP	−7.33	−7.49	2.14	−13.43	−13.60	2.24

Over forested areas, a good agreement between SNAP and GAMMA VV backscatter was observed with mean values of approximately −6 dB in all cases and an RMSE of 2.71 to 3.1. For VH, a better agreement was observed with a mean backscatter of around −12.5 dB and slightly lower RMSE values between 2.18 and 2.81. RMSE values on steeper slopes were recorded to be higher by about 0.4 for both polarizations. In descending orbit, backscatter of both polarizations was slightly higher on steep slopes than on flat areas. The opposite was observed for the ascending orbit.

RMSE values over grassland were lower than over forest with values of 1.81 to 2.59 for VV and 1.59 to 2.43 for VH. Backscatter over flat grassland was lower than that over forest by about 0.7 to 1.6 dB. The same trend of lower VH backscatter in ascending orbit and higher backscatter in descending orbit observed over forest was observed over grassland. For VV however, higher backscatter was observed in ascending mode and in descending orbit, backscatter processed by GAMMA was slightly lower, but the equivalent from SNAP much higher.

5. Discussion

5.1. Software Usability

During this study and preceding work on pyroSAR, effort was invested into making the two processing software packages as easy to use as possible. In the case of GAMMA, this meant creating an API that wraps the command line interface into modularized Python workflows including a suite of convenience functions. In SNAP, effort was invested into creating flexible and reliable workflows delivering consistent results while improving the overall processing speed.

For SAR processing in any software, numerous parameters can be set whose influence on the results require many years of training and expertise in interpreting SAR data. By providing easy-to-use workflows and a Jupyter notebook for reproducing the created results, it is intended to further lower the entry barrier to utilizing SAR imagery. Although simplifying SAR workflows usually comes with a reduction in parameterization flexibility, great care is taken to keep the workflows and their configuration as flexible as possible and demonstrate the impact of different processing choices in the Jupyter notebook.

Particularly, the use of SNAP required a large effort during this study since a bug prevented the use of external DEMs for terrain flattening in SNAP6, and initial processing was held back by the much longer processing time in comparison to GAMMA. In addition to this, SNAP required a lot of memory, thus initial processing was not possible on a 16GB local machine. Processing on a large server cluster with 500GB of RAM and 48 logical CPUs was still slower than using GAMMA on a local laptop. Although it is recognized that SNAP's workflow chaining in memory, without the need to create intermediate products, is theoretically beneficial due to fewer read and write operations of intermediate products, apparently more development time is needed to fully implement this philosophy and actually gain speed in processing and memory efficiency.

The large need for resources of SNAP was drastically reduced by executing each processing node individually while writing intermediate products. This way, processing time was reduced by a factor of seven and the memory consumption was reduced so that processing on a local machine became possible. The isolated execution of single nodes was also highly beneficial in error tracing. This feature is currently only available in pyroSAR and not in SNAP itself; therefore, the large resource footprint and long processing time are likely limiting wider adoption of the software. Naturally, it is of interest to communicate these findings with the SNAP developers and community to contribute to the improvement of the software.

In addition to the processing speed, testing SNAP with different parameterizations at the beginning of this study was oftentimes held back by difficulties in interpreting the short ambiguous error messages. This way, a lot of time had to be invested in repeated trial runs in order to find the origin of generic error messages. pyroSAR tries to mitigate this problem by providing more verbose error messages and reacting to those that can be interpreted. For example, a mechanism was implemented during this study to automatically remove certain parameters from a processing node in case the currently used SNAP version does not accept this particular parameter, write the modified workflow to a new XML file and rerun the processing. This was, for example, necessary to be able to use SNAP6 workflows in SNAP5 since, in the Terrain-Flattening node, two new parameters were introduced in the newer version.

Throughout this study, GAMMA was found to work very reliably yet the authors benefited largely from several years of know-how built into pyroSAR in order to reduce the complexity of this rather difficult-to-use software.

If a large data cube with long SAR time series is to be built, software continuity is essential to ensure that all data have been processed in the same way. Although this is certainly not fully possible due to several changes in the internal IPF processor during the lifetime of Sentinel-1, a user is advised to use only one version of SNAP or GAMMA for building larger data sets and continuously expanding them with newly acquired data. For this, pyroSAR offers the modularized SNAP processing scheme described earlier, giving a user the option to selectively assign processor versions to specific nodes.

This way, only critical nodes can be executed in newer versions of SNAP. For example, in early 2018, the border noise removal became obsolete with a new IPF version, which caused an error in older versions of SNAP. This particular step could thus be executed by the newer version, leaving all other nodes working unchanged. Optimally, this kind of mechanism would directly be implemented in SNAP so that a user could operate multiple subversions of SNAP without explicitly installing them. A similar mechanism has not yet been developed for GAMMA.

Ideally, a user could also order data processed with a specific IPF version directly from ESA to also exclude optimizations made in newer versions for the sake of data continuity in case newer versions introduced changes, reducing the comparability with older scenes. While it is clear that a large data volume such as the entire S1 archive cannot be re-processed, an on-demand service coupled to the rolling archive could have a user select the specific IPF version of his or her choice.

5.2. Use of DEM Products

In this study, the suitability of four openly available DEM options for SAR processing was investigated. The extent to which two newer options, the ALOS World 30 m (AW3D30) DEM and the TanDEM-X 90 m DEM, stand up against the well-established SRTM variants in 1 and 3 arc seconds, was assessed. These four products represent three very different approaches of creating DEMs. The AW3D30 DEM was created photogrammetrically from optical ALOS PRISM data, the SRTM DEM from a bistatic C-Band SAR space shuttle mission and the TDX DEM from a bistatic X-Band SAR two-satellite constellation. Each having their advantages and disadvantages, large discrepancies between the four were observed, reflected in deviations of up to 2500 m from the median of all four in the Alps and even up to 15,000 m over Fiji. These large deviations need to be considered for SAR data processing, which otherwise would result in topographic normalization of low quality.

In the Alps, the SRTM 1 arcsec DEM was selected as the best option for processing based on general usability not requiring additional pre-processing such as masking, an overall low magnitude of deviations from the median, and the higher resolution in comparison to the 3 arcsec variant and the available TDX DEM. The AW3D30 DEM was found to contain several outliers of high magnitude around Lake Garda, which would have to be masked out prior to processing using an ancillary mask. The TDX DEM was found to contain several regions with artifacts, which are likely to be reduced in future updates of the products with improvements to the processor and/or manual corrections.

In the case of Fiji, no DEM fulfilled all criteria of being able to be used as provided, having a high resolution and containing a low number of high deviations. Both the SRTM 3 arcsec and TDX DEM required a re-coding of no data values over water so that SAR backscatter over water is not masked in the output products. The product of the highest quality was the low-resolution SRTM 3 arcsec DEM with only a few maximum deviations, all being of low magnitude. The TDX DEM, although also containing only deviations of similarly low magnitude, did not include a water mask of high quality, resulting in noise over water areas and a high omission of water around the coastlines of the Fiji Islands. Utilizing the TDX water indication mask delivered together with the current DEM product did not sufficiently mask out water bodies either. The SRTM 3 arcsec water mask was used instead. The SRTM 1 arcsec and AW3D30 contained several artifacts of very high magnitude above 15,000 m. Several of those found in the SRTM DEM were also found in the ALOS DEM. Since the ALOS DEM utilizes the SRTM DEM in areas where low accuracy was achieved with the photogrammetric approach, these artifacts were likely copied without further visual analysis. Because of the differences in water masking, the SRTM 3 arcsec water mask was applied to the other three DEMs as well since it was found to be the most accurate one. An overview of the findings in subsection "Quantification of Outliers" in Sections 4.2.1 and 4.2.2 is given in Table 18.

Table 18. Comparison of the 95th percentile (p95) of DEM deviation magnitude and the frequency of maximum outlier occurrence (%) for both test sites. Summary of Tables 7 and 8.

	p95		%	
	Alps	**Fiji**	**Alps**	**Fiji**
AW3D30	1197.80	14913.00	5.90	37.07
SRTM-1Sec-HGT	526.48	10275.38	3.69	49.41
SRTM-3Sec	298.47	296.20	4.74	8.81
TDX90m	530.46	170.16	85.67	4.71

5.3. Quality of Topographic Normalization

To compare different DEMs and their impact on SAR image topographic normalization, SNAP5 Terrain Flattening had to be used as the use of external DEMs was not possible in the considered version of SNAP6. Major improvements of the terrain flattening procedure were introduced in SNAP6, which is shown in Figures 9 and 18 of this study, where the SRTM 1 arcsec *auto-download* option was used to show the benefit over SNAP5. Further improvements to this procedure are expected for SNAP7, which is due to be released in summer 2019. The SNAP results of topographic normalization quality can thus not objectively be compared to those of GAMMA knowing that they show inferior results to the latest version of SNAP. By providing the Jupyter notebook with this publication, it is intended to begin establishing an open testing framework, that can easily be adjusted to future processing software updates or even extended to additional software solutions not yet considered in this study.

For summarizing the findings of subsection "Comparison of Single Image Processing Results" in Sections 4.2.1 and 4.2.2, Tables 19 and 20 compare the results of the GAMMA and SNAP processing results, respectively.

Table 19. Slope and coefficient of variation from the comparisons of GAMMA-processed RTC backscatter over forested regions with local incident angle made in Figures 13 and 17.

	Slope		CV	
	Alps	**Fiji**	**Alps**	**Fiji**
AW3D30	0.00	−0.02	−0.2080	−0.2652
SRTM-1Sec-HGT	0.00	−0.02	−0.1965	−0.2519
SRTM-3Sec	−0.01	−0.04	−0.2506	−0.2830
TDX90m	−0.01	−0.03	−0.2706	−0.2845

Table 20. Slope and coefficient of variation from the comparisons of SNAP-processed RTC backscatter over forested regions with local incident angle made in Figures 13 and 17.

	Slope		CV	
	Alps	**Fiji**	**Alps**	**Fiji**
AW3D30	−0.05	−0.07	−0.3635	−0.3104
SRTM-1Sec-HGT	−0.05	−0.07	−0.3512	−0.3161
SRTM-3Sec	−0.04	−0.06	−0.3152	−0.2942
TDX90m	−0.05	−0.06	−0.3378	−0.2938

The use of the SRTM 1 arcsec DEM resulted in the highest quality of backscatter normalization using GAMMA, both in the Alps and in Fiji. This DEM was also found to contain the smallest number of maximum deviations from the median across the whole Alps scene and low magnitudes for these deviations, second in ranking only to the SRTM 3 arcsec DEM. In contradiction, this DEM contained the most outliers in Fiji, where several artifacts of high DEM deviation were observed. The SRTM 3 arcsec DEM performed worse with similar results for the Alps and Fiji, although showing a low number of DEM outliers with low magnitude. The TDX DEM performed similarly to the SRTM 3 arcsec DEM, albeit containing nearly 86% or maximum deviations in the Alps. Higher slopes and variations were observed in Fiji for all DEMs.

In contradiction to the GAMMA results, the SRTM 3 arcsec and TDX DEMs performed best for SNAP processing and thus better reflect the findings of the DEM outlier assessment summarized in Table 18. The former, which consistently contained few outliers and low deviation magnitude in both test sites, also performed best in the SAR processing comparison. The high number of maximum deviations in the TDX DEM over the Alps did not have a noticeable influence on the SAR processing result.

It is concluded that the DEM comparisons performed in this study show valuable findings in assessing the suitability for SAR processing by highlighting the differences between them in regions of high deviations but are not directly suited to fully assess the resulting quality of the topographic normalization. Since this study focused on analyzing DEM outliers above 100 m, only very few points across the whole images were taken into consideration. However, although rarely occurring, these extreme outliers present in the data will occasionally have a large impact on the processing results and should be removed prior to processing to avoid misinterpretation. This becomes even more critical in a data cube environment where the analysis of individual images loses importance when several hundreds of images are being analyzed.

Large differences in the RMSE between SNAP and GAMMA products of about 2 dB in flat terrain and up to 4.5 dB on steep slopes were observed, which can only partially be explained by the differences in topographic normalization quality. Additional analyses are necessary to further investigate the differences between the two processing software packages and assess whether their results can be aligned more closely. The largest conceptual differences between the two processing workflows are GAMMA's conversion to slant range prior to normalization, while SNAP operates entirely in ground range, and the choice of resampling methods during geocoding.

5.4. Time Series Analysis

Section 4.4 of this study—Time Series Analysis—investigated the temporal interoperability of SNAP and GAMMA processed gamma0 for two land cover classes; annually consistent evergreen tropical forest, and seasonal grassland, taking into consideration the influence of both internal and external factors in the overall consistency of the output backscatter. The evaluation considered the influence of orbit direction, choice of software and topography.

The software analysis for the backscatter time series highlighted little discrepancies, and thus good interoperability, for the analyzed land covers, the different acquisition geometries and for small slopes. Contrariwise, large slopes yielded large discrepancies. Nevertheless, a useful property highlighting a consistent gamma0 processing chain is the intra-software interoperability among different viewing geometries at all slopes.

In view of the construction of a Sentinel-1 data cube, it can be concluded that both the commercial and open-source software workflows presented in this study do provide reliable gamma0 products, with the general recommendation of including a geometrical distortion (layover, shadow) map and the local incident angle as an auxiliary data cube product. The quality of the generated time series benefited from the custom border noise removal presented in Section 3.3. After processing, a data cube can aid in temporally exploiting and analyzing the time series in terms of backscatter and geophysical parameters, or simple change detection be performed. However, it needs to be kept in mind that changes in backscatter originating from DEM artifacts and insufficient topographic normalization might still be present in the data and care is thus to be taken when analyzing pixel time series out the spatial image context.

6. Conclusions and Further Recommendations

The choice of DEM as an input for creating gamma0 RTC backscatter is a major influencing factor for the correction of topographic effects such as foreshortening. Hence, it was assessed whether particular DEMs are generally better suited for the task and whether regional differences require site-specific selection. As DEM options, the two SRTM variants in 1 and 3 arcsecs, the ALOS 30m World DEM and the TanDEM-X 90m DEM were selected. In a first analysis, the four DEMs were compared, assessing their direct similarity and quantity of outliers. A second test focused on their influence on the quality of the resulting terrain correction. As factors for selecting between the four options, the need for additional pre-processing and thus overall usability, the frequency of outliers and the spatial resolution were considered. The highest overall quality was found in the two SRTM DEMs, containing accurate handling of water bodies and a low number of outliers, thus reducing the need for additional processing. The lowest quality was found in the TanDEM-X DEM containing an inaccurate water mask and also several large artifacts in mountainous regions. However, neither DEM was found to fulfill all criteria, with results differing between the two test sites. Hence, it is necessary to perform the presented comparisons prior to processing data for a test site in order to prevent the introduction of systematic backscatter errors that can be particularly difficult to assess in a data cube time series analysis scheme that potentially incorporates hundreds of scenes. Optimally, differences in DEM quality and their influence on SAR processing could be further quantified so that the development of an automated method is made possible that locally detects and corrects inconsistencies while taking actual terrain changes into consideration.

To assess the influence of the DEM quality on the actual processing result, the SAR images were compared with the local incident angle maps. This way, insufficiently corrected dependencies on the terrain were quantified both across the whole image, represented through a slope in the linear regression equation, and locally through the amount of variation around the mean backscatter. The findings of this analysis could not be predicted by those of the preceding DEM comparison and thus present an additional direct method to help select an optimal DEM.

A direct comparison of SNAP and GAMMA was not conducted as comprehensively as anticipated, as the use of external DEMs was not possible in the SNAP version available at the time of writing. Large improvements were confirmed between SNAP5 and SNAP6, with the latter performing similar to GAMMA. Large differences were found in areas of steep shadowed slopes, yet no assessment was made as to whether SNAP6 or GAMMA delivered better quality. Most remaining differences between the result of SNAP6 and GAMMA could not be explained by the quality of the terrain correction and are thus more likely the result of different methods used for resampling and interpolating the images during geocoding.

One step that is missing in both solutions is the accurate removal of Sentinel-1 border noise. While SNAP, as opposed to GAMMA, offers an implementation, this method was found to not sufficiently remove this noise. Alternatively, a custom implementation in pyroSAR was used, which was shown to reliably remove these artifacts and was thus applied prior to processing with either software.

In terms of the overall usability of the used processing software solutions, advantages and disadvantages can be attested to for both. SNAP offers a convenient, user-friendly graphical user interface and is open source, but its usage is often hindered by difficult-to-interpret error and log messages. Furthermore, the complex Java structure of the toolbox complicates the search and fix of encountered bugs.

GAMMA, on the other hand, offers a very basic command line interface requiring a lot of experience to use in order to develop workflows as flexible and easy to use as those of SNAP. However, once this is achieved, it works very robustly and fast.

The final choice of software is to be left to the user depending on the requirement and available resources. The study outlined a generally good interoperability of products except for extreme slopes tilted away from the sensor. However, this study focused on forests and grassland only and it is thus recommended to specifically test product interoperability for the land cover classes of interest. Furthermore, it is recommended to mask backscatter values in areas of incident angles greater 60° as an increase of RMSE between the products of SNAP and GAMMA was observed, which was attributed to a decrease in topographic normalization quality. This threshold value might change with future software versions and improved normalization routines.

In order to directly encompass the results of this study, a Jupyter notebook is provided. This way, the analyses presented here can easily be reproduced in new study sites, with new versions of the software solutions discussed, and also, with additional processing software options not yet taken into consideration. Additionally, the increased transparency is anticipated to contribute to an open discussion of best practices for producing ARD data and thus, generally accelerate progress in this field. The accompanying Jupyter notebook (see Supplementary Materials) together with pyroSAR is envisaged as an open test bed for extending ARD assessments towards SAR data domains, such as polarimetry and interferometry. In this context, the authors acknowledge the CEOS Analysis-Ready Data for Land (CARDL4L) guidelines as the best foundation for formally describing and standardizing the term ARD. No formal assessment was made of how far the workflows presented here meet the requirements of this guideline. This is considered as a future goal such that a user might eventually be able to select a certain level of analysis "readiness" in either software and conveniently process data accordingly.

As a data cube option to analyze the processed data, the Open Data Cube (ODC) technology was selected. Functionality to ingest data into this environment is presented in the accompanying Jupyter notebook and thus know-how is provided for lowering the entry border to working with such an environment. However, setting up such a cube and ingesting data into it is anticipated to require little effort only in comparison to ensuring a consistently high level of data quality and workflow transparency. The authors thus encourage application of the developed analyses provided in the Jupyter notebook and welcome additions to its online GitHub repository and the pyroSAR framework in order to improve the quality of ARD backscatter data and thus further lower the entry barrier to using SAR data. The ODC, in general, is seen as an interesting option for providing not only the SAR data itself but also the implementation of algorithms exploiting the data, further leveraging open science.

Supplementary Materials: A Jupyter notebook and auxiliary code for reproducing this study are available online at https://github.com/johntruckenbrodt/S1_ARD and http://www.mdpi.com/2306-5729/4/3/93/s1.

Author Contributions: Conceptualization: J.T., T.F., T.J., C.D., C.T. and A.S.; Data Curation: J.T., T.F., C.W., T.J. and D.S.; Formal Analysis: J.T. and C.W.; Investigation: J.T., C.W., T.J., D.S., C.D. and C.T.; Methodology: J.T., C.W., D.S., T.F., T.J., C.D., C.T. and C.R.; Software: J.T., C.W. and T.J.; Supervision: J.T., D.S., C.D., C.T. and C.R.; Validation: C.W., D.S. and C.R.; Visualization: J.T., C.W. and T.J.; Writing—Original Draft Preparation: J.T., C.W., T.F., C.R., T.J., A.S. and G.G.; Writing—Review and Editing: J.T., T.F., C.W., D.S., C.D., C.T., C.R. and T.J.

Funding: This project was partially funded through DFG (German Research Foundation) project HyperSense (grant No. TH 1435/4-1). This research was supported by the Common Sensing project funded by the UK Space Agency's International Partnership Programme (http://commonsensing.org.gridhosted.co.uk/).

Acknowledgments: The authors wish to extend their gratitude to the Fiji Ministry of Lands for providing data to support this study.

Conflicts of Interest: The authors declare no conflict of interest.

References

1. Onoda, M.; Young, O.R. *Satellite Earth Observations and Their Impact on Society and Policy*; Springer: Singapore, 2017.
2. Anderson, K.; Ryan, B.; Sonntag, W.; Kavvada, A.; Friedl, L. Earth observation in service of the 2030 agenda for sustainable development. *Geo-Spat. Inf. Sci.* **2017**, *20*, 77–96. [CrossRef]
3. Wulder, M.A.; Coops, N.C. Satellites: Make earth observations open access. *Nature* **2014**, *513*, 30–31. [CrossRef] [PubMed]
4. COPE-SERCO. *Sentinel Data Access Annual Report 2019*; COPE-SERCO-RP-19-0389; COPE-SERCO: Frascati, Italy, 3 May 2019.
5. Giuliani, G.; Chatenoux, B.; De Bono, A.; Rodila, D.; Richard, J.-P.; Allenbach, K.; Dao, H.; Peduzzi, P. Building an earth observations data cube: Lessons learned from the swiss data cube (sdc) on generating analysis ready data (ard). *Big Earth Data* **2017**, *1*, 100–117. [CrossRef]

6. Lewis, A.; Oliver, S.; Lymburner, L.; Evans, B.; Wyborn, L.; Mueller, N.; Raevksi, G.; Hooke, J.; Woodcock, R.; Sixsmith, J.; et al. The australian geoscience data cube — foundations and lessons learned. *Remote Sens. Environ.* **2017**, *202*, 276–292. [CrossRef]

7. Swiss Data Cube. First Sentinel-1 Analysis Ready Data Ingested. Available online: https://www.swissdatacube.org/index.php/2018/12/05/first-sentinel-1-analysis-ready-data-ingested/ (accessed on 9 April 2019).

8. Haarpaintner, J.; Killough, B.; Ofori-Ampofo, S.; Boamah, E.O. Advanced sentinel-1 analysis ready data for the ghana open data cube and environmental monitoring. In Proceedings of the International Workshop on Retrieval of Bio- & Geo-physical Parameters from SAR Data for Land Applications, Oberpfaffenhofen, Germany, 5 November 2018.

9. Dhu, T.; Dunn, B.; Lewis, B.; Lymburner, L.; Mueller, N.; Telfer, E.; Lewis, A.; McIntyre, A.; Minchin, S.; Phillips, C. Digital earth australia – unlocking new value from earth observation data. *Big Earth Data* **2017**, *1*, 64–74. [CrossRef]

10. Veci, L.; Lu, J.; Foumelis, M.; Engdahl, M. Esa's multi-mission sentinel-1 toolbox. In Proceedings of the EGU, Vienna, Austria, 23–28 April 2017.

11. Geudtner, D.; Torres, R.; Snoeij, P.; Davidson, M.; Rommen, B. Sentinel-1 system capabilities and applications. In Proceedings of the 2014 IEEE Geoscience and Remote Sensing Symposium, Quebec City, QC, Canada, 13–18 July 2014; pp. 1457–1460.

12. Ariza-Porras, C.; Bravo, G.; Villamizar, M.; Moreno, A.; Castro, H.; Galindo, G.; Cabera, E.; Valbuena, S.; Lozano, P. Cdcol: A geoscience data cube that meets colombian needs. In Proceedings of the Colombian Conference on Computing, Cali, Colombia, 19–22 September 2017; Springer International Publishing: Cham, Switzerland, 2017; pp. 87–99.

13. Baumann, P.; Rossi, A.P.; Bell, B.; Clements, O.; Evans, B.; Hoenig, H.; Hogan, P.; Kakaletris, G.; Koltsida, P.; Mantovani, S.; et al. Fostering cross-disciplinary earth science through datacube analytics. In *Earth Observation Open Science and Innovation*; Mathieu, P.-P., Aubrecht, C., Eds.; Springer International Publishing: Cham, Switzerland, 2018; pp. 91–119.

14. Kreiser, Z.; Killough, B.; Rizvi, S.R. Water across synthetic aperture radar data (wasard): Sar water body classification for the open data cube. In Proceedings of the IGARSS 2018-2018 IEEE International Geoscience and Remote Sensing Symposium, Valencia, Spain, 22–27 July 2018; pp. 437–440.

15. CEOS. *Analysis Ready Data for Land: Normalized Radar Backscatter*; CEOS: Reston, VA, USA, 14 December 2018.

16. Small, D. Flattening gamma: Radiometric terrain correction for sar imagery. *IEEE Trans. Geosci. Remote Sens.* **2011**, *49*, 3081–3093. [CrossRef]

17. Wicks, D.; Jones, T.; Rossi, C. Testing the interoperability of sentinel 1 analysis ready data over the united kingdom. In Proceedings of the IEEE International Geoscience and Remote Sensing Symposium, Valencia, Spain, 22–27 July 2018; pp. 8655–8658.

18. Giuliani, G.; Chatenoux, B.; Honeck, E.; Richard, J.-P. Towards sentinel-2 analysis ready data: A swiss data cube perspective. In Proceedings of the IEEE International Geoscience and Remote Sensing Symposium, Valencia, Spain, 22–27 July 2018.

19. Killough, B. Overview of the open data cube initiative. In Proceedings of the IEEE International Geoscience and Remote Sensing Symposium, Valencia, Spain, 22–27 July 2018; pp. 8629–8632.

20. Frau, L.; Rizvi, S.R.; Chatenoux, B.; Poussin, C.; Richard, J.; Giuliani, G. Snow observations from space: An approach to map snow cover from three decades of landsat imagery across switzerland. In Proceedings of the IEEE International Geoscience and Remote Sensing Symposium, Valencia, Spain, 22–27 July 2018; pp. 8663–8666.

21. Small, D.; Miranda, N.; Ewen, T.; Jonas, T. Reliably flattened backscatter for wet snow mapping from wide-swath sensors. In Proceedings of the ESA Living Planet Symposium, Edinburgh, UK, 9–13 September 2013.

22. Rüetschi, M.; Schaepman, M.E.; Small, D. Using multitemporal sentinel-1 c-band backscatter to monitor phenology and classify deciduous and coniferous forests in northern switzerland. *Remote Sens.* **2017**, *10*.

23. Rüetschi, M.; Small, D.; Waser, L. Rapid detection of windthrows using sentinel-1 c-band sar data. *Remote Sens.* **2019**, *11*.

24. Howell, S.E.L.; Small, D.; Rohner, C.; Mahmud, M.S.; Yackel, J.J.; Brady, M. Estimating melt onset over arctic sea ice from time series multi-sensor sentinel-1 and radarsat-2 backscatter. *Remote Sens. Environ.* **2019**, *229*, 48–59. [CrossRef]

25. Truckenbrodt, J.; Cremer, F.; Baris, I.; Eberle, J. Pyrosar: A framework for large-scale sar satellite data proessing. In Proceedings of the Big Data from Space, Munich, Germany, 19–20 February 2019; Soille, P., Loekken, S., Albani, S., Eds.; Publications Office of the European Union: Munich, Germany, 2019; pp. 197–200. [CrossRef]

26. Truckenbrodt, J.; Baris, I.; Cremer, F.; Kidd, R. Pyrosar Version 0.9.1 Online Documentation. Available online: https://pyrosar.readthedocs.io/en/v0.9.1/ (accessed on 5 July 2019).

27. Truckenbrodt, J.; Baris, I.; Cremer, F. Spatialist: A Python Module for Spatial Data Handling. Available online: https://github.com/johntruckenbrodt/spatialist (accessed on 9 April 2019).

28. ESA. Snap—Esa Sentinel Application Platform. Available online: http://step.esa.int/ (accessed on 9 April 2019).

29. Gamma Remote Sensing. Gamma Software. Available online: https://www.gamma-rs.ch/ (accessed on 9 April 2019).

30. GDAL/OGR Contributors. Gdal/ogr Geospatial Data Abstraction Software Library. Available online: http://gdal.org (accessed on 9 April 2019).

31. Barrilero, O.; Peters, M.; Cara, C.; Veci, L.; Engdahl, M.; Ramoino, F.; Volden, E. Evolutions and roadmap of snap and the sentinel toolboxes. In Proceedings of the ESA Living Planet Symposium, Milan, Italy, 13–17 May 2019.

32. Kluyver, T.; Ragan-Kelley, B.; Perez, F.; Granger, B.; Brussonnier, M.; Frederic, J.; Kelley, K.; Hamrick, J.; Grout, J.; Corlay, S.; et al. Jupyter notebooks—A publishing format for reproducible computational workflows. In Proceedings of the International Conference on Electronic Publishing, Göttingen, Germany, 7–9 June 2016.

33. van der Walt, S.; Colbert, S.C.; Varoquaux, G. The numpy array: A structure for efficient numerical computation. *Comput. Sci. Eng.* **2011**, *13*, 22–30. [CrossRef]

34. Hunter, J.D. Matplotlib: A 2d graphics environment. *Comput. Sci. Eng.* **2007**, *9*, 90–95. [CrossRef]

35. Jones, E.; Oliphant, T.; Peterson, P. Scipy: Open Source Scientific Tools for Python. Available online: http://www.scipy.org/ (accessed on 14 April 2019).

36. Robitaille, T.P.; Tollerud, E.J.; Greenfield, P.; Droettboom, M.; Bray, E.; Aldcroft, T.; Davis, M.; Ginsburg, A.; Price-Whelan, A.M.; Kerzendorf, W.E.; et al. Astropy: A community python package for astronomy. *Astron. Astrophys.* **2013**, *558*, A33.

37. Pedregosa, F.; Varoquaux, G.; Gramfort, A.; Michel, V.; Thirion, B.; Grisel, O.; Blondel, M.; Prettenhofer, P.; Weiss, R.; Dubourg, V.; et al. Scikit-learn: Machine learning in python. *J. Mach. Learn. Res.* **2011**, *12*, 2825–2830.

38. Reuter, H.I.; Nelson, A.; Jarvis, A. An evaluation of void-filling interpolation methods for srtm data. *Int. J. Geogr. Inf. Sci.* **2007**, *21*, 983–1008. [CrossRef]

39. Slater, J.A.; Garvey, G.; Johnston, C.; Haase, J.; Heady, B.; Kroenung, G.; Little, J. The srtm data "finishing" process and products. *Photogramm. Eng. Remote Sens.* **2006**, *72*, 237–247. [CrossRef]

40. Farr, T.G.; Rosen, P.A.; Caro, E.; Crippen, R.; Duren, R.; Hensley, S.; Kobrick, M.; Paller, M.; Rodriguez, E.; Roth, L.; et al. The shuttle radar topography mission. *Rev. Geophys.* **2007**, *45*. [CrossRef]

41. JAXA. *Alos Global Digital Surface Model (DSM) Product Description*; JAXA: Tokyo, Japan, April 2019.

42. DLR. *Tandem-x Ground Segment Dem Products Specification Document*; DLR: Cologne, Germany, 7 May 2018.

43. Wegmüller, U.; Werner, C.; Magnard, C. *Geocode_Back; Gamma Diff&Geo: Reference Manual*; Guemligen, Switzerland, 1 December 2017.

44. Miranda, N.; Hajduch, G. *Masking "no-value" Pixels on Grd Products Generated by the Sentinel-1 Esa Ipf*; CLS: New York, NY, USA, 29 January 2018.

45. Visvalingam, M.; Whyatt, J.D. Line generalisation by repeated elimination of points. *Cartogr. J.* **1993**, *30*, 46–51. [CrossRef]

46. Meier, E.; Frei, U.; Nüesch, D. Precise terrain corrected geocoded images. In *Sar Geocoding: Data and Systems*; Schreier, G., Ed.; Herbert Wichmann Verlag GmbH: Karlsruhe, Germany, 1993.

47. Büttner, G.; Kosztra, B.; Soukup, T.; Sousa, A.; Langanke, T. *Clc2018 Technical Guidelines*; European Environment Agency: Copenhagen, Denmark, 25 October 2017.

48. Schmitt, A.; Wendleder, A.; Hinz, S. The kennaugh element framework for multi-scale, multi-polarized, multi-temporal and multi-frequency sar image preparation. *ISPRS J. Photogramm. Remote Sens.* **2015**, *102*, 122–139. [CrossRef]
49. Vreugdenhil, M.; Wagner, W.; Bauer-Marschallinger, B.; Pfeil, I.; Teubner, I.; Rüdiger, C.; Strauss, P. Sensitivity of sentinel-1 backscatter to vegetation dynamics: An austrian case study. *Remote Sens.* **2018**, *10*, 1396. [CrossRef]

Article

Building a SAR-Enabled Data Cube Capability in Australia Using SAR Analysis Ready Data

Catherine Ticehurst [1,*], Zheng-Shu Zhou [2], Eric Lehmann [3], Fang Yuan [4], Medhavy Thankappan [4], Ake Rosenqvist [5], Ben Lewis [4] and Matt Paget [1]

[1] Commonwealth Scientific and Industrial Research Organisation (CSIRO) Land & Water, Canberra ACT 2601, Australia
[2] CSIRO Data61, Floreat WA 6014, Australia
[3] CSIRO Data61, Canberra ACT 2601, Australia
[4] Geoscience Australia, GPO Box 378, Canberra ACT 2601, Australia
[5] solo Earth Observation (soloEO), Tokyo 104-0054, Japan
* Correspondence: Catherine.Ticehurst@csiro.au

Received: 28 May 2019; Accepted: 12 July 2019; Published: 15 July 2019

Abstract: A research alliance between the Commonwealth Scientific and Industrial Research Organization and Geoscience Australia was established in relation to Digital Earth Australia, to develop a Synthetic Aperture Radar (SAR)-enabled Data Cube capability for Australia. This project has been developing SAR analysis ready data (ARD) products, including normalized radar backscatter (gamma nought, γ^0), eigenvector-based dual-polarization decomposition and interferometric coherence, all generated from the European Space Agency (ESA) Sentinel-1 interferometric wide swath mode data available on the Copernicus Australasia Regional Data Hub. These are produced using the open source ESA SNAP toolbox. The processing workflows are described, along with a comparison of the γ^0 backscatter and interferometric coherence ARD produced using SNAP and the proprietary software GAMMA. This comparison also evaluates the effects on γ^0 backscatter due to variations related to: Near- and far-range look angles; SNAP's default Shuttle Radar Topography Mission (SRTM) DEM and a refined Australia-wide DEM; as well as terrain. The agreement between SNAP and GAMMA is generally good, but also presents some systematic geometric and radiometric differences. The difference between SNAP's default SRTM DEM and the refined DEM showed a small geometric shift along the radar view direction. The systematic geometric and radiometric issues detected can however be expected to have negligible effects on analysis, provided products from the two processors and two DEMs are used separately and not mixed within the same analysis. The results lead to the conclusion that the SNAP toolbox is suitable for producing the Sentinel-1 ARD products.

Keywords: Sentinel-1; Synthetic Aperture Radar; Data Cube; dual-polarimetric decomposition; interferometric coherence; Digital Earth Australia

---

1. Introduction

1.1. Background

Synthetic Aperture Radar (SAR) data have been shown to provide different and complementary information to the more common optical remote sensing data. Radar backscatter response is a function of topography, land cover structure, orientation, and moisture characteristics—including vegetation biomass—and the radar signal is able to penetrate clouds, providing information about the earth's surface where optical sensors cannot. Despite these advantages, it is not used as extensively or operationally as optical data. Reasons for this have included the traditionally high cost of SAR data acquisition and the relatively complex and specialized processing methods [1].

The release of freely available European Copernicus programme data, especially the routinely acquired global coverage of Sentinel-1 SAR data, has opened up opportunities for greater exploration and application of SAR data globally. The Sentinel-1A and 1B SAR satellites have been operating since 2014 and 2016 respectively, and have been building up an archive of dual polarized C-band data, including extensive, wall-to-wall acquisitions over Australia since December 2016 at the spatial resolution of ~3 m × 22 m in the default acquisition mode of interferometric wide (IW) swath [2].

Digital Earth Australia (DEA) is an analysis platform for observations of all forms [3], but particularly those captured from satellites which have unique potential and pose particular challenges for their full exploitation. DEA uses images and information recorded by satellites orbiting our planet to detect physical changes across Australia. DEA was originally built upon the extensive Landsat archive processed into an analysis ready data (ARD) product (including atmospheric correction to surface reflectance, co-registration, and associated cloud/cloud shadow masks), and is being developed to feature other satellite datasets including SAR. A research alliance between the Commonwealth Scientific and Industrial Research Organization (CSIRO) and Geoscience Australia (GA) was established to develop SAR capability for DEA (referred to as the Australian SAR Data Cube project).

GA and the CSIRO are also both partners in the Open Data Cube (ODC) initiative, in which ODC platforms [4] aim to enable easier access to satellite ARD, as they remove the need for the user to pre-process Earth observation datasets, and provide access to archived remotely sensed data in a format ready for use. The definition of ARD with respect to SAR data is being actively developed through the Committee on Earth Observation Satellites (CEOS) analysis ready data for land (CARD4L) framework [5]. The Australian SAR Data Cube project has been utilizing this information in developing SAR ARD products for Australia, which currently include radar backscatter (gamma nought, γ^0), eigenvector-based dual-polarization decomposition and interferometric coherence, all generated from the Sentinel-1 IW swath mode data available through the Copernicus Australasia Regional Data Hub [6]. These three products have been selected since their processing methods are relatively well advanced, and they have already been used for environmental and agricultural applications within Australia [7,8]. Another reason for the selection of these three SAR ARD products for Australia is due to the availability of the dual-polarized Sentinel-1 SAR imagery by the European Space Agency (ESA).

The processing workflows for producing these SAR ARD products make use of ESA's free Sentinel-1 Toolbox within the Sentinel Application Platform (SNAP). SNAP is an open source platform, allowing easy access and sharing of processing workflows with the capability of batch processing through its graph processing tool (GPT). To evaluate the performance of SNAP for producing SAR ARD products, selected scenes are also processed to radar backscatter and interferometric coherence using the proprietary software GAMMA [9] for comparison.

This manuscript first describes the three SAR ARD products being developed for the Australian SAR Data Cube project and the applications they have been used for, with particular emphasis on Sentinel-1. It then details the workflows used to produce the SAR ARD products using the SNAP toolbox and gives an example of how the three selected products provide complementary information about the landscape. We then evaluate the outputs from the SNAP toolbox and compare them to outputs from the proprietary GAMMA software.

1.2. SAR ARD Products

The Australian SAR ARD products are currently produced using Sentinel-1 data in IW swath mode, which has been acquired systematically since October 2014 (for Sentinel-1A). It covers a swath of 250 km at a spatial resolution of ~3 m × 22 m (single look complex–SLC), or 20 m × 22 m for its ground range detected (GRD) high resolution class (HR) sampled to 10 m × 10 m, allowing regional coverage at a pixel size compatible with optical sensors such as the Landsat data series. Each Sentinel-1 satellite carries a dual polarization C-band SAR sensor (i.e., switchable H or V transmitter and parallel H and V receivers). Over land it is typically configured to acquire VH (vertical transmit–horizontal receive) and VV (vertical transmit–vertical receive) polarizations in IW mode [2]. Sentinel-1A in conjunction with Sentinel-1B have been routinely acquiring the dual-polarimetric IW products across the whole of Australia every 12 days since December 2016.

The SAR ARD products being developed for an Australian SAR data cube are normalized radar backscatter, dual-polarization decomposition, and interferometric coherence. These products are currently being tested for integration into the DEA data cube (which is part of the ODC initiative), which involves indexing of the products and retaining relevant metadata information to meet CARD4L standards. The normalized radar backscatter is the most widely used, however there are advantages to including the dual-polarization decomposition and interferometric coherence in a range of applications. These will be demonstrated in the following sections, based on information available in the literature, as well as an example of some SAR ARD data for Australia.

1.2.1. Radar Backscatter

Radar backscatter is the most widely used of the SAR products due to it being the simplest to produce and understand. It typically gives the proportion of radar signal backscattered to the receiver as amplitude (or intensity). Radar backscatter is dependent on the characteristics of the surface it is interacting with including its dielectric properties, orientation, and structure [10].

Applications based on SAR backscatter have appeared extensively for decades including a range of applications such as mangrove monitoring [11], forest biomass [12], and flood extent mapping [13,14]. However, the availability of free SAR archive data has increased the use of SAR for multi-temporal analysis, often improving results compared to single-date scenes [8]. Multi-temporal Sentinel-1 SAR can be used to identify patches of deforestation based on the detection of radar shadows from two viewing angles (using Sentinel-1 in ascending and descending mode) [15]. It has also been used to map fire scars in areas where persistent cloud-cover hampered efforts with optical remote sensing technology [16], as well as for the identification of irrigated agriculture [17].

1.2.2. Dual-Polarimetric Decomposition

For a fully polarimetric SAR system, quad-polarimetric decompositions enable the scattering mechanisms to be extracted as a single scattering matrix from the averaged Mueller matrix, decomposed into the sum of elementary matrices from the coherent scattering matrix, or characterized into physical scattering mechanisms by eigenvector-based decompositions of the coherency or covariance matrix [18]. These methods are used to distinguish land cover types exhibiting different scattering behaviors. Since the default imaging mode (IW) of Sentinel-1 works with selectable dual polarization, the quad-polarizations are not available. However, eigenvector-based dual-polarimetric decomposition can be applied to characterize the behavior of the scatterers to a certain extent, resulting in entropy, anisotropy, and alpha parameters [19].

This information is useful in applications such as land cover classification and change detection analysis. Zhou et al. [8] demonstrate how including the dual-polarimetric decomposition bands (entropy, anisotropy, and alpha) along with the normalized radar backscatter of multi-temporal Sentinel-1 data improves the discrimination of dryland crop type in the Wheatbelt of Victoria, Australia, as well as detecting the growth stage of an irrigated rice region in New South Wales. Cloude [19] demonstrates the ability of entropy/alpha in discriminating forest from non-forest, and in highlighting the complex scattering behavior of urban environments.

1.2.3. Multi-Temporal Coherence

Multi-temporal (or interferometric) coherence is a by-product when generating interferograms for applications such as deformation monitoring. However interferometric coherence can also be useful for determining whether the scattering properties of a surface change through time. This can be related to land-cover change or vegetation growth. The coherence between two images reduces over time as the land surface changes, which is more pronounced for the shorter wavelengths (such as C-band). This is particularly so for vegetation cover. However, objects that do not change through time (or change very slowly) can have a high coherence value between two multi-temporal images. In particular, buildings or bare ground (with constant soil moisture) can have a high coherence compared to their surroundings [20]. Sentinel-1 interferometric coherence is of interest to the ESA as demonstrated through the SINCOHMAP project [21], which is developing methods for land cover and vegetation mapping. Tamm et al. [22] found it was feasible to use Sentinel-1 12-day repeat pass interferometric coherence for identifying the dates that grasslands have been mown. However ploughed fields and remnant grass created confusion. One of the challenges of using interferometric coherence is that precipitation can cause temporal decorrelation [22].

2. SNAP Graph Processing Tool Workflow

2.1. SNAP Processing

The code developed in the SAR data cube project comprises a collection of shell scripts and python code to enable batch processing of the SNAP graph processing tool (GPT) XML files [23]. To produce a SAR ARD product, the list of available Sentinel-1 files is first extracted through the Sentinel Australasia Regional Access (SARA) portal [24]. SARA's web API allows queries based on area of interest, date range, Sentinel-1 level-1 data type (GRD or SLC), and sensor mode (in this case IW is used). The GPT executable is then run using the GPT graph XML files on the list of Sentinel-1 zip files. This process is currently run on the Australian National Computational Infrastructure [25], the same facility hosting SARA, so no data transfer is required for Sentinel-1 level-1 data access.

While each SAR ARD product requires its own processing workflow, some common steps and parameters are used for all. For IW mode, Sentinel-1 acquires data in three sub-swaths (and numerous bursts that are synchronized between passes) using the TOPSAR (terrain observation with progressive scans SAR) method [26]. To form a complete image from the SLC data, each sub-swath needs to be processed, including 'debursting' to remove the gaps between each burst, and then merged together [2]. Precise orbit file correction is applied to each SAR ARD product to ensure best geo-positional accuracy. All SAR ARD products are geometrically corrected using the 'SRTM 1Sec HGT' option available in the SNAP toolbox, which is automatically downloaded within the processing, or using the pre-downloaded and/or refined SRTM 1 arc-second DEM as a local DEM. The Australian Albers equal area projection was selected for the SAR ARDs, ensuring interoperability with the extensively used Landsat ARD products in DEA. An output pixel size of 25 m × 25 m is used for compatibility with the DEA Landsat series. The current output format is the BEAM-DIMAP flat binary image file format, as it is the native output by SNAP. However, GeoTIFF image file format, with internal compressed tiling enabled, is preferred as it improves compatibility with other software tools and improves performance (e.g., windowed reading). Conversion from BEAM-DIMAP to compressed GeoTIFF file format is performed after the initial SAR ARD product is created.

Each SAR ARD workflow is now described in more detail as it is processed in SNAP. Each of the processing steps, as built in the SNAP graph builder tool, is shown in Figure 1. All processing parameters are left as their default option unless specified.

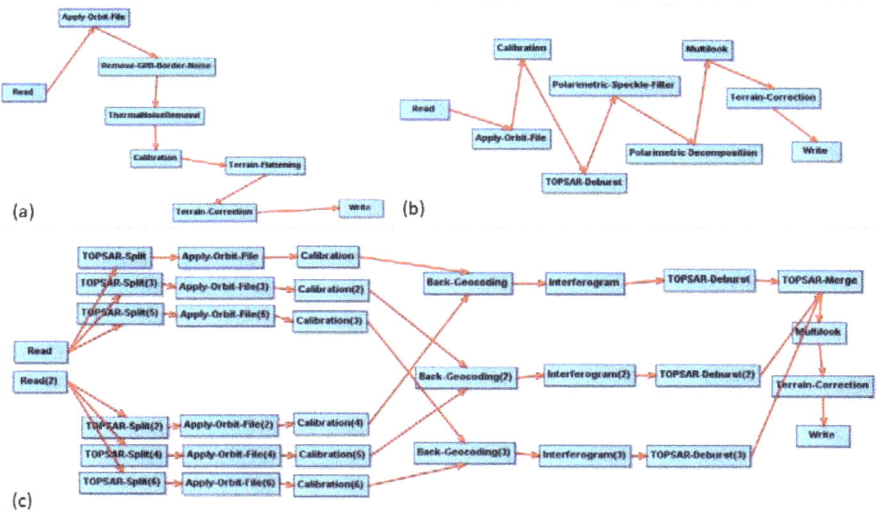

Figure 1. Workflows in the SNAP graph builder tool for producing Synthetic Aperture Radar (SAR) anlysis ready data (ARD) products (**a**) gamma nought radar backscatter, (**b**) dual-polarimetric decomposition, (**c**) interferometric coherence.

The normalized radar backscatter ARD is gamma nought (γ^0) as recommended in the CARD4L report [27], since it accounts for topographic variation better than the more traditional sigma naught (σ^0) [28]. This ARD is generated from the ESA's Sentinel-1 GRD product, rather than the SLC product, to save on processing time since debursting and multi-looking have already been applied [29]. It requires removal of border noise along the edge of some scenes, where the 'borderLimit' is set to 800 pixels, and the 'trimThreshold' to 10.0, to correct for some of the early Sentinel-1 scenes that had a wide strip of border noise with relatively high intensity values. Thermal noise removal is applied, before the image is calibrated to γ^0 backscatter. Radiometric terrain flattening is then applied using the 'SRTM 1Sec HGT' product available in SNAP, followed by the range Doppler terrain correction to orthorectify the image. The local incidence angle image is also output to meet CARD4L recommendations. No speckle filtering is applied to the radar backscatter SAR ARD product, as the type and parameters for speckle filtering (if one is required) is determined by its application.

The dual-polarization decomposition SAR ARD is generated from ESA's Sentinel-1 SLC product. The image is calibrated, but kept in complex format, and then deburst. The improved Lee Sigma filter is applied as it results in a smoother output image, before the H-alpha dual-polarization decomposition step to generate the alpha, anisotropy, and entropy bands. Each band is then multi-looked to create square pixels, before range Doppler terrain correction to orthorectify the image.

Processing the Sentinel-1 data into an interferometric coherence product first requires the matching of suitable interferometric scene pairs. For the interferometric coherence ARD product, scenes are considered as a suitable pair for processing if they have the same relative orbit number, and their acquisitions are a maximum of 12 days apart. Once the image pairs are defined and precise orbit correction applied, radiometric calibration is applied, followed by back-geocoding to each sub-swath before an interferogram is generated. The window size used for estimating coherence when generating the interferogram is set at 30 pixels in the range direction, and 9 pixels in azimuth, to produce a smoother output image. The flat-earth (reference) phase is subtracted in this step to remove the effects of the earth's curvature. The sub-swaths are then deburst and merged into a single image, followed by multi-looking and range Doppler terrain correction to orthorectify the image.

The batch processing of Sentinel-1 data to ARD for backscatter, dual-polarimetric decomposition, and interferometric coherence has been automated for execution on the Raijin super-computer at the National Computational Infrastructure (NCI, http://nci.org.au/). The code used to process the Sentinel-1 data on the NCI is available on the GitHub: https://github.com/opendatacube/radar. One complication of running the processing on high-performance facilities at the NCI is that the compute nodes do not have an external network interface. In the SNAP processing sequences discussed above, this creates issues when some of the steps are set to automatically download data from the ESA servers. This mainly applies to the apply-orbit-file, terrain-correction and back-geocoding steps, which require access to external orbit files or tiles of digital elevation model (DEM) data. To circumvent this issue, the necessary files of external data are pre-downloaded to the NCI file system prior to processing, and the SNAP workflow is made to use these files during execution.

Python code is used to submit jobs to Raijin to process the Sentinel-1 files as per the user's specifications (date range, spatial extents, etc.). Processing of the XML workflow using the GPT occurs in a multi-threaded way, allowing each job to be executed on multiple computer processing units (CPUs). Typical processing times ('walltime', using 8 CPUs) required by the tasks of interest are:

- Backscatter: 30–40 min per scene with typical input scene sizes of 0.5 Gb to 1 Gb.
- Dual-polarimetric decomposition: 80–85 min per scene with typical input scene sizes of 4.5 Gb.
- Interferometric coherence: 55–65 min per pair of Sentinel scenes with typical input scene sizes of 4.5 Gb.

2.2. Demonstration of SAR ARD Products

Normalized radar backscatter is often used in SAR applications, while dual-polarimetric decomposition and interferometric coherence are less common. However, depending on the application, there are benefits in using more than one of the ARD products. The benefits of utilizing the normalized radar backscatter, dual-polarimetric decomposition, and interferometric coherence are shown within the coastal zone of the Fitzroy River catchment in Western Australia.

Sentinel-1A data were processed to all three ARD products described in Section 2.1 for the Fitzroy River catchment (Figure 2) for December 2016 to April 2017. Sentinel-2 data were also used for a similar date (based on cloud cover) to identify land cover. The radar backscatter also had a Lee Sigma speckle filter applied to reduce speckle effects.

Figure 2 shows the γ^0 VV, γ^0 VH, entropy and interferometric coherence bands for three dates, along with a cloud-free Sentinel-2 scene. This scene shows a tidal inlet surrounded by mudflats and coastal mangroves. The middle of the scene consists of mangroves within the intertidal zone with larger mangroves adjacent to narrow river channels and the smaller mangroves and hypersaline areas further inland. The eastern side of the scene is outside the intertidal zone and consists of scattered trees among the grassland. The VH backscatter (Figure 2c) contrasts the very low backscatter of the mudflats with the high backscatter of the mangroves along the narrow channels. The VV backscatter (Figure 2b) shows further detail within the mudflats (blue areas along the lower-middle of the scene) as well as over water. The entropy band (Figur 2e) provides additional information with flooded wetlands and vegetated sand banks having a high entropy value, and the exposed non-vegetated sand banks having a low entropy value. This contrasts with the VV backscatter which has a low backscatter over water, but a high backscatter for the vegetated sand banks. A temporally flooded area towards the south-east of the scene (green area in Figure 2e) is not visible at all in the radar backscatter. Figure 2f demonstrates how the multi-temporal interferometric coherence can provide additional information in an environment as complex as a mangrove coastal zone. The information shown in the interferometric coherence image (Figure 2f) is very different to that available in the backscatter images (Figure 2b,c). High interferometric coherence values are observed over the slightly elevated bare areas within the coastal zone that aren't subject to tidal flooding.

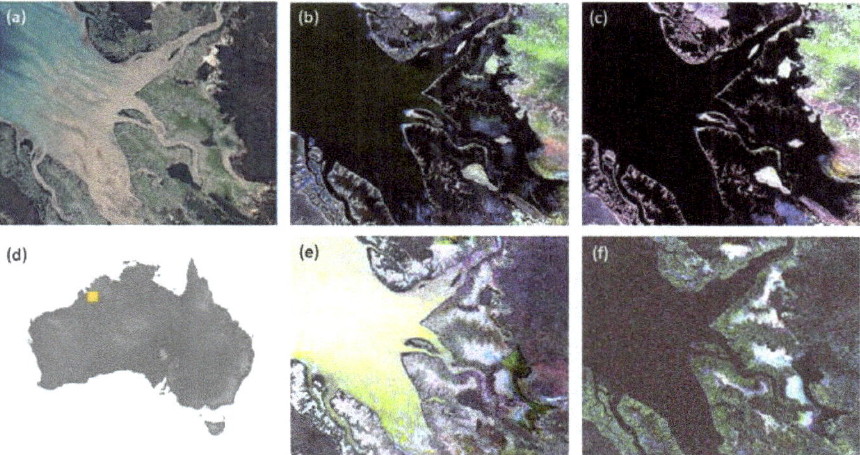

Figure 2. (**a**) Sentinel-2 true color image for 1 April 2017; (**b**) VV (vertical transmit – vertical receive) gamma nought backscatter with red-green-blue (RGB) as 30 Dec 2016, 11 January 2017, and 23 January 2017; (**c**) VH (vertical transmit – horizontal receive) gamma nought backscatter with RGB dates same as (**b**); (**d**) location of lower Fitzroy River catchment; (**e**) entropy band with RGB dates same as (**b**); (**f**) VV interferometric coherence with RGB as 18–30 December 2016, 30 December 2016–11 January 2017, 11–23 January 2017.

3. Assessment of Suitability of SNAP Toolbox for Australian SAR Data Cube Applications

To test how robust the proposed method of producing Sentinel-1 SAR ARD data is, an evaluation was done comparing the proposed workflow (SNAP toolbox and its default SRTM DEM from NASA's Shuttle Radar Topography Mission [30]) to some of the best available options (proprietary GAMMA software and a refined DEM). The SNAP toolbox has the advantage in that it is open source with a relatively simple processing workflow, while the proprietary GAMMA software is widely used for specialized SAR processing. The standard SRTM DEM available in the SNAP toolbox is openly available and automatically downloaded during the processing workflow, however the refined DEM [31] is a refined version of the SRTM DEM with Australia-wide coverage. The refined DEM has void filling, vegetation removal, and smoothing applied to reduce noise associated with low relief areas. The aims of these comparisons were to:

- Test how well the SNAP processing software compares to the proprietary GAMMA software (often considered to be one of the most reputable and industry best SAR software) when producing Sentinel-1 γ^0 backscatter.
- Compare how the standard SRTM DEM available in the SNAP processing software (referred to as the SRTM_DEM) and the refined, Australia-wide DEM (referred to as the GA_DEM [31]) influence γ^0 backscatter.
- Compare γ^0 backscatter from SNAP and GAMMA in an area of relatively steep topography and an area of relatively flat terrain.
- Compare the effects of look angle from a far-range and near-range image over the same area on the Sentinel-1 γ^0 backscatter.
- Evaluate the absolute geometric accuracy of the γ^0 images based on the location of a corner reflector within a scene.
- Compare the interferometric coherence ARD product generated from SNAP to the one generated from GAMMA.

Two study sites used for these comparisons are:

- Lake Eucumbene (Figure 3a), an area in the alpine region of southeast Australia containing relatively steep topography (height differences of ~1120 to 1550 m AHD) where the orbit paths overlap to give a near- and far-range image; and
- The Surat Basin (Figure 3b), a topographically flat (height differences of ~320 to 480 m AHD) dryland agricultural area in eastern Australia. This site is also used for radiometric and geometric calibration of SAR satellites including the Sentinel-1 constellation, through a permanent corner reflector array deployed there [32].

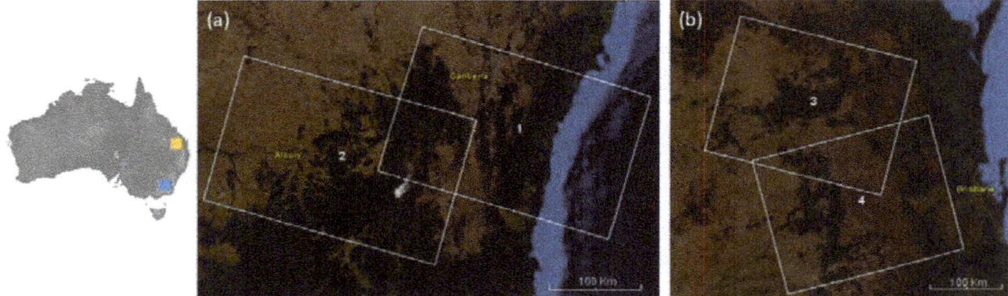

Figure 3. Location of Sentinel-1 images for (**a**) Lake Eucumbene far range (1) and near range (2), (**b**) Surat Basin for descending (3) and ascending mode (4).

The Sentinel-1 images processed to γ^0 backscatter for these two study sites are shown in Table 1. Two Sentinel-1 image pairs were used for generating the interferometric coherence (shown in Table 2). The processing steps and parameters used in the GAMMA software were selected to be the same as those in SNAP. There may however be slight differences due to the internal settings for some functions, which are not visible to the user. Note that a comparison could not be performed for the dual-polarimetric decomposition method, as it is only available within SNAP, not GAMMA.

The assessment is divided into four sections: Comparison of the γ^0 image from the SNAP toolbox and GAMMA software; comparison of the γ^0 image from the SRTM_DEM and GA_DEM; comparison of the γ^0 image from the near range and far range; and comparison of the interferometric coherence image from the SNAP toolbox and GAMMA software. These sections evaluate the radiometric and geometric consistencies of the different software, DEMs, and viewing angles. Note that all Sentinel-1 scenes are VV, with the exception of a scene in the Surat Basin, which is HH. This is a Sentinel-1B scene in ascending mode, which was selected to assess the γ^0 image's absolute geometric accuracy due to the availability of an accurately characterized corner reflector within this scene [32].

Table 1. Sentinel-1 ground range detected (GRD) images used for comparison of gamma nought images.

Filename	Ascending/Descending	Date	Look Angle	Polarization	Site Name
S1A_IW_GRDH_1SDV_20190110T191608_20190110T191633_025419_02D0D3_9C65.zip	Descending	1 Janurary 2019	Far range	VV	Lake Eucumbene
S1A_IW_GRDH_1SDV_20190115T192426_20190115T192451_025492_02D371_B5AA.zip	Descending	15 Janurary 2019	Near range	VV	Lake Eucumbene
S1B_IW_GRDH_1SDV_20180911T192100_20180911T192129_012671_01761D_377D.zip	Descending	11 September 2018	Mid range	VV	Surat Basin
S1B_IW_GRDH_1SSH_20180916T083200_20180916T083229_012737_017833_7C98.zip	Ascending	16 September 2018	Mid range	HH	Surat Basin

Table 2. Sentinel-1 single look complex (SLC) images used for comparison of interferometric coherence images.

Filename	Ascending/Descending	Date	Look Angle	Polarization	Site Name
S1A_IW_SLC__1SDV_20190115T192425_20190115T192452_025492_02D371_101B.zip	Descending	15 January 2019	Near range	VV	Lake Eucumbene
S1A_IW_SLC__1SDV_20190127T192425_20190127T192452_025667_02D9DB_A9BD.zip	Descending	27 Janurary 2019	Near range	VV	Lake Eucumbene
S1B_IW_SLC__1SDV_20180911T192100_20180911T192130_012671_01761D_7DDC.zip	Descending	11 September 2019	Near range	VV	Surat Basin
S1B_IW_SLC__1SDV_20180923T192100_20180923T192130_012846_01_B7E_7767.zip	Descending	23 September 2019	Near range	VV	Surat Basin

3.1. Comparison of SNAP and GAMMA Output for Radar Backscatter

GAMMA and SNAP software were used to produce radiometrically terrain corrected Sentinel-1 γ^0 backscatter products for Lake Eucumbene and Surat Basin study sites, both using the same input parameters and the same DEM (the refined GA_DEM). Figure 4 shows the γ^0 backscatter output for Lake Eucumbene from the far-range (top row) and near-range (middle row) images, as well as the Surat Basin (bottom two rows with HH along the bottom row) as produced from the GAMMA software and SNAP toolbox. The first column of Figure 4 shows the γ^0 images from GAMMA in the red band and SNAP in the blue/green bands (i.e., cyan). The grey color of these images illustrates there is generally similar radiometric and geometric agreement between the GAMMA and SNAP γ^0 backscatter products. The middle column of Figure 4 shows the GAMMA minus SNAP difference images. The geometric features visible in the difference images indicate systematic (rather than random) differences in image geometry. These differences are most prominent in areas with steep terrain and layover/radar shadowing effects (as illustrated in the GAMMA–SNAP difference images of Figure 4b,f), with shifts of 3–4 pixels detected. In terrain with moderate topography displacements in the order of 1–2 pixels were observed. In both cases the directions of the displacements furthermore differ systematically depending on the slope aspects relative to the radar, indicating different radiometric terrain correction approaches between GAMMA and SNAP. On slopes facing away from the radar a backscatter difference of 0.1 to 0.4 dB (with GAMMA values greater than SNAP) was observed compared to no detectable difference on the sides facing the radar. The magnitude of the effect was larger in the far-range image pair. The open farmlands to the east of Lake Eucumbene have backscatter intensity values within 0.1 dB. The flat agriculture areas of the Surat Basin study site show good agreement, with minor differences over vegetated areas (Figure 4m).

Figure 4c,g show a close up of the near range and far-range images respectively, including part of the lake. This area contains relatively steep terrain leading to the water's edge. Most of this area shows good radiometric and geometric agreement, except along small sections of the water's edge where SNAP γ^0 backscatter values are higher (showing in cyan tones). The near-range image in Figure 4g includes minor differences where γ^0 backscatter from GAMMA is higher than from SNAP (red tones). This occurs within a forested area adjacent to the banks of the lake. In flat open farmland, isolated buildings act as point scatterers as seen in the far range and near range (Figure 4d,h respectively), illustrating a small but systematic geometric difference between the GAMMA and SNAP outputs of less than one pixel in the north-south (azimuth) direction.

In the descending image over the Surat Basin study site (Figure 4i), small systematic differences between the GAMMA and SNAP γ^0 backscatter values are visible (Figure 4j). A close-up of a local dam and forest areas (Figure 4k) show no visible difference. The ascending image (HH polarization) over the Surat Basin study site (Figure 4l) shows minor differences (Figure 4m) particularly along the vegetated river bank. A close-up of the corner reflector in Figure 4n shows that the geometric difference between the GAMMA and SNAP scenes is approximately one pixel in the east-west (range) direction. The absolute geometric accuracy of the γ^0 HH image agrees to 0.8 arc-second (which is within a Sentinel-1 pixel) when compared to the coordinates of the corner reflector for both GAMMA and SNAP.

The frequency histograms of the backscatter intensity values (in dB) from Figure 4 are shown in Figure 5: For Lake Eucumbene for far range (Figure 5a), near range (Figure 5b), and the Surat Basin study site descending (Figure 5c) and ascending (Figure 5d) both of which have a mid-range look angle. Overall the results are similar, however the GAMMA output (red line) has a lower number of mid-range backscatter intensity pixels ($\sim$ −15 to −10 dB) compared to the SNAP output products (blue line). This effect occurs irrespective of radar look angle, however the differences are lower in the Surat Basin images (Figure 5c,d).

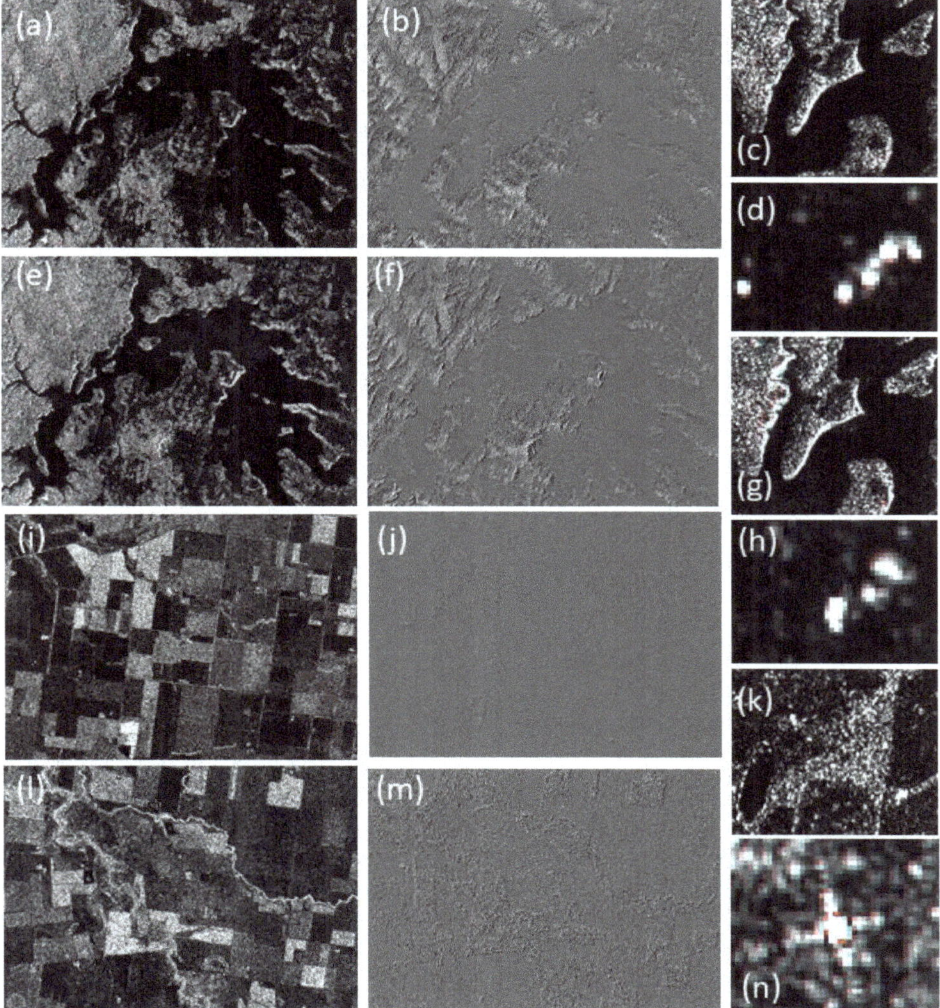

Figure 4. Gamma nought comparisons of GAMMA and SNAP (**a**) Lake Eucumbene far range (red = GAMMA with GA_DEM, cyan = SNAP with GA_DEM); (**b**) GAMMA minus SNAP difference image of (**a**); (**c**) close up of (**a**); (**d**) close up of (**a**); (**e**) Lake Eucumbene near range (red = GAMMA with GA_DEM, cyan = SNAP with GA_DEM); (**f**) GAMMA minus SNAP difference image of (**e**); (**g**) close up of (**e**); (**h**) close up of (**e**); (**i**) Surat Basin descending (red = GAMMA with GA_DEM, cyan = SNAP with GA_DEM); (**j**) GAMMA minus SNAP difference of (**i**); (**k**) close up of (**i**); (**l**) Surat Basin ascending (red = GAMMA with GA_DEM, cyan = SNAP with GA_DEM); (**m**) GAMMA minus SNAP difference of (**l**); (**n**) close up of (**l**). Note: all GAMMA minus SNAP difference images are stretched to the same grey-scale range: −0.3 to 0.3 intensity.

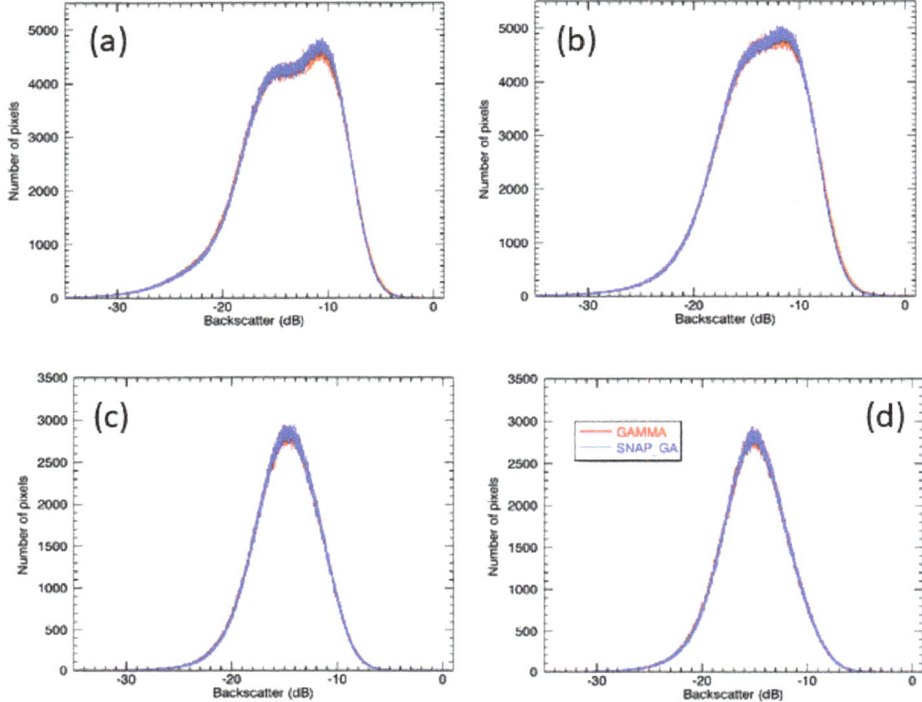

Figure 5. Frequency histograms of backscatter intensity (in dB) from the GAMMA and SNAP gamma nought images for Lake Eucumbene Sentinel-1 study site (**a**) far range, (**b**) near range, and Surat Basin study sites (**c**) descending and (**d**) ascending (HH).

The GAMMA–SNAP difference images (middle column of Figure 4) have low mean absolute values varying from 0.013×10^{-3} to 1.3×10^{-3} intensity, with the SNAP image lower than the GAMMA processed image except for the near-range image of Lake Eucumbene (Table 3). (Note that a small number of pixels (<0.005%) behaved as strong point scatterers resulting in very high backscatter intensity values. These were masked as they incorrectly influenced overall image statistics). The standard deviation of the difference image is lowest for the Surat Basin study site in descending mode and highest for the far-range difference image of Lake Eucumbene.

Table 3. Mean and standard deviations (as intensity) of gamma nought difference images from the GAMMA and SNAP comparisons for Lake Eucumbene and Surat Basin study sites.

Comparison	Study Site	Look Angle	Mean	Standard Deviation
GAMMA–SNAP	Lake Eucumbene	Far range	-0.2×10^{-3}	0.04
	Lake Eucumbene	Near range	1.3×10^{-3}	0.03
	Surat	Mid range	-0.01×10^{-3}	0.01
	Surat HH	Mid range	-0.13×10^{-3}	0.03

3.2. Comparison of SNAP Output Using SRTM_DEM and GA_DEM for Radar Backscatter

SNAP software was used to produce Sentinel-1 γ^0 images using the SRTM_DEM and GA_DEM for Lake Eucumbene and Surat Basin study sites; all used the same input parameters except for the DEMs. Figure 6 shows the γ^0 backscatter output for Lake Eucumbene from the far-range (top row) and near-range (middle row) images, as well as the Surat Basin site (bottom two rows with HH along the bottom row) as produced from the SNAP toolbox with the different DEMs. The first column of Figure 6 shows the γ^0 images produced using the SRTM_DEM in the red band and using the GA_DEM in the blue/green bands (i.e., cyan). The grey colour of these images illustrate similar γ^0 backscatter products, both in radiometric and geometric quality. The middle column of Figure 6 shows the difference images as produced from SNAP using the SRTM_DEM and GA_DEM. The relief patterns visible in Figure 6b,f illustrate a systematic shift between the two DEMs of 1–2 pixels in the east–west (range) direction and 0.5–1 pixels in the north–south (azimuth) direction within the moderate and steep terrain. Across the lake surface and immediate shorelines, the two DEMs are identical and consequently, there are no observable differences in backscatter.

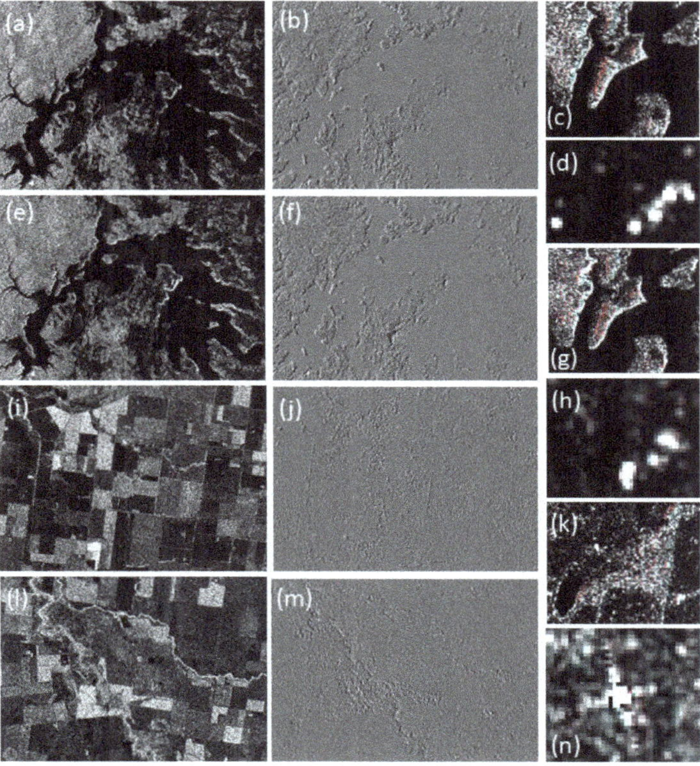

Figure 6. Gamma nought comparisons of SRTM_DEM and GA_DEM (**a**) Lake Eucumbene Far range (red = SNAP with SRTM_DEM, cyan = SNAP with GA_DEM); (**b**) SNAP SRTM_DEM minus GA_DEM difference image of (**a**); (**c**) close up of (**a**); (**d**) close up of (**a**); (**e**) Lake Eucumbene near range (red = SNAP with SRTM_DEM, cyan = SNAP with GA_DEM); (**f**) SNAP SRTM_DEM minus GA_DEM difference image of (**e**); (**g**) close up of (**e**); (**h**) close up of (**e**); (**i**) Surat Basin descending (red = SNAP with SRTM_DEM, cyan = SNAP with GA_DEM); (**j**) SNAP SRTM_DEM minus GA_DEM difference image of (**i**); (**k**) close up of (**i**); (**l**) Surat Basin ascending (red = SNAP with SRTM_DEM, cyan = SNAP with GA_DEM); (**m**) SNAP SRTM_DEM minus GA_DEM difference image of (**l**); (**n**) close up of (**l**). Note: all SNAP SRTM_DEM minus GA_DEM difference images are stretched to the same grey-scale range: −0.3 to 0.3 intensity.

Figure 6c,g show the same close-up view around the edge of Lake Eucumbene as Figure 4c,g. As expected, the main difference in backscatter occurs in areas where the two DEMs are different. This is mostly along the edge of forested areas next to the lake, where the refined GA_DEM has been corrected to ground level (and hence tall vegetation has been removed). As a direct effect from the geometric displacement between the DEMs, the γ^0 backscatter produced using the SRTM_DEM has a higher backscatter intensity along the edges of the forest facing towards the radar. The close-up view of flat open farmland near Lake Eucumbene (Figure 6d,h) shows that the geometric agreement between the SRTM_DEM and GA_DEM are within a pixel.

For the Surat Basin study site in descending mode, the main differences occur along the edges of forests (Figure 6k), similar to the Lake Eucumbene area. A close-up view of the corner reflector in the ascending image (bottom-left of Figure 6n) shows the geometric difference between the SRTM_DEM and GA_DEM is within a pixel.

The SRTM–GA DEM difference images (second column of Figure 6) have absolute mean values close to zero, with small variations observed (0.1×10^{-3} intensity or less, Table 4) being a direct a consequence of the geometric shift between the SRTM and GA DEMs.

Table 4. Mean and standard deviations (as intensity) of gamma nought difference images from the SNAP toolbox from the SRTM_DEM and GA_DEM comparisons for the Lake Eucumbene and Surat Basin study sites.

Comparison	Study Site	Look Angle	Mean	Standard Deviation
	Lake Eucumbene	Far range	0.09×10^{-3}	0.03
SRTM–GA DEM	Lake Eucumbene	Near range	0.07×10^{-3}	0.04
	Surat	Mid range	-0.1×10^{-3}	0.01
	Surat HH	Mid range	0.08×10^{-3}	0.01

3.3. Comparison of Near-Range and Far-Range Effects for Radar Backscatter

Another important factor to consider when using multi-temporal Sentinel-1 ARD products relates to radiometric and geometric variation between the near-range and far-range images processed with the same processor. Figure 7a shows Lake Eucumbene with far range in red and near range in cyan, and Figure 7b shows the difference image. Differences on land appear random and caused by actual incidence angle differences, with no systematic geometric shifts detected. The forested areas of relatively steep terrain show higher backscatter intensity values in the far-range compared to the near-range image. Patterns are visible on the lake (with the near range having higher backscatter intensity values), which are caused by surface condition variations between the two acquisition dates (an interval of 5 days). The mean and standard deviation in intensity for this difference image are -2.5×10^{-3} and 0.07 respectively, which is a larger standard deviation than seen in the GAMMA–SNAP and DEM comparisons. However, the geometric accuracy between the two images is within a pixel.

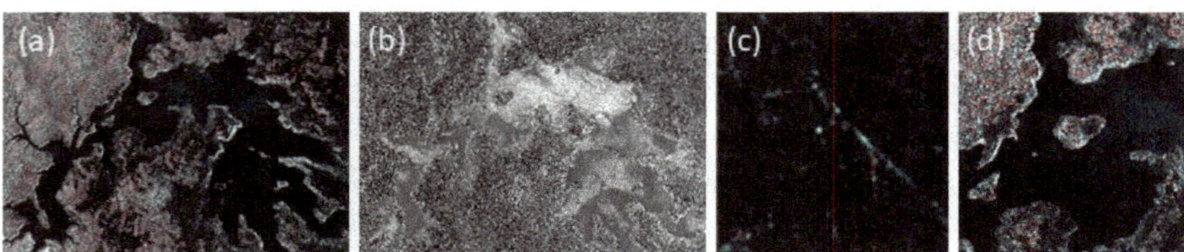

Figure 7. (**a**) Lake Eucumbene study site; (**b**) near range minus far range difference image of (**a**) (**c**) flat farm land with bitumen road; (**c**) close-up view of Lake Eucumbene. (Colors in (**a**), (**c**) and (**d**) are red = far-range gamma nought from 10 January 2019, cyan = near-range gamma nought from 15 January 2019).

Figure 7c shows a closeup of the backscatter from far-range (red) and near-range (cyan) images for an area of relatively flat farmland. The bitumen road is easily visible in the near-range (cyan), but is not visible in the far-range image due to the low backscatter return from specular scattering. There are radiometric differences between the far-range and near-range images, most likely due to the different look angles resulting in different interactions between the radar and land surface.

This difference has a direct impact on surface water mapping. Figure 7d shows a close-up view of Lake Eucumbene. There is good geometric agreement between the two dates and look angles, since the boundary between water and land match well. The blue tones within the water indicate higher backscatter in the near range over water. This can lead to confusion when discriminating water from non-water based on threshold values. Figure 8 shows histograms generated from multiple pairs of near- and far-range Sentinel-1 observations, where water and non-water pixels are identified using Landsat 8 spectral classifications (the water observations from space [33]) acquired within 5 days from corresponding Sentinel-1 scenes. The figure illustrates consistently better separation between water and non-water at larger look angles, regardless of variation caused by weather conditions (e.g., waves).

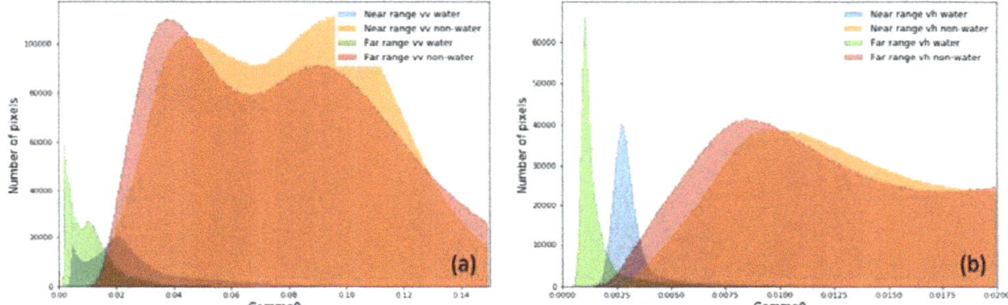

Figure 8. Histograms of water and non-water pixels for the near-range and far-range Sentinel-1 scenes over Lake Eucumbene for (**a**) VV and (**b**) VH. Water pixels are identified using nearby Landsat 8 spectral classifications (see text). Four pairs of SAR images are included to show how varying weather conditions can broaden the backscatter distributions over water but backscatter values over water are consistently lower in the far range.

Other factors influencing the use of overlapping Sentinel-1 images relate to changes in the land surface (e.g., soil moisture) or atmospheric effects (such as heavy rain which can influence the C-band wavelength). In this study, we have particularly selected two pairs of data acquired in the dry season with lower chance of meteorology impact, however these factors need to be considered when using multi-temporal SAR ARD for environmental and agricultural applications.

3.4. Comparison of SNAP and GAMMA Output for Interferometric Coherence

GAMMA and SNAP software were used to produce Sentinel-1 interferometric coherence products for the Lake Eucumbene and Surat Basin study sites, both using the same input parameters and the same DEM (the SRTM_DEM). For SNAP, the method used to generate interferometric coherence is shown in Section 2.1. The left-hand column in Figure 9 shows the two interferometric coherence images with SNAP in red and GAMMA in cyan, with Lake Eucumbene along the top row and the Surat Basin along the bottom. The SNAP–GAMMA coherence difference images are shown in the middle column.

For the Lake Eucumbene scene (Figure 9a), bare areas have a high coherence (white) along the eastern side of the lake. The shoreline around the lake also has a high coherence as it is void of vegetation. Coherence within the lake is very low due to decorrelation of water surface over time. There is a 0.5–1 pixel shift between the SNAP and GAMMA coherence images, with the GAMMA coherence image to the northeast of the SNAP coherence image. The greatest difference in coherence between SNAP and GAMMA occurs in steep terrain due to this systematic pixel shift and the effects of layover. This difference is greatest along the steep barren slopes around the edge of the lake, possibly due to misalignment, and forested slopes along the western side of Lake Eucumbene (Figure 9c), where gradients up to 50% result in radar shadow and layover effects.

For the Surat Basin, bare fields have a high coherence (white), and the higher-biomass crops have a very low coherence (Figure 9d). There is an approximate 1 pixel shift in this scene image with the GAMMA coherence image northeast of the SNAP coherence image. This effect is visible in the SNAP–GAMMA interferometric difference image (Figure 9e), where the greatest difference occurs along the crop edges. The bare fields have a difference in coherence of −0.01 to −0.02 with the SNAP coherence being lower. This difference is more variable where there is crop growth.

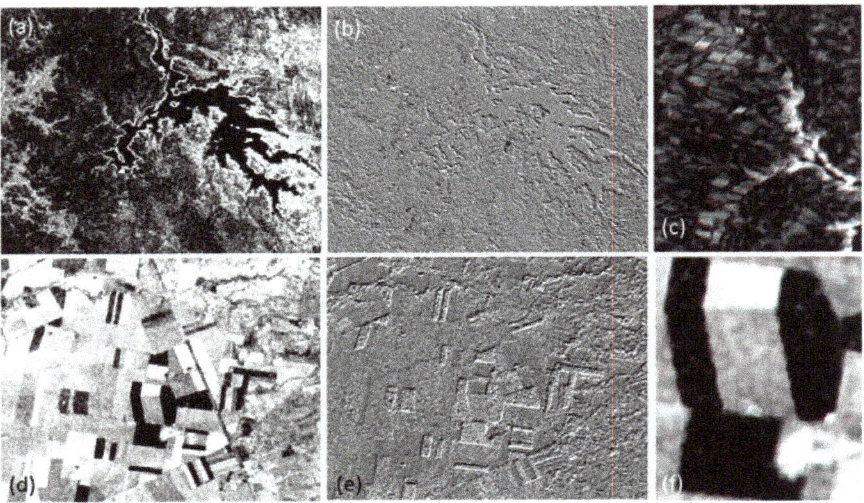

Figure 9. Interferometric coherence comparisons of SNAP and GAMMA (**a**) Lake Eucumbene (red = SNAP, cyan = GAMMA); (**b**) SNAP minus GAMMA difference image of (**a**); (**c**) close up of (**a**); (**d**) Surat Basin (red = SNAP, cyan = GAMMA); (**e**) SNAP minus GAMMA difference image of (**d**), (**f**) close up of (**d**).

The frequency histograms of the interferometric coherence values from Figure 9 are shown in Figure 10 for Lake Eucumbene (Figure 10a) and the Surat Basin (Figure 10b) from SNAP (blue) and GAMMA (red). The frequency distribution of interferometric coherence values between SNAP and GAMMA are similar, with some variation in the lower values (<0.15) for the Lake Eucumbene scene and in the higher values (>0.6) for the Surat Basin scene.

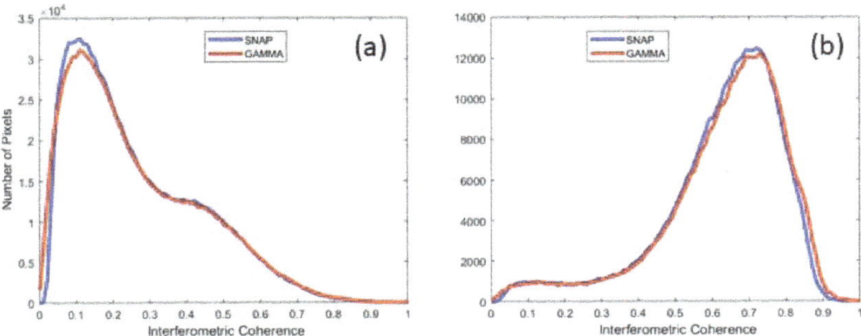

Figure 10. Frequency histograms of interferometric coherence from images processed by GAMMA and SNAP for (**a**) Lake Eucumbene study site, (**b**) Surat Basin study site.

The SNAP–GAMMA coherence difference images (second column of Figure 9) have absolute mean values of 0.003 and −0.007 for Lake Eucumbene and the Surat Basin respectively, with similar standard deviations from both study sites (0.031 and 0.024, respectively). These minor differences may be related to the different internal processing methods used to produce interferometric coherence in SNAP and GAMMA, which is apparent within the cropping area where coherence changes abruptly along crop edges, or along steep slopes.

4. Conclusions

Three SAR ARD products have been developed for the Australian SAR Data Cube project due to their potential benefits in environmental and agricultural applications. The characteristics of the output ARD products are designed to use freely available SAR data with processing workflows based on open source software, in particular the Sentinel-1 SAR and the SNAP processing toolbox. The methods used to produce the SAR ARD products with SNAP have been described along with an example of the different information they can provide in a coastal environment.

To evaluate the quality of the SAR ARD products generated using the SNAP toolbox, they were compared to equivalent products generated from one of the industry-best proprietary software, GAMMA. These comparisons also investigated the effects on γ^0 backscatter due to variations related to: Near- and far-range look angles; SNAP's default Shuttle Radar Topography Mission (SRTM) DEM and a refined Australia-wide DEM; as well as different terrain. A comparison of the interferometric coherence produced from SNAP and GAMMA was also performed. The GAMMA software does not provide dual-polarimetric decomposition, at the time of writing this article, so this product comparison could not be made.

The γ^0 images are geometrically aligned to within a pixel in the evaluation images from the near-range and far-range images of Lake Eucumbene, and the Surat Basin study site over flat terrain. The ascending HH image of the Surat Basin study site had an absolute geometric accuracy (0.8 arc-second) well within a Sentinel-1 pixel when compared to a well-characterized corner reflector within the scene.

Based on the radiometric and geometric assessment, comparisons show the γ^0 images produced from the SNAP and GAMMA software packages have small but systematic differences, due to different radiometric terrain correction algorithms, however we are unable to conclude which one is better. These differences increase with slope with largest differences being in terrain affected by layover and shadowing, and is also influenced by slope aspect (i.e., orientation of slope relative to the radar). The overall difference (based on standard deviations) between the GAMMA–SNAP difference images is larger in the far-range image compared to the near-range image.

Comparison between the γ^0 images produced from SNAP using the (NASA) SRTM_DEM and GA_DEM revealed systematic geometric displacements between the products in moderate and steep terrain, possibly because the refined GA_DEM is corrected to the ground surface, whereas the SRTM_DEM is a digital surface model, and hence is still influenced by the height of the trees. These small systematic differences were most noticeable in forest areas.

Based on the results evaluated in this study, the greatest difference in the γ^0 backscatter is between the overlapping near-range and far-range images, rather than processing software or DEM.

Comparison of the interferometric coherence images from SNAP and GAMMA showed a 0.5–1 pixel shift resulting in a small difference between products most notable in areas of steep terrain and crop edges, however overall image statistics were very similar.

Overall the geometric differences were minor, and the radiometric differences were most likely related to different viewing geometries that are not fully corrected for in the processing. Co-registration to a standard spatial layer would likely reduce the geometric differences between SNAP and GAMMA, particularly in flat terrain. The results from this study indicate that the SNAP Sentinel-1 Toolbox can be considered acceptable for processing Sentinel-1 data into ARD products, in terms of radiometric and geometric requirements. The systematic geometric and radiometric issues detected can be expected to have negligible effects on analysis, provided the products from the two processors are used separately and not mixed within the same analysis.

However, some factors still require consideration for operational applications and for scaling to regional- and national-scale extents. Scenes are currently processed individually, so there may be radiometric inconsistencies between overlapping swaths due to different viewing geometries. Appropriate co-registration should be implemented in multi-temporal Sentinel-1 data processing for better alignment. Further quantitative and multi-temporal analysis will be needed for a more accurate assessment, particularly for different environments within the Australian landscape. Future work will also include further testing of the SAR ARD processing workflows and development of applications based on these products. While the code is currently designed to operate on the National Computational Infrastructure, it is also being developed to enable processing on local computing facilities, or Amazon web services. Integration of the radar backscatter, dual-polarimetric decomposition, and interferometic coherence SAR ARD products into the Digital Earth Australia data cube is currently being tested. The software developed by this project is provided as open source tools via GitHub (GitHub, 2019).

Author Contributions: Writing—original draft preparation, C.T.; writing—review and editing, C.T.; Z.-S.Z.; E.L.; F.Y.; M.T.; and A.R.; software, E.L.; F.Y. and B.L.; methodology, formal analysis, Z.-S.Z.; F.Y. and A.R.; conceptualization, M.P.

Funding: This research received no external funding.

Acknowledgments: The authors wish to thank Biswajit Bala (CSIRO), Norman Mueller (GA) and Sean Chua (GA) for providing initial reviews.

Conflicts of Interest: The authors declare no conflict of interest.

References

1. Reiche, J.; Lucas, R.; Mitchell, A.L.; Verbesselt, J.; Hoekman, D.H.; Haarpaintner, J.; Kellndorfer, J.M.; Rosenqvist, A.; Lehmann, E.A.; Woodcock, C.E.; et al. Combining satellite data for better tropical forest monitoring. *Nat. Clim. Chang.* **2016**, *6*, 120–122. [CrossRef]

2. European Space Agency (ESA) Sentinel Online–Sentinel-1. Available online: https://sentinels.copernicus.eu/web/sentinel/missions/sentinel-1 (accessed on 28 April 2019).

3. Geoscience Australia Digital Earth Australia. Available online: http://www.ga.gov.au/about/projects/geographic/digital-earth-australia (accessed on 28 April 2019).

4. OpenDataCube–An Open Source Geospatial Data Management & Analysis Platform. Available online: https://www.opendatacube.org/ (accessed on 28 April 2019).

5. Committee on Earth Observation Satellites–CEOS Analysis Ready Data. Available online: http://ceos.org/ard/ (accessed on 28 April 2019).

6. Copernicus Australasia. Commonwealth of Australia. Available online: http://www.copernicus.gov.au/ (accessed on 28 April 2019).

7. Lehmann, E.A.; Caccetta, P.; Lowell, K.E.; Mitchell, A.; Zhou, Z-S.; Held, A.; Milne, T.; Tapley, I. SAR and optical remote sensing: assessment of interoperability and complementarity in the context of a large-scale operational forest monitoring system. *Remote Sens. Environ.* **2015**, *156*, 335–348. [CrossRef]

8. Zhou, Z-S.; Caccetta, P.; Devereux, D.; Caccetta, M.; Woodcock, R.; Paget, M.; Held, A. Preparation of analysis ready POLSAR data for the Australian Geoscience Data Cube. In Proceedings of the IEEE International Geoscience and Remote Sensing Symposium (IGARSS), Fort Worth, TX, USA, 23–28 July 2017. [CrossRef]

9. Gamma Remote Sensing. Gamma Remote Sensing—GAMMA Software. Available online: http://www.gamma-rs.ch/software (accessed on 28 April 2019).

10. Earth Online, Radar Course 2. European Space Agency. Available online: https://earth.esa.int/web/guest/missions/esa-operational-eo-missions/ers/instruments/sar/applications/radar-courses/content-2/-/asset_publisher/qIBc6NYRXfnG/content/radar-course-2-parameters-affecting-radar-backscatter (accessed on 20 March 2019).

11. Lucas, R.M.; Mitchell, A.L.; Rosenqvist, A.; Proisy, C.; Melius, A.; Ticehurst, C. The potential of L-band SAR for quantifying mangrove characteristics and change: Case studies from the tropics. *Aquat. Conserv. Mar. Freshw. Ecosyst.* **2007**, *17*, 245–264. [CrossRef]

12. Quegan, S.; Le Toan, T.; Chave, J.; Dall, J.; Exbrayat, J-F.; Ho Tong Minh, D.; Lomas, M.; D-Alessandro, M.M.; Paillou, P.; Papathanassiou, K.; et al. The European Space Agency BIOMASS mission: Measuring forest above-ground biomass from space. *Remote Sens. Environ.* **2019**, *227*, 44–60. [CrossRef]

13. Smith, L. Satellite remote sensing of river inundation area, stage, and discharge: A review. *Hydrol. Process.* **1997**, *11*, 1427–1439. [CrossRef]

14. Tsyganskaya, V.; Martinis, S.; Marzahn, P.; Ludwig, R. SAR-based detection of flooded vegetation–review of characteristics and approaches. *Int. J. Remote Sens.* **2018**, *39*, 2255–2293. [CrossRef]

15. Bouvet, A.; Mermoz, S.; Ballère, M.; Koleck, T.; Le Toan, T. Use of the SAR shadowing effect for deforestation detection with Sentinel-1 time series. *Remote Sens.* **2018**, *10*, 1250–1268. [CrossRef]

16. Lohberger, S.; Stängel, M.; Atwood, E.C.; Siegert, F. Spatial evaluation of Indonesia's 2015 fire-affected area and estimated carbon emissions using Sentinel-1. *Glob. Chang. Biol.* **2017**, *24*, 644–654. [CrossRef] [PubMed]

17. Gao, Q.; Zribi, M.; Escorihuela, M.J.; Baghdadi, N.; Segui, P.Q. Irrigation Mapping Using Sentinel-1 Time Series at Field Scale. *Remote Sens.* **2018**, *10*, 1495–1512. [CrossRef]

18. Lee, J.S.; Pottier, E. *Polarimetric Radar Imaging: From Basics To Applications*; CRC Press: Boca Raton, FL, USA, 2009.

19. Cloude, S.R. The dual polarisation Entropy/alpha decomposition: A PALSAR case study. In Proceedings of the dual polarisation Entropy/alpha decomposition: A PALSAR case study, Frascati, Italy, 22–26 January 2007.

20. Plank, S. Rapid damage assessment by means of multi-temporal SAR—A comprehensive review and outlook to Sentinel-1. *Remote Sens.* **2014**, *6*, 4870–4906. [CrossRef]

21. SinCohMap–Sentinel-1 Interferometric Coherence for Vegetation and Mapping. Available online: http://www.sincohmap.org/ (accessed on 28 April 2019).

22. Tamm, T.; Zalite, K.; Voomansik, K.; Talgre, L. Relating Sentinel-1 Interferometric Coherence to mowing events on grasslands. *Remote Sens.* **2016**, *8*, 802–820. [CrossRef]

23. GitHub. Opendatacube/radar. Available online: https://github.com/opendatacube/radar (accessed on 28 April 2019).

24. Commonwealth of Australia SARA Sentinel Australasia Regional Access. Available online: https://copernicus.nci.org.au/sara.client/#/home (accessed on 28 April 2019).

25. NCI. National Computational Infrastructure. 2019. Available online: http://nci.org.au/ (accessed on 28 April 2019).

26. De Zan, F.; Guarnieri, A.M. TOPSAR: Terrain Observation by Progressive Scans. *IEEE Trans. Geosci. Remote Sens.* **2006**, *44*, 2352–2360. [CrossRef]

27. Analysis Ready Data for Land–Normalised Radar Backscatter. Committee on Earth Observation Satellites. Available online: http://ceos.org/ard/files/CARD4L_Product_Specification-Backscatter-v3.2.pdf (accessed on 4 April 2019).

28. Small, D. Flattening Gamma: Radiometric Terrain Correction for SAR Imagery. *IEEE Trans. Geosci. Remote Sens.* **2011**, *49*, 3081–3093. [CrossRef]

29. European Space Agency (ESA) Sentinel Online–Data Products. Available online: https://sentinel.esa.int/web/sentinel/missions/sentinel-1/data-products (accessed on 28 April 2019).

30. Farr, T.G.; Rosen, P.A.; Caro, E.; Crippen, R.; Duren, R.; Hensley, S.; Kobrick, M.; Paller, M.; Rodriguez, M.; Roth, L.; et al. The Shuttle Radar Topography Mission. *Rev. Geophy.* **2007**, *45*. [CrossRef]

31. Gallant, J.C.; Dowling, T.I.; Read, A.M.; Wilson, N.; Tickel, P.; Inskeep, C. 1 second SRTM Derived Digital Elevation Models User Guide. 2011; Geoscience Australia. Available online: www.ga.gov.au/topographic-mapping/digital-elevation-data.html (accessed on 28 April 2019).

32. Garthwaite, M.C.; Nancarrow, S.; Hislop, A.; Thankappan, M.; Dawson, J.H.; Lawrie, S. Design of Radar Corner Reflectors for the Australian Geophysical Observing System. *Geosci. Aust.* **2015**, *3*. [CrossRef]

33. Mueller, N.; Lewis, A.; Roberts, D.; Ring, S.; Melrose, R.; Sixsmith, J.; Lymburner, L.; McIntyre, A.; Tan, P.; Curnow, S.; et al. Water observations from space: Mapping surface water from 25 years of Landsat imagery across Australia. *Remote Sens. Environ.* **2016**, *174*, 341–352. [CrossRef]

 *data*

Article

Dynamic Data Citation Service—Subset Tool for Operational Data Management

Chris Schubert [1],* , **Georg Seyerl [1] and Katharina Sack [2]**

1   Data Centre—Climate Change Centre Austria, 1190 Vienna, Austria
2   Institute for Economic Policy and Industrial Economics, WU—Vienna University of Economics and Business, 1020 Vienna, Austria
*   Correspondence: chris.schubert@ccca.ac.at

Received: 31 May 2019; Accepted: 30 July 2019; Published: 1 August 2019

Abstract: In earth observation and climatological sciences, data and their data services grow on a daily basis in a large spatial extent due to the high coverage rate of satellite sensors, model calculations, but also by continuous meteorological in situ observations. In order to reuse such data, especially data fragments as well as their data services in a collaborative and reproducible manner by citing the origin source, data analysts, e.g., researchers or impact modelers, need a possibility to identify the exact version, precise time information, parameter, and names of the dataset used. A manual process would make the citation of data fragments as a subset of an entire dataset rather complex and imprecise to obtain. Data in climate research are in most cases multidimensional, structured grid data that can change partially over time. The citation of such evolving content requires the approach of "dynamic data citation". The applied approach is based on associating queries with persistent identifiers. These queries contain the subsetting parameters, e.g., the spatial coordinates of the desired study area or the time frame with a start and end date, which are automatically included in the metadata of the newly generated subset and thus represent the information about the data history, the data provenance, which has to be established in data repository ecosystems. The Research Data Alliance Data Citation Working Group (RDA Data Citation WG) summarized the scientific status quo as well as the state of the art from existing citation and data management concepts and developed the scalable dynamic data citation methodology of evolving data. The Data Centre at the Climate Change Centre Austria (CCCA) has implemented the given recommendations and offers since 2017 an operational service on dynamic data citation on climate scenario data. With the consciousness that the objective of this topic brings a lot of dependencies on bibliographic citation research which is still under discussion, the CCCA service on Dynamic Data Citation focused on the climate domain specific issues, like characteristics of data, formats, software environment, and usage behavior. The current effort beyond spreading made experiences will be the scalability of the implementation, e.g., towards the potential of an Open Data Cube solution.

Keywords: dynamic data citation; subset; data curation; persistent identifier; data provenance; metadata; versioning; query store; data sharing; FAIR principles

---

1. Summary

Data with a spatial reference, so-called geospatial data, e.g., on land use, demographic statistics, geology, or air quality, are made accessible by interoperable and standardized web services. This means that data, whether stored in a database or file-based systems, are transformed into Open Geospatial Consortium (OGC) [1] conformal data services. These include catalog services for searching and identifying data via their metadata, view services for visualizing information in the internet browser, and more comprehensive services such as the Web Feature and Web Coverage Service (WCS) [2], which allow access to more complex data structures, such as multidimensional parameters. Data and data services are becoming more sophisticated, more dynamic, and more complex due to their fine-grained information and consume more and more storage space. The high dynamics of the content offered can be explained by data updates, which take place at less and less frequent intervals, and the increasing number of new available sensors.

According to the objective and strategy of GEO—the Group on Earth Observation [3]—even more people, not only scientific domain experts, get access to climate, earth observation and in situ measures to extract information on their own.

Due to increasing interoperable and technologically "simplified" data access, the citation of newly created data derivatives and their data sources becomes essential for data analyses, such as the intersection of different data sources. The description of entire process chains with regard to information extraction, including the methods and algorithms applied, will become essential in the practice of data reproducibility [4]. In order to obtain this information in a structured system, the concept of data provenance [5,6] was defined, which describes the sequence of how data were generated.

It is common practice that behavior related to data usage goes away from downloading and using desktop tools to web-based analysis. The Open Data Cube (ODC) [7,8] as an open source framework for geospatial data management and effective web based data analysis for earth observation data. There is a growing number of implementations of ODC on national and regional level. Therefore, precise citation processes [9] should be considered in available data infrastructures.

For proper data management, data citation and evidence as robust information of data provenance in relation to the core principles on data curation [10–12] will be relevant. Each data object should be citable, referenceable, and verifiable regarding its creators, the exact file name, from which repositories it originates from, as well as the last access time.

The requirements [13] for citation of data should take into account: (i) the precise identification and time stamp of access to data, (ii) the persistence of data, and (iii) the provision of persistent identifiers and interoperable metadata schemes that reflect the completeness of the source information. These are the basic pillars of data citation, reflected in the Joint Declaration of Data Citation Principles [10] and the FAIR (Findable, Accessible, Interoperable, Reusable on Data Sharing) Principles [13].

These were considered in the Research Data Alliance Data Citation Working Group (RDA Data Citation WG) [14,15] and summarized as 14 recommendations of the document "Data Citation of Evolving Data: Recommendations of the Working Group on Data Citation" (WGDC) [9,10]. This outcome forms the basis for the concept of dynamic data citation. Nevertheless, there are still barriers within the sophisticated offer on huge widespread characteristics on syntactical data formats and scientific domain issues. The Earth Observation domain is handling data curation in different principles than the climate model domain. Stockhause et al. [16] give a detailed overview of the evolving data characteristics and compare the different approaches.

As a recently established research data infrastructure, the Data Centre at the Climate Change Centre Austria (CCCA) started with a dynamic data citation pilot concept focused on NetCDF Data for the RDA working group in 2016 and implemented completely the recommendation so that since 2017, operational service can be provided for regional and global atmospheric datasets.

The current development efforts are scaling up techniques with the aim to extend our coverage on existing services especially towards the objective to cover the requirements of scientific domain on Open Earth Observation and the Open Data Cube environment and to offer the technical approach as an extension for the domain.

The overall objective of this article was to demonstrate the technical implementation and to provide the future potential of benefits regarding the RDA recommendations, with operational service offered as evidence, such as sustainable storage consumption using the Query Store for the data subset, and automatic adaptation into interoperable metadata description to keep the data provenance information.

2. Introduction on Dynamic Data Citation

Citing datasets in an appropriate manner is agreed upon as good scientific praxis and well established. Data citation as a collection of text snippets provides information about the creator of the data, the title, the version, the repository, a time stamp, and a persistent identifier (PID) for persistent data access. These citation principles can easily be applied in a data repository to static data. If only a fragment of a dataset is requested, which is served by subset functionalities, a more or less dynamic citation is required [9]. The consideration is to identify exactly those parts, subsets, of the data that are actually needed for research studies or reports, even if the original data source evolves with new versions, e.g., by corrections or revisions.

With data-driven web services, the data used are not always static, especially in collaborative iteration and creation cycles [14]. This is particularly valid for climatological research, where different data sources and models serve as input for new data as derivatives, e.g., climate indices like calculation of the number of tropical nights, which is based on different climate model ensembles. From a data quality point of view, it is preferable that such derivatives also be affected and updated automatically by the performed correction chain. Such changes in consideration on dependencies in data creation should be communicated as automatically as possible. A research data infrastructure should be able to provide an environment for dynamic data. With the reproducibility of results in mind, it is essential to be able to accurately verify a particular dataset, its exact version, or the creation of data fragments. The reproducibility of the data fragments and their relationship to their originals is essential if data processing has to be repeated.

Creating subsets is a common procedure for setting up needed customized data extraction for experiments or studies. Either only specific areas of interest or only a certain time interval are needed, but also particular information layers, such as the distribution of the mean surface temperature, can be of interest for a further effective processing. However, it is also a known fact that the storage of subsets created cannot scale with increasing amounts of data [8]. This implies that subsets are always copies of the original, and redundant storage consumption is not an economical option for capacity reasons (storage costs). The objective on considerations of the RDA—WGDC is to store only the criteria that create the subsets as arguments in a query store. In general, these are only few kilobytes compared to mega- to gigabytes with a subset of, e.g., Austrian climate scenarios. Such a query can be executed again, and the subset will be created on demand for a needed use.

To ensure that the stored queries are available for long-term use, to be executed again, and created subsets are available to other users, they are assigned to unique persistence identifiers and verification techniques. These are the core concepts of the RDA recommendations on dynamic data citation.

With such implementation, an operator of data infrastructures or service provider has to allocate only temporary storage for access to a subset. For the aforementioned OGC-compliant web services, the storage plays a minor role too, as the mechanisms for the provision of data fragments are very similar to subset services, such as browser-controlled zooming in by controlling the bounding box parameters. However, for such web applications, the RDA recommendations provide the targeted added value that queries are provided by a persistent identifier and thus enable delivering information about the data origin, which is reflected in the inheritance and adaptation of metadata to newly generated data fragments.

The 14 RDA recommendations for the creation of reproducible subsets in a context of easy and precise identification for dynamic data is a very demanding but pragmatic guidance. The RDA—Recommendations for the Scalable Dynamic Data Citation Methodology serves as a guideline with technical requirements for implementation, which are underpinned with practical examples in an understandable manner.

The 14 RDA Recommendation on Dynamic Data Citation

The recommendations for creating reproducible subsets reflecting the results of expert discussion, which served as a guideline on how to identify dynamic subsets from existing data sources. Short core messages on each recommendation are given, based on Rauber et al. [14,15].

Four pillars for structuring the recommendations were identified, see Figure 1:

- Framework on preparing the data and a query store;
- Identifying specific data in a persistent manner;
- Resolving PID and retrieving data; and
- Guaranteeing modifications and adaptability for data infrastructures as well as changes in software environments.

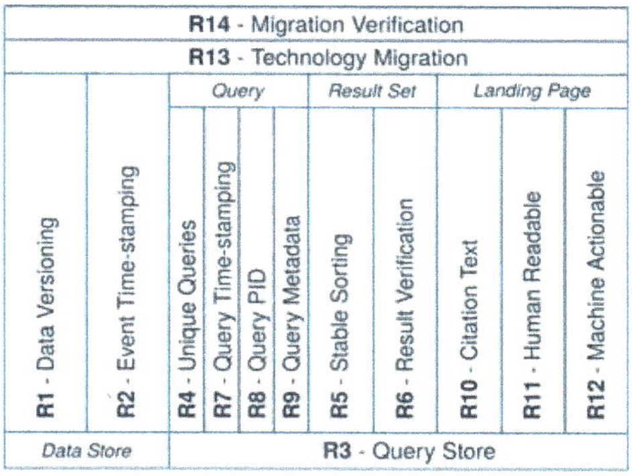

R14 - Migration Verification			
R13 - Technology Migration			
	Query	*Result Set*	*Landing Page*
R1 - Data Versioning / R2 - Event Time-stamping	R4 - Unique Queries / R7 - Query Time-stamping / R8 - Query PID / R9 - Query Metadata	R5 - Stable Sorting / R6 - Result Verification	R10 - Citation Text / R11 - Human Readable / R12 - Machine Actionable
Data Store	R3 - Query Store		

Figure 1. A structured order for the Research Data Alliance (RDA) recommendation on dynamic data citation.

The recommendations in detail are summarized and adapted according to the implementation at the CCCA Data Centre as listed below. More information equipped with practical examples can be found in Rauber et al. [10].

R1—Data Versioning: Versioning ensures the former states of available datasets which can be retrieved. This information about this version is described within the metadata and the URI, which directs to the query store.

R2—Timestamping: Ensuring that all operations on data get timestamps is part of each data repository or database. The timestamp is provided in metadata.

R3—Query Store Facilities: Enabling a query store is an essential building block, with queries and associated metadata in order to enable re-execution in the future. The [UNI DATA] subset service (NCSS) provides a catalogue of subset arguments which are prepared in URIs.

R4—Query Uniqueness: Detecting identical queries and its arguments, e.g., by a normalized form and its comparison.

R5—Stable Sorting: Ensuring a stable sorting of the records in the dataset is unambiguous and reproducible. Executed queries are available in a query library, and if R4—Query Uniqueness response is a positive true result, the user has to apply still existing ones.

R6—Result Set Verification: Computing a kind of checksum generates a hash key as fixity information on the query result to ensure the verification of the correctness of re-execution. The check sum algorithm runs on each created subset and its execution.

R7—Query Time-stamping: A timestamp is assigned to the query, based on the last update to the entire database.

R8—Query PID: Each new query with a purpose of republishing is assigned a new handle identifier as a PID.

R9—Store the Query: Storing query and all related arguments, e.g., check-sum, timestamp, superset PID, and relation, based on R3—Query Store Facilities.

R10-Automated Citation Texts: Generating citation texts based on snippets of authors, title, date, version and repository information. It lowers the barrier for citing and sharing the data.

R11-Landing Page: PIDs resolve to a human readable landing page that provides the data and metadata, including the relation to the superset (PID of the data source) and citation text snippet. The metadata are held in DCAT-AP Schema, adapted by the European Commission [17]

R12-Machine Actionability: Providing an API/machine actionable interface to access metadata and data via the provided ckan API. The query re-execution creates a new download link which is available for 72 h.

R13-Technology Migration: When data are migrated to a new infrastructure environment (e.g., new database system), ensuring the migration of queries and associated fixity information.

R14-Migration Verification: Verify successful data and query migration, ensuring that queries can be re-executed correctly.

3. Purpose of Implementation and Development Tasks

The CCCA—Data Centre operates a research data infrastructure for Austria with highly available server cluster, storage capacity, and linked to high-performance computing facilities of the Vienna Scientific Cluster and the Central Institute for Meteorology and Geodynamics (ZAMG), the national weather service. The main portfolio of CCCA Services is to enable a central access point of Austrian research institutions and the Greater Alpine Region for storing and distributing scientific data and information in an open and interoperable manner regarding FAIR principles.

The CCCA—Data Centre developed a web-based tool for dynamic data citation. The main motivation in 2015 was simply to have a technical solution to providing a persistent identifier and an automatically generated citation text. At this point, the issue of what happens with evolving data and its version concept arises. Consequently, this led to the incentive to provide proper components for an appropriate data lifecycle and assign a dynamically persistent identifier (PID) for all associated data derivatives. With the RDA recommendations, the approach of a query store was convincing, and an appropriate decision base to follow this concept on identifying uniquely queries which can be executed again when needed was created. With the CCCA—Data Centre's task to provide large file sizes like climate scenarios, the argumentation to reduce redundancies for the storage consumption was the most convincing argument for the planned implementation at this time.

In cooperation with the Data Citation Working Group, a concept for a technical pilot implementation was accompanied.

This pilot implementation on Dynamic Data Citation at the CCCA Data Centre focused on CF standard [18] compliant NetCDF data to manage high-resolution climate scenarios for Austria in a time range from 1965 until 2100. NetCDF is an open standard and machine-independent data format for structured and multidimensional data. It includes attributes, dimensions, and variables. For example, for the Austrian Climate Scenarios, calculated temperature records on daily basis are available in 1×1 km gridded, geo-referenced data in multiple single files. The scenarios include different "representative concentration pathways" (RCPs) [19], ensembles of different GCM (general circulation models) and RCM (regional climate model) runs, for high-resolution conclusions, which are combined with statistical methods for the integration of in situ observations. The open accessible entire data package includes, for Austria, over 1200 files with a size up to 16 GB per file. Due to user requirements, in particular for the development of data-driven climate services and the characteristics of the climate scenarios provided, a subset service, Figure 2, was required.

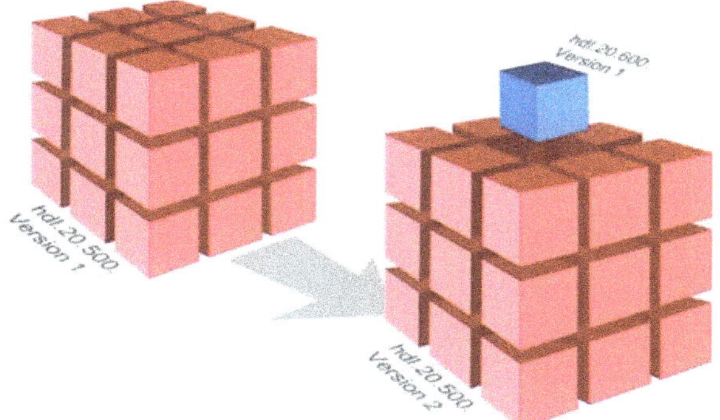

Figure 2. Schematic draft of subset needs, which includes the control on versioning and the alignment with the persistent identifier (PID), here handle.NET identifier—hdl. For the fragmented subset (blue cube), a new identifier is aligned, coupled with its own version number.

Especially for such large files, the first argument is decreasing the download rate, and the second is again storing the subsets on a desktop workstation. The continuous process chain on data fragments will be broken. Normally, GIS or data analytic tools are used to intersect the individual 'area of interest' or choosing a separate, distinguished layer or simply selecting a given time frame is still a common behavior. In a case of republishing, to offer a reuse or reproducibility study, all metadata and siblings' relation to the origin and different version would be lost and have to be described again. To do this manually is time-consuming, while describing the processes with all arguments for the intersection procedure will be imprecise. The CCCA Data Centre wants to overcome these troublesome processes mostly related with complex data structures especially for the climate services.

The overall approach on the CCCA-DC software environment was to set up a system which follows open source licenses. All developments and modules are available on the CCCA GitHub [20]. The data in the storage system, which are embedded in a highly available Linux Server Cluster, are managed by the ckan [21] software packages as a Python application server. This collaborative development framework is specialized in data management and catalogue systems, which is used as a central system component. For ckan, many extensions especially for the geospatial scientific domain are available, which brings a lot of synergies and benefits in its own modular software developments. One essential component for a provided catalogue of services is the flexible metadata scheme functionality. The Data Catalog Vocabulary DCAT [22] as a ckan default metadata profile was extended by the DCAT-AP and GeoDCAT-AP [23], a development by the Joint Research Centre of European Commission, which meets interoperable requirements for data exchange between distributed data servers. With this solution, heterogeneous data formats can be described with a common core schema for metadata and enable a uniform transformation into other profiles, such as Dublin Core, INSPIRE, and ISO 19115 metadata for geographical information.

The graphical user interface of the CCCA data server is based on the ckan web server and includes all functionalities, such as catalog and search functions, view services for web-based visualization of data content, as well as the implemented subset service. A Python API interface is also provided via ckan, which enables machine-to-machine communication for automatically steered processes.

For the unique identification of a data object, persistent identifiers (PIDs) are used, see Figure 2, and its registry guarantees uniqueness according to the specifications of internet identifiers to other data objects. For the CCCA, the Handle.NET® Registry Server was used for PID assignment. The advantage of Handle is the unlimited and instant assignment of identifiers, the technical coherence on standards, and encoding, which is essential for each newly created query.

The primary component for processing and creating data fragments is the Unidata Thredds Data Server (TDS) [24]. This server is responsible for processing NetCDF data, such as visualizing the data. In addition to TDS, the NetCDF Subset Services (NCSS) was embedded. NCSS provides a catalog of subsetting parameters that allows creating data fragments while retaining the original resolution and characteristics of the original data. These parameters include geographic coordinates, date ranges, and multidimensional variables. NCSS uses "HTTP GET" [25] in the following structure:

http://{host}/{context}/{service}/{dataset}[/dataset.xml | /dataset.html | {?query}]

where elements proposed as:

{host}—server name
{context}—"thredds" (usually)
{service}—"ncss" (always)
{dataset}—logical path for the dataset, obtained from the catalog
dataset.xml—to get the dataset description in xml
dataset.html—to get the human-readable web form
datasetBoundaries.xml—to get a human-readable description of the bounding boxes
{?query}—to describe the subset that you want.

The subsetting parameters for the element {?query} allow a combination of different parameters, like the name of variables, the location points or bounding box, arguments which specify a time range, the vertical levels, and the returned format.

Figure 3 illustrates the implemented components and gives an overview about the relationships between requests (blue arrows) and responses (orange arrows) between the server. The application server takes the requests via the Web server and generates URL-based (HTTP GET) requests with the subsetting parameters (subset requests). These requests are stored in the query store and are assigned with the Handle identifier.

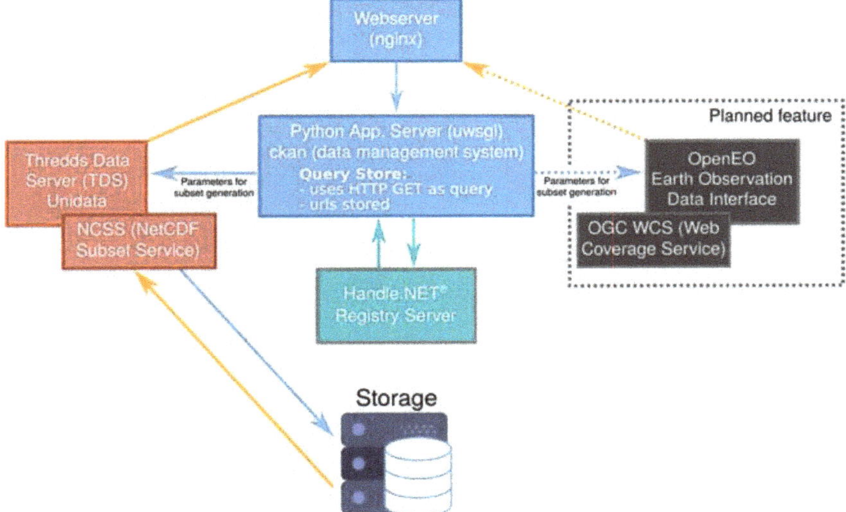

Figure 3. Simplified structure of server and hardware components for dynamic data citation within the CCCA Data Centre environment: (i) ckan web server, (ii) the application server for access, data management used as query store, (iii) Handle.NET® Registry Server for PID allocation, and (iv) the Unidata Thredds Data Server (TDS), NCSS Subset Service and planned features on Open EO support.

Within the ckan data management system, the required meta information for the subset dataset is compiled from the original meta data via adaptation and inheritance and tagged with the necessary description of the relationship as well as versions as supplementary meta data elements. The metadata of the newly created data subset also contain the original metadata elements, such as a short description, the data creator, licenses, etc. The supplementary elements are based on the query arguments and the meta information from the application server, which are automatically adapted. These are the title of the subsets, the selected parameters, the new spatial extent, and the changed time interval. In addition, there is the contact of the subset creator, the time of creation, the check-sum to verify if it is the same result if the request is repeated, the file size, and then the relationship to other records and their version.

The Thredds server retrieves the defined arguments from the query store via NCCS and thus creates the subset directly from the data store in which the original NetCDF data are contained. The data format is again NetCDF; other formats like comma-separated values (CSV) are also supported and return them to the web server. There, the subset is available as a resource for download, but also as a view service (OGC-WMS) for web-based visualization.

4. User Interface of the Application on Dynamic Citation Service

The Subset and Dynamic Data Citation Service at the CCCA Data Server is accessible for everyone. Due to performance reasons via Thredds, only registered users get access, Figure 4, for the comprehensible functionality on defining and republishing the subset at the data server.

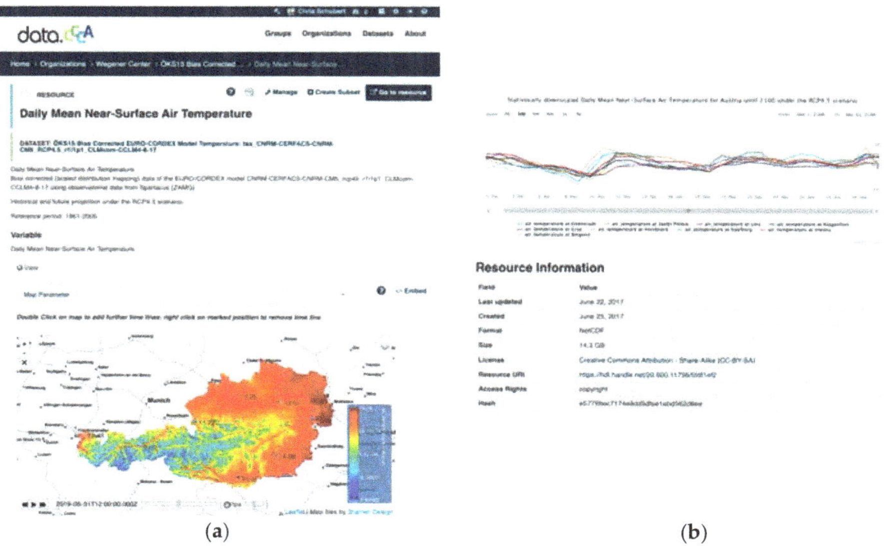

(a)   (b)

Figure 4. The general landing page of a data resource, after the personalized login: the general landing page of a dataset resource after login, where the subset can be created (on top): (**a**) The visualization is a view service (WMS), created by Thredds, and it allows the user by activating the time control to visualize each time step up to 2100; (**b**) additionally, it shows a timeline diagram after a point of interest on the map window is created.

After creating the subset, Figure 5, the user immediately receives a dynamically generated citation text containing the original author, the name of the subset, version, selected parameters, and the persistent identifier. This citation proposal can be used for correct reference in studies, publications, etc. and is clearly assignable to the entire research community. For a newly created and published subset, all metadata are inherited from the original data and supplemented by the defined arguments, such as the customized bounding box and the name of the creator, as well as the relation as a first step for data provenance information.

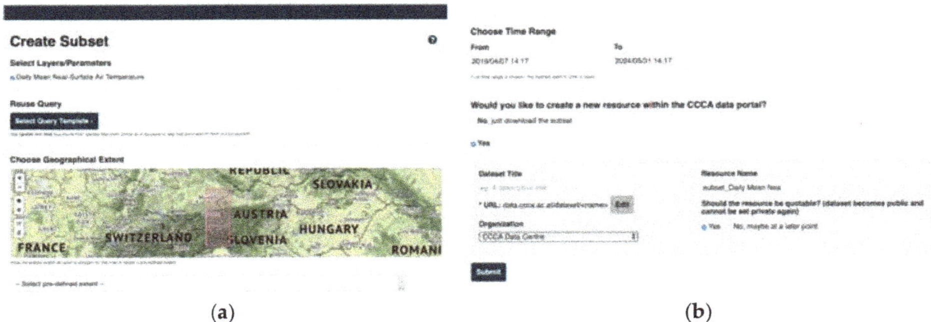

(**a**)   (**b**)

Figure 5. GUI of the subset creation function: (**a**) The upper part of web page for defining the parameter, or reuse of a still existing query, defining a bounding box either by polygon or predefined administrative units, (**b**) allows choosing a time range for other datasets like the globally available radio occultation data packages, a fourth dimension—e.g., the Potential High was introduced and can choose.

Versioning is used to ensure that previous states of records are maintained and made retrievable. Being able to refer to previous versions of datasets is important for reproducibility of simulation, calculations, and methods in general. The given Handle PID resolves into the landing page of the subset resource, where detailed metadata are provided. The web application generates automated citation texts. It includes predefined text snippets like the title, author, publishing date, version, and the data repository. For subsets, the aforementioned filter arguments based on queries were used and provided as text information, see Figure 6. The generated citation texts are in a form that lowers barriers for data sharing and reusability with proper credits.

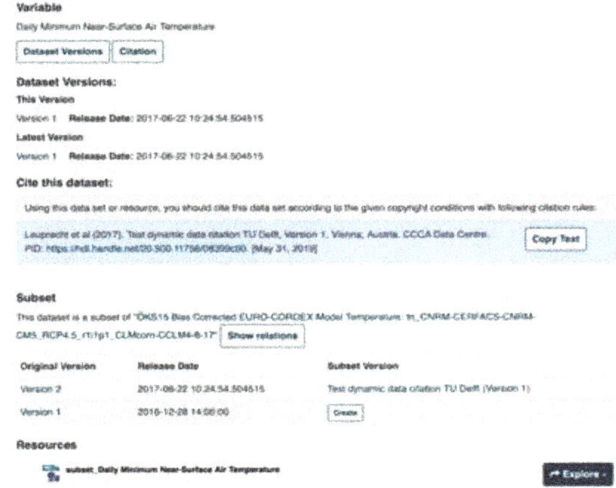

Figure 6. The screenshot gives an impression of what versions, relations, and the suggested text for citation looks like. In addition, the user could create, with the same arguments, a subset based on oldest versions but normally on a new version published. If new versions are available, a notification will be sent to the subset creator, which is part of the metadata profile.

5. Discussion and Next Steps

The implementation of the CCCA Data Centre's Dynamic Subsetting of evolving data shows its feasibility on a Pilot for NetCDF software and data processing environment. Nevertheless, limitations exist and can be seen both in the particular scope of the data format and in the lack of hardware configurations that enable interfaces and connectivity to other data infrastructures. The given requirements for CCCA data only lie in the CF conformity. Thus, all described functionalities are automatically available to the data providers. Due to the performance of the NetCDF format, the system independence and the multidimensional structured description of geospatial content, this format is used as an ingest and transfer format for the Open Data Cube. Integrated Python libraries allow a seamless transformation of data formats that are commonly used in the Earth Observation sector, such as GeoTIFF. Open Data Cube is a Python-based software framework that allows analyzing and processing the entire data package as a Data Cube to generate new earth observation products and services. Further considerations for the described dynamic citation implementation consist oφ setting up the data management software components with regard to the linkage with PID and the automated extraction of metadata on local Open Data Cube implementation in order to apply exactly this gap of the dynamic data citation within the Data Cubes. A first showcase within the framework of the Austrian Data Cube in cooperation with the Vienna University of Technology and the EODC—Earth Observation Data Center in Austria is currently in the conception phase, see Figure 3.

Another potential field of application is seen in the direction of OGC-compliant Web Services. The focus of these techniques is more on the interoperable web-based provision of data. The Web Coverage Service (WCS) describes the effective handling of subset generation and data fragments for effective further processing. The aspect to the requirements in the direction of dynamic data citation is taken into account but is not implemented so consistently in data infrastructures. This gap is not the aim of OGC standards themselves, but data infrastructure operators as well as their users should be guided towards these needs.

With this demonstrated implementation, an effort is undoubtedly made from a technical as well as a development as well as maintenance cost perspective. The big advantage, which can be shown here, is the avoidance of redundancies for storage consumption of generated subsets, whether locally or via cloud storage systems, and the exact citation to such individually created subsets so that they can be made accessible for other users.

The considered reflection and implementation regrettably go only in one direction, that is, from a dataset to its own data fragment. Inheriting the meta-information from an original to its subsets is not a dialectical challenge. What needs to come next in data curation and data management science is a method for how to deal with grouping of data ensembles and the merging of metainformation and contrary metadata elements.

6. Conclusions

The citation of data, which are mostly static, serves the description of the origin, the credits on authorship, and a link for accessing and downloading an entire dataset. In many research environments, data grow dynamically and through updating, which is a challenge for research data repositories. New versions can be created continuously through corrections; this can be done regularly, for example, on a monthly basis, but also quite agilely at irregular intervals and helps to improve data quality.

When data are used as the basis for a study or calculation, it can be ensured that the exact data version is available for verification in a study. This is especially the case for data derivatives where new algorithms are applied to the original data at a given point in time, e.g., the calculation of climate indices based on different climate models. The citation of the data should make it possible to identify the data fragment in a reliable and efficient process for all aspects of reproducibility of research and published studies.

The RDA recommendations of the Working Group on Data Citation (WGDC) enable researchers and data infrastructures to identify and cite data they are using. The recommendations support a dynamic, query-centric view of the data and enable precise identification by associating the queries to the subsets that are generated.

The Subset and Dynamic Data Citation Service of the CCCA was one of the first operational adaptations of the RDA Citation Working Group recommendations. This implementation is also listed as an RDA Adoption Story [26] as a factsheet, which also contains some useful information about the development effort required for implementation and acceptance.

This ongoing operational service for subset creation and dynamic data citation is evidence of the applicable approach of the RDA Recommendation.

Nevertheless, the observation of user behavior shows that there are still obstacles to republishing the created subsets on the CCCA server. Reasons for this behavior could be the minor number of users in Austria, especially for the climate scenario scope. In order to expand the user community, the implemented subset service was applied to datasets with a global 5-dimensional atmospheric dataset. An extension was also made by providing climate scenarios for the Western Balkan region in Europe, where institutions, such as their national weather services, can create their scenarios covering the national territories as subsets.

The additional strategy for expanding the user community is to extend the service to the scientific field of satellite-based Earth observation, such as through the Open EO approach and the Open Data Cube environment. The RDA is supporting this planned activity at CCCA through the RDA Adoption Grant Program for the next 12 months.

With the present implementation of the dynamic data citation of evolving data, the feasibility is given on the one hand, while on the other hand, experiences as well as software developments can be passed on in order to obtain a more exact estimation of efforts for future implementation for other data infrastructures in order to realize mechanisms for proper data management.

Author Contributions: All authors were involved in the conceptualization of this paper. Software used for subsetting and dynamic data citation at the CCCA Data Centre was developed by K.S. and G.S., the Server Configuration was done by G.S.

Funding: This CCCA Data Centre Infrastructure has received funding from the Austrian Federal Ministry of Education, Science and Research (bmbwf) under the Research Investment Programme (HRSM) within the project CCCA and GEOCLIM as well as in-kind contribution by ZAMG. The planned activities for the Earth Observation Data extension as option of RDA outputs will be funded by the "RDA Europe 4.0" Project. Contracts and agreements are under preparation at the time of publishing this article.

Acknowledgments: In addition to the developer team at the CCCA Data Centre, I am very grateful to Andreas Rauber from the Technical University of Vienna and Chair the RDA Citation Working Group for his constructive discussion and his curiosity regarding the climate science as well as the geospatial domain, providing concrete consultancy. In addition, I would like to express my passionate thanks to the RDA, the fruitful discussion on plenaries and presentation, but in particular Marieke Willems for the support and marketing actions by RDA.

Conflicts of Interest: The authors declare no conflict of interest.

References

1. OGC. The Open Geospatial Consortium. Available online: https://www.opengeospatial.org (accessed on 29 July 2019).
2. Baumann, P. OGC Web Coverage Service (WCS) 2.1 Interface Standard—Core. 2018. Available online: http://docs.opengeospatial.org/is/17-089r1/17-089r1.html (accessed on 29 July 2019).
3. GEO Group on Earth Observation. GEO Strategic Plan 2016–2025: Implementing GEOSS. Available online: https://www.earthobservations.org/documents/GEO_Strategic_Plan_2016_2025_Implementing_GEOSS.pdf (accessed on 25 May 2019).
4. Craglia, M.; Nativi, S. Mind the Gap: Big Data vs. Interoperability and Reproducibility of Science. In *Earth Observation Open Science and Innovation*; Pierre-Philippe, M., Christoph, A., Eds.; Springer: Cham, Switzerland, 2018; pp. 121–141. [CrossRef]
5. Lawrence, B.; Jones, C.; Matthews, B.; Pepler, S.; Callaghan, S. Citation and peer review of data: Moving towards formal data publication. *Int. J. Digit. Curation* **2011**, *6*, 4–37. [CrossRef]

6. Pröll, S.; Rauber, A. Scalable data citation in dynamic, large databases: Model and reference implementation. In Proceedings of the 2013 IEEE International Conference on Big Data, Santa Clara, CA, USA, 6–9 October 2013; IEEE Press: Piscataway, NJ, USA, 2013; pp. 307–312.

7. Open Data Cube Initiative. Open Data Cube Whitepaper. Open Data Cube Partners. 2017. Available online: https://docs.wixstatic.com/ugd/f9d4ea_1aea90c5bb7149c8a730890c0f791496.pdf (accessed on 26 January 2018).

8. Sudmanns, M.; Tiede, D.; Lang, S.; Bergstedt, H.; Trost, G.; Augustin, H.; Baraldi, A.; Blaschke, T. Big Earth data: Disruptive changes in Earth observation data management and analysis. *Int. J. Digit. Earth* **2019**. [CrossRef]

9. Buneman, P.; Davidson, S.B.; Frew, J. Why data citation is a computational problem. *Commun. ACM* **2016**, *59*, 50–57. [CrossRef] [PubMed]

10. CODATA-ICSTI Task Group on Data Citation Standards and Practices. Out of cite, out of mind: The current state of practice, policy, and technology for the citation of data. *Data Sci. J.* **2013**, *12*, 1–67.

11. FORCE11 Data Citation Synthesis Group. *Joint Declaration of Data Citation Principles*; Martone, M., Ed.; FORCE11 Data Citation Synthesis Group: San Diego, CA, USA, 2014; Available online: http://www.force11.org/datacitation (accessed on 15 June 2019).

12. FORCE11 FAIR Data Publishing Group. FAIR Guiding Principles. 2017. Available online: https://www.force11.org/fairprinciples (accessed on 15 June 2019).

13. Silvello, G. Theory and Practice of Data Citation. *J. Assoc. Inf. Sci. Technol.* **2018**, *69*, 6–20. [CrossRef]

14. Rauber, A.; Asmi, A.; van Uytvanck, D.; Pröll, S. Data Citation of Evolving Data: Recommendations of the Working Group on Data Citation (WGDC). Result of the RDA Data Citation WG. 20 October 2015. Available online: http://rd-alliance.org/system/files/documents/RDA-DC-Recommendations_151020.pdf (accessed on 25 May 2019).

15. Rauber, A.; Asmi, A.; van Uytvanck, D.; Pröll, S. Identification of Reproducible Subsets for Data Citation, Sharing and Re-Use. *Bull. IEEE Tech. Committe Digit. Libr.* **2016**. Available online: http://www.ieee-tcdl.org/Bulletin/v12n1/papers/IEEE-TCDL-DC-2016_paper_1.pdf (accessed on 25 May 2019).

16. Stockhause, M.; Lautenschlager, M. CMIP6 Data Citation of Evolving Data. *Data Sci. J.* **2017**, *16*, 1–13. [CrossRef]

17. DCAT Application Profile for Data Portals in Europe (DCAT-AP). Available online: https://joinup.ec.europa.eu/solution/dcat-application-profile-data-portals-europe/releases;http://data.europa.eu/w21/c319d97f-19bf-4365-af0b-eadcd3256293 (accessed on 29 July 2019).

18. CF Conventions and Metadata. Available online: https://cfconventions.org/latest.html (accessed on 29 July 2019).

19. Representative Concentration Pathways (RCP's). Available online: https://www.aimes.ucar.edu/docs/IPCC.meetingreport.final.pdf (accessed on 25 May 2019).

20. CCCA Server Software. Available online: https://github.com/ccca-dc (accessed on 29 July 2019).

21. CKAN. Open Source Data Management System. Available online: https://ckan.org (accessed on 29 July 2019).

22. W3C. Data Catalog Vocabulary (DCAT). Available online: http://www.w3.org/TR/2014/REC-vocab-dcat-20140116/ (accessed on 29 July 2019).

23. GeoDCAT-AP Is an Extension to the "DCAT Application Profile for European Data Portals" (DCAT-AP) for the Representation of Geographic Metadata. Available online: https://inspire.ec.europa.eu/documents/geodcat-ap (accessed on 15 June 2019).

24. Unidata Thredds Data Server. Available online: https://www.unidata.ucar.edu/software/thredds/current/tds/ (accessed on 29 July 2019).

25. NetCDF Subset Service. Available online: https://www.unidata.ucar.edu/software/tds/current/reference/NetcdfSubsetServiceReference.html (accessed on 29 July 2019).

26. RDA Adoption Stories. Available online: https://www.rd-alliance.org/dynamic-data-citation-frequently-modifying-high-resolution-climate-data (accessed on 29 July 2019).

Article

A Topology Based Spatio-Temporal Map Algebra for Big Data Analysis

Sören Gebbert [1,*], Thomas Leppelt [2] and Edzer Pebesma [1]

[1] Institute for Geoinformatics, University of Münster, Heisenbergstraße 2, 48149 Münster, Germany; edzer.pebesma@uni-muenster.de
[2] Deutscher Wetterdienst, Frankfurter Straße 135, 63067 Offenbach am Main, Germany; thomas.leppelt@gmail.com
* Correspondence: soerengebbert@googlemail.com

Received: 14 April 2019; Accepted: 27 May 2019; Published: 18 June 2019

Abstract: Continental and global datasets based on earth observations or computational models challenge the existing map algebra approaches. The available datasets differ in their spatio-temporal extents and their spatio-temporal granularity, which makes it difficult to process them as time series data in map algebra expressions. To address this issue we introduce a new map algebra approach that is topology based. This topology based map algebra uses spatio-temporal topological operators (STTOP and STTCOP) to specify spatio-temporal operations between topological related map layers of different time-series data. We have implemented several topology based map algebra tools in the open source geoinformation system GRASS GIS and its open source cloud processing engine actinia. We demonstrate the application of our topology based map algebra by solving real world big data problems using a single algebraic expression. This included the massively parallel computation of the NDVI from a series of 100 Sentinel2A scenes organized as earth observation data cubes. The processing was performed and benchmarked on a many core computer setup and in a distributed container environment. The design of our topology based map algebra allows us to deploy it as a standardized service in the EU Horizon 2020 project openEO.

Keywords: topology based map algebra; data cubes; big data; map algebra; earth oberservation; GRASS GIS

1. Introduction

Continental and global time series data from earth observation satellites [1–3] or computational simulations with arbitrary spatio-temporal granularities require very sophisticated tools for efficient analysis and processing. The NASA Landsat mission produces a large time series of earth observation data using different spectral bands that differ in their geographical locations, spatial resolution, spatial extents and their sensing time. The same is true for the ESA Copernicus mission, that includes a wide range of earth observation satellites. The publicly available NASA and ESA earth observation archives contain multiple petabytes of data, growing by several petabytes each year. There are several public global and continental climate datasets [4] available with high spatial- and temporal resolution.

A rapidly changing global environment, the global climate change, its continental and global effects on agriculture or natural hazards raise the requirement to analyse these large time series data and their relations to each other. A major challenge from the perspective of data analysis is how to process this kind of data altogether, handling the different spatio-temporal extents and their different spatio-temporal granularities.

In this research we develop a topology based map algebra to process large scale time series datasets with different spatio-temporal granularities and extents using algebraic expressions. We show how to apply topological algebraic expressions to Landsat8, Sentinel2A and climate time series data to compute vegetation indices and hydro-thermal coefficients. We demonstrate the big data analysis capabilities of the topology based algebra by computing the NDVI from 100 Sentinel2A scenes using the tools that we developed on a many core computer system and a distributed docker container environment. Our research is based on the spatio-temporal capabilities of the temporal enabled GRASS GIS [5]. It makes use of spatio-temporal topological features of the GRASS GIS Temporal Framework [6] to formulate algebraic expressions with spatio-temporal topological operators.

2. Related Work

The book [7] from Dana Tomlin introduced the concept of map algebra as a general language in geographical information systems (GIS) for analysis and processing of raster based geographic data with two dimensions. This data is mainly referenced as raster layers in GIS. By integrating time as the third dimension by [8,9] a new class of algebra was introduced into the GIS world. The new map algebra approach works on space-cubes, time-cubes and hyper-cubes that have two to three spatial- and one temporal dimension. Space-, time- and hyper-cubes have by definition equidistant spatial dimensions and require periodic time intervals that are equidistant. The computational spatial region must be equal for all layers in these cubes.

The map algebra approach of Dana Tomlin are available in many GIS applications. The GIS software systems GRASS GIS, ArcGIS and ERDAS Imagine integrate map algebra concepts. The Google Earth Engine [10] framework supports two different approaches to apply mathematical operations on images[1]. It is possible to use algebraic expressions with spatial, comparison and logical operators or nested functions on image objects. However, mathematical operations can not be applied on times-series data[2] directly. The user must implement code to iterate over an image collection or map an algorithm to apply mathematical operations for each image in the collection.

2.1. GRASS GIS

We chose GRASS GIS to implement the topology based map algebra, because it is a full-featured, free and open source temporal geographical information system [5,6,11]. GRASS GIS has been used in numerous environmental scientific applications by [12–16]. A comprehensive overview of its application in environmental modelling is given in [17].

The temporal enabled GRASS GIS can efficiently manage, analyse and process continental- and global-scale time-series raster, 3D raster or vector data sets. Being free and open source allows users to modify it, which enabled us to integrate the topology based map algebra as a main feature.

Our topology based map algebra was implemented as three new modules in the temporal enabled GRASS GIS version 7 [5]. It utilises the GRASS GIS Temporal Framework [6], PyGRASS [18] and the GRASS GIS modules g.copy[3], r.mapcalc[4] and r3.mapcalc[5].

2.2. Actinia

Actinia [19] is an open source REST API for scalable, distributed, high performance processing of geographical data that uses GRASS GIS for computational tasks. It provides a REST API to process satellite images, time series of satellite images, arbitrary raster data with geographical relations and vector data. We improved Actinia in context of this work to support massive parallel processing of topology based map algebra expressions in a distributed cloud environment. Actinia and GRASS GIS are software components of the EU Horizon 2020[6] openEO project [20], that is an open source

---

1. https://developers.google.com/earth-engine/image_math, June 2018.
2. Time-series data is called image collection in Google Earth Engine.
3. https://grass.osgeo.org/grass76/manuals/g.copy.html.
4. https://grass.osgeo.org/grass76/manuals/r.mapcalc.html.
5. https://grass.osgeo.org/grass76/manuals/r3.mapcalc.html.
6. H2020 grant 776242.

interface between earth observation data infrastructures and front-end applications. Actinia is actively developed and deployed on different cloud platforms from cloud provides like Amazon, Google, Deutsche Telekom and others.

2.2.1. Time in GRASS GIS

The temporal enabled GRASS GIS uses the concept of linear, discrete time represented by time instances and time intervals. Time intervals and time instances represent the time stamps of map layers. Time intervals are left closed and right open. The end time is not part of the time interval but represents the start time of a potential successor. Time intervals can overlap or contain each other. The temporal model supports gaps and time instances. Time in GRASS GIS is described in detail in [5].

Time is measured using the Gregorian calender time, also called absolute time, conform to ISO 8601[7] and as relative time defined by an integer and a unit of type year, month, day, hour, minute or second. The smallest supported temporal granule is one second. The definition of absolute and relative time follows the temporal database concepts collected in [21].

2.2.2. Map Algebra in GRASS GIS

The GRASS GIS module r.mapcalc and r3.mapcalc implement a subset of the map algebra functionality described in [7–9], especially the local and focal algebraic methods. Zonal algebraic methods are performed by dedicated GRASS GIS modules like r.series, r.neighbors, r.univar, r3.univar, r.stats, r3.stats and several others.

With the integration of time in GRASS GIS by [5] the space-cube map algebra module r3.mapcalc was enabled to perform spatio-temporal operations on time-cubes using the existing map algebra syntax.

2.2.3. Data Cubes in GRASS GIS

A data cube is an aggregation concept from relational databases introduced by [22] that works on attribute data organized in a N-dimensional cube. We use the concept of earth observation data cubes, that organize earth observations like satellite images in a multi-dimensional cube. Dimensions are spatial and temporal coordinates, attributes are pixel values.

Space-time datasets (STDS) are the spatio-temporal data types in GRASS GIS to manage series of time-stamped map layers and were introduced in [5]. They can be interpreted as sparse data cubes with irregular time dimensions. There are three spatio-temporal GRASS GIS data types:

- Space-Time Raster Datasets (STRDS) that manage time series of raster map layers. These are sparse raster data cubes in which each time stamped slice stores only pixels from the area of interest in this slice, that can be different from any other slice and is a subset of the spatial extent of the whole data cube.
- Space-Time Raster 3D Datasets (STR3DS) that manage time series of 3D raster map layers. These are sparse voxel data cubes, using the same storage approach as the STRDS.
- Space-Time Vector Datasets (STVDS) that manage time series of vector map layers. STVDS can be interpreted as vector data cubes.

An arbitrary number of time-stamped map layers can be registered in a single space-time dataset. A single time-stamped map layer can be registered in several different space-time datasets at the same time. Space-time datasets have a dedicated temporal type. Therefore a STDS can have either absolute time or relative time. The same is true for time-stamped map layers.

STDS can contain map layers that have time intervals or time instances attached as time stamps. Intervals and instances of time can be mixed in a space-time dataset. The spatial extents and spatial

---

[7] http://en.wikipedia.org/wiki/ISO_8601.

resolutions of associated raster or 3D raster map layers in a STDS can be different. See [5,6] for more details.

2.3. Temporal Granularity and Topological Relations

A calendar has multiple granularities that can be described using a temporal hierarchy, see [23]. We use the temporal Gregorian calendar hierarchy of years, months, days, hours, minutes and seconds. A year is composed of 12 months, a month contains between 28 and 31 days, a day has 24 h, a hour has 60 min and one minute has 60 s. A glossary about temporal granularity is available in [24]. The temporal granularity is defined in the GRASS GIS Temporal Framework as the largest common divider of time intervals and gaps between intervals or instances from all time-stamped map layers that are collected in a space-time dataset (STDS). It supports space-time datasets that have complex temporal structures. See [6] for more details. Temporal relations between time-stamped map layers follow temporal interval logic defined by [25], see Figure 1.

The GRASS GIS Temporal Framework can compute topological relations between the spatial extents in two and three dimensions based on [26], see Figure 2. Spatial extents are represented as axis aligned bounding boxes.

	A in relation to B	B in relation to A
A — B —	equivalent	equivalent
A — B —	follows/adjacent	precedes/adjacent
A — B —	overlaps	overlapped
A — B —	after	before
A — B —	during	contains
A — B —	starts	started
A — B —	finishes	finished

Figure 1. Temporal relations between time intervals after [25].

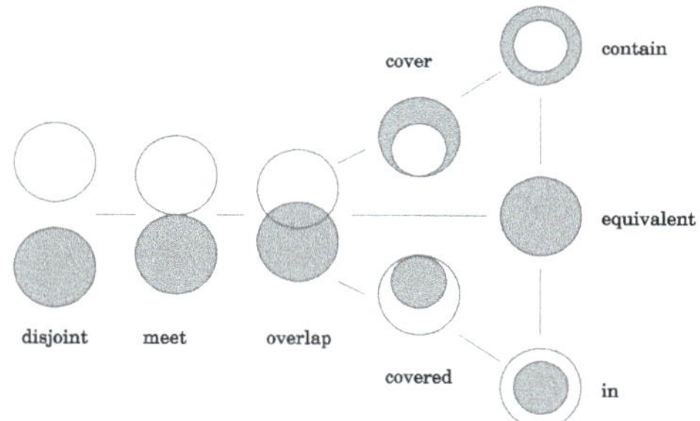

Figure 2. Spatial topological relationships visualised after [26].

2.4. Spatio-Temporal Operations

Spatio-temporal operations are based on the approaches implement in the GRASS GIS Temporal Framework by [6]. It provides boolean operations like intersection, union, disjoint union on time intervals and time instances. It support spatial boolean operations between spatial extents. As boolean operations, spatial intersection, spatial union and spatial disjoint union are available.

3. Topology Based Spatio-Temporal Map Algebra

We designed our topology based map algebra to perform two different tasks. The first task is the creation of subsets from space-time datasets based on algebraic expressions. That allows us to extract map layers from different space-time datasets based on their spatio-temporal topological relations. We implemented the dedicated GRASS GIS module t.select to perform such tasks. This module supports as input all STDS types that can also be mixed in a single expression. This algebraic approach is also part of the second task.

The second task is to perform spatio-temporal operations between STDS of the same type. We implemented this approach for STRDS and STR3DS as GRASS GIS modules t.rast.algebra and t.rast3d.algebra.

The syntax of the topology based map algebra is similar to the syntax of r.mapcalc [27]. The result of an algebraic operation is a new STDS. Two GRASS GIS modules r.mapcalc and r3.mapcalc are used to perform the spatial operations on raster layers, space- and time-cubes in our topology based map algebra.

A spatio-temporal expression of our topology based map algebra has the following form:

$$STDS \; = \; expression$$

A new space-time dataset is created by an expression that contains other space-time datasets, raster or 3D raster map layers, scalars, spatio-temporal operators as well as spatial and temporal functions.

An important feature of our topology based map algebra is the application of computational regions based on the spatial-extents of raster map layers that are involved in a single spatial operation. Hence, time-stamped raster map layers in a STRDS can have different spatial extents. This mode is activated by using the -s flag in the GRASS GIS algebra module t.rast.algebra. It assures that spatial operations that are applied to spatio-temporal related raster map layers are using the disjoint union of all spatial extents and the smallest raster cell size of all involved raster map layers for this operation.

3.1. The Spatio-Temporal Topological Operator STTOP

A core feature of our topology based map algebra is the spatio-temporal topological operator (STTOP). This operator defines what spatial and temporal operations should be performed between two entities. An expression involving the STTOP has always the following form:

$$left \; entity \; \{STTOP\} \; right \; entity$$

The spatio-temporal topology of an entity is arbitrary, it can contain components with equal, overlapping or including time stamps and spatial extents. The following entities are supported in an expression:

$$
\begin{aligned}
expression := \; & STDS \; \{STTOP\} \; STDS \\
:= \; & STDS \; \{STTOP\} \; map \; layer \\
:= \; & STDS \; \{STTOP\} \; scalar \\
:= \; & STDS \; \{STTOP\} \; expression \\
:= \; & expression \; \{STTOP\} \; STDS \\
:= \; & expression \; \{STTOP\} \; map \; layer \\
:= \; & expression \; \{STTOP\} \; scalar \\
:= \; & expression \; \{STTOP\} \; expression
\end{aligned}
$$

Expressions can be nested. The result of an expression evaluation is an entity that always contains a series of time-stamped components of type map layer or scalar. The temporal topology is evaluated between the time-stamped components. Spatial topology evaluation is performed between time-stamped map layers.

A single map layer as well as a scalar can be defined as entity on the right side of the STTOP. In this case the single map layer or the scalar on the right side are applied to all map layers on the left side of the operator.

An operation between STDS A and B that result in a new STDS C can be expressed as:

$$C = A \{STTOP\} B$$

Operation between several STDS A, B, C, D and E that result in a new STDS Z can be expressed as follows:

$$Z = A \{STTOP\} ((B \{STTOP\} C \{STTOP\} D) \{STTOP\} E)$$

Braces are used to specify non-default operator precedence's, otherwise the operator evaluation is performed from the left to the right. Intermediate results are specified as $STDS\ B^*$, B^{**} and B^{***} that represent a series of time-stamped components after expression evaluation. The evaluation and therefore the operator precedence of the expression above is shown below:

$$B^* = B \{STTOP\} C$$
$$B^{**} = B^* \{STTOP\} D$$
$$B^{***} = B^{**} \{STTOP\} E$$
$$Z = A \{STTOP\} B^{***}$$

The STTOP can specify a selection operation or a spatial operation combined with spatio-temporal topological relations and temporal operators. It is shown in Figure 3 with all available sub-operators.

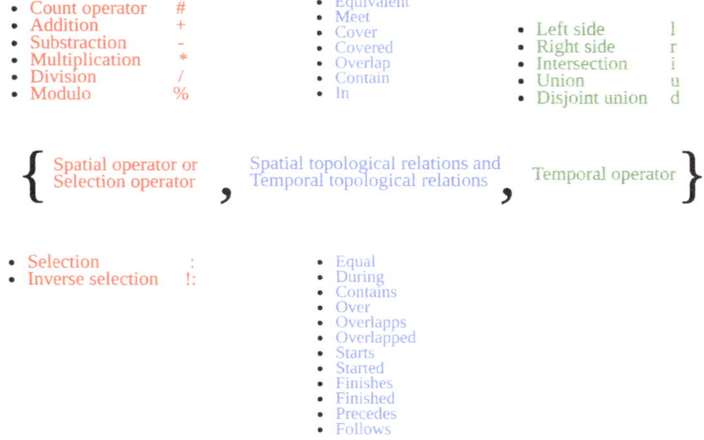

Figure 3. The spatio-temporal topological operator. The operator is specified in curled braces that contains the spatial and selection operators in red, the spatio-topological relations in blue and the temporal operators in green.

The STTOP is not commutative, the result of $A \{STTOP\} B$ may not be equal to $B \{STTOP\} A$.

3.1.1. Spatial Operators

Spatial operations are performed between raster map layers or between 3D raster map layers that have spatio-temporal topological relations to each other. A single scalar or series of time-stamped scalars can be used in a spatial operation as well. However, a series of time-stamped scalars is the result of an expression evaluation so that only single scalars can be specified directly in an expression.

Spatial operators are defined as the first sub-operators in the STTOP. If spatial operations are applied, new raster or 3D raster map layer will be created and registered in a new STRDS or STR3DS.

A special operator is the *count* operator that returns the number of time-stamped components on the right side of the operator that have adjacent temporal-topological relations to the time-stamped map layers on the left side of the operator.

All spatial operators are listed in Table 1.

Table 1. Spatial operators and their precedence in the topology based map algebra.

Symbol	Description	Precedence
#	count	1
%	modulus	1
/	division	1
*	multiplication	1
+	addition	2
−	subtraction	2

Spatial operators can be used directly in an expression, without the specification of temporal topological relations or operations. Then only time-stamped components with equal temporal topological relations are used in spatial operations. Equal spatial topological relations are not enforced because of the GRASS GIS computational region concept. For example, the creation of a new STRDS C that contains the sum of raster map layers from STRDS A and STRDS B that have equal time stamps, can be expressed as:

$$C = A + B$$

3.1.2. Temporal Selection Operators

Temporal selection operators were introduced to select and extract map layers of different STDS that have certain spatio-temporal topological relations to each other. This operator is defined as the first sub-operator in the STTOP. The selection operator does not create new map layers. It selects map layers based on their spatio-temporal topological relations and registers them in a new STDS. If the selection operation performs temporal operations, like temporal extent intersection, then the resulting map layers are copies of the original map layers and new time stamps, based on the temporal operations, are assigned. The original map layer must be copied, since GRASS GIS does not support multiple time stamps for single map layers. The type of the resulting STDS is defined by the type of the most left STDS in a selection expression. All supported selection operators are listed in Table 2. Examples for temporal selections are available in Appendix C.

Table 2. Temporal selection operators and their precedence in the topology based map algebra.

Symbol	Description	Precedence
:	Selection Operator	1
! :	Inverse Selection Operator	1

3.1.3. Spatio-Temporal Topological Relations

Spatio-temporal topological relations are defined as the second sub-operators in the STTOP. Operations between two series of time-stamped components are based on their spatio-temporal topological relations to each other. Using topological relations is a convenient way to handle time series of components that have arbitrary spatio-temporal topologies. Spatio-temporal relations as shown in Figures 1 and 2 are used to decide what operations should be performed between time-stamped components that can be map layers or scalars.

One or several topological relations can be specified as a single operator. Specifying several temporal topological relations will be interpreted using a boolean OR operation. The temporal topological relation is valid if one topological relation out of the specified list of relations is fulfilled. If several topological relations are fulfilled then a single map layer at the left side can be topologically related to many time-stamped components on the right side of the STTOP and vice verse. Specifying temporal or spatial topological relations can result in one to many relations between two series of time-stamped components. If one to many relations occur from the left side to right then an implicit aggregation approach is applied. A detailed example of implicit aggregation is described in Figure A1 in Appendix A.

Spatial and temporal topological relations can be combined in the second sub-operator. Temporal relations have OR relationships to each other. Spatial relations have also OR relationships to each other. However, temporal and spatial relationships are connected with an AND relation. This is important to distinct map layers that have equal time stamps but different spatial extents. If spatial and temporal topological relations are specified together as sub-operator, then one of the temporal and one of the spatial relationships must be fulfilled. The delimiter of a list of topological operators is the the logical OR |:

$$expression\ \{+,\ equals\ |\ during\ |\ equivalent,\ l\}\ expression$$

3.1.4. Temporal Operators

Temporal operators are located at the third position in the STTOP. They are performed between the temporal extents of two or more spatio-temporal related time-stamped components. The boolean temporal operations intersect, union and disjoint union are supported as well as decision operators left side and right side. The decision operators specifies if the time interval or time instance of a time-stamped map layer on the left side or of a time-stamped component on the right side of the STTOP should be used for the resulting time-stamped components of a selection or spatial operation. All supported temporal operations are shown in Figure 4.

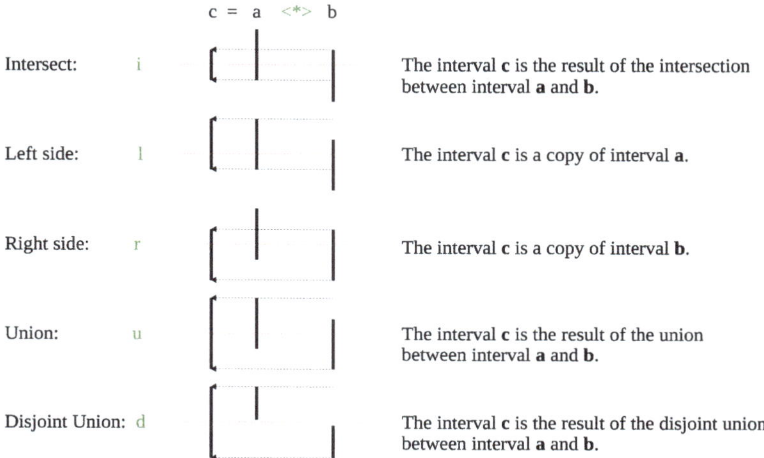

Figure 4. Temporal operations between time intervals implemented in the topology based map algebra.

3.2. Conditional Expressions

Conditional expression are required to formulate comparison and logical decision statements in our topology based map algebra. The syntax of conditional expressions is similar to [27] with the exception that topological relations must be specified. A conditional expression consists of an *if* statement that specifies the comparison operations between STDS or time-stamped components and the *then* and *else* expressions. The result of a conditional expression is a series of time-stamped map layers.

$$expression := if(\{topological\ relations\},$$
$$if\ statement,$$
$$then\ expression,$$
$$else\ expression)$$

3.2.1. The Spatio-Temporal Topological Comparison Operator STTCOP

One key component of a conditional expression is the spatio-temporal topological comparison operator (STTCOP) that can only be specified in the *if* statement. The spatio-temporal topological comparison operator must be used to specify comparison operations between time-stamped components like map layers, boolean values, scalars, date and time definitions. This operator expects a series of time-stamped components on the left and the right side. The result of the evaluation of a STTCOP is a series of time-stamped components. A single component can be of type map layer, scalar, date, datetime or boolean. Boolean, date and datetime component types are used in comparison operations. They cannot be used in spatial operations. The result of the evaluation of an *if* statement is either a list of time-stamped boolean values or a list of spatial comparison operations. Topological relations defined before the *if*, *then*, *else* expressions and time-stamped boolean values resulting from the evaluation of the *if* statement are used to select the spatial operations performed in the *then* or *else* expression. Spatial operations of the *then* expressions are performed if time-stamped boolean values are *true* and have a valid topological relation to the resulting map layers in the *then* expression. Spatial operations in the *else* expressions are performed if no topological relation to the *if* statement exists or if they are false. This is demonstrated in Figures A2 and A3 in Appendix A.

Spatial comparison operations are handled differently. They are applied to *then* and *else* expressions, if topological relations exist. Spatial comparison operations are then integrated in the resulting spatial operations. This is visualised in Figure A4 in Appendix A.

The STTCOP is not commutative, the result of $A\ \{STTOP\}\ B$ may not be equal to $B\ \{STTOP\}\ A$.

The STTCOP is described in Figure 5 with all available sub-operators. The STTCOP is build upon 4 sub-operators:

1. The comparison operator;
2. The spatio-temporal topological operator;
3. The aggregation operator;
4. The temporal operator.

The spatio-temporal topological operator and the temporal operator are similar to the STTOP. Specific for the STTCOP are the comparison, boolean and aggregation operators. The comparison and boolean operators are shown in Table 3. The aggregation operator is specified using the boolean AND & and OR | symbols. The cause of existence of these operators is the requirement to decide what kind of logical operation should be performed in a one to many relationship. If a time interval on the left side contains multiple time intervals with boolean values from the right side, the question arises how these values should be aggregated in a boolean operation? The aggregation operator describes the logical operations that should be performed between all boolean values on the right side of the STTCOP. An example of this operation is demonstrated in Figure A2.

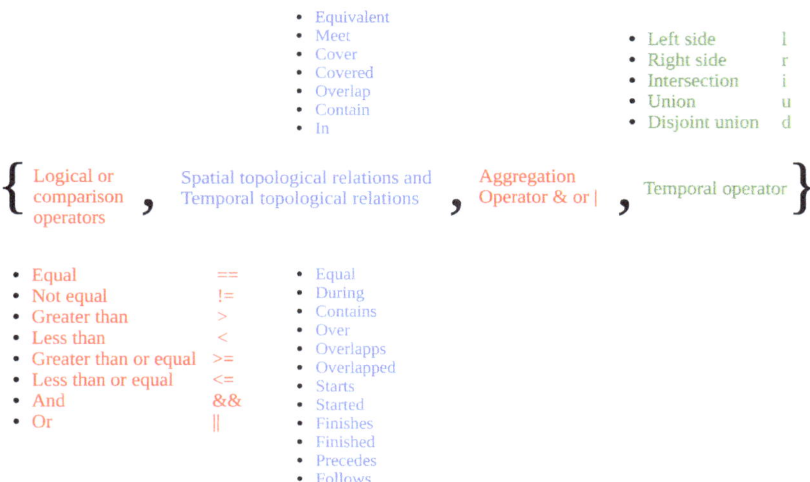

Figure 5. The spatio-temporal topological comparison operator (STTCOP). The operator is specified in curled braces that contain the logical and comparison operator in red, the spatio-topological relations in blue, the aggregation operator in red and the temporal operator in green.

Table 3. Comparison and boolean operators of the STTCOP.

Symbol	Description	Precedence
==	equal	1
!=	not equal	1
>	greater than	1
>=	greater than or equal	1
<	less than	1
<=	less than or equal	1
&&	Boolean and	2
||	Boolean or	2

The STTCOP can be simplified in the same way as the STTOP. Boolean and comparison operators can be used without specifying topological relations and temporal operations, if only equal temporal topological relationships are required. For example:

$$\{\&\&, equals, \&, l\} \;\rightarrow\; \&\&$$

$$\{>=, equals, l\} \;\rightarrow\; >=$$

3.3. Spatio-Temporal Functions

We implemented several spatio-temporal functions that include the following functionalities:

- Temporal buffering of time instances and time intervals;
- Temporal topological operation like shifting and snapping of time intervals;
- Mathematical functions like $log()$, $sin()$, $sqrt()$, $null()$ and many more;
- Date- and time functions like $start\_time()$, $start\_year()$, $start\_doy()$ and many more;
- Special functions to use different spatio-temporal datatypes in a single expression like $map()$ and $tmap()$.

A complete list of spatio-temporal functions is available in Appendix B.

3.4. Neighbourhood Operations

The topology based map algebra supports neighbourhood operations based on indices. The syntax is similar to [27] with the difference that the temporal dimension was added. STRDS neighbourhood operations can be temporal $A[t]$ or spatio-temporal $A[t, x, y]$. The temporal index is based on a nearest neighbour approach rather then temporal topological relations like precedes or follows. This has the advantage that neighbourhood operations can also be performed on raster map layers that have time intervals or time instances and no adjacent temporal topological relations to their temporal neighbours. The difference between raster and 3D raster algebra is, that the 3D raster algebra is based on space time 3D raster datasets and a four dimensional neighbourhood operator $[t, x, y, z]$. Several examples are available in Appendix C.

3.5. Temporal Granularity Algebra Approach

The granularity mode in our topology based map algebra was introduced to handle topological well aligned STDS in a convenient way. In granularity mode all spatio-temporal operators in our topology based map algebra imply equal temporal topological relations between time-stamped map layers.

The GRASS GIS Temporal Framework provides methods to compute the temporal granularity of space time datasets with valid temporal topology based on the Gregorian calendar hierarchy [6]. It can determine if the granularity has the unit years, months, days, hours, minutes or seconds. Different space time datasets may have different temporal granularities. To perform spatio-temporal operations between them, one must know what their common temporal granularity is. Hence, the GRASS GIS Temporal Framework was extended to compute the common temporal granularity of series of time-stamped map layer, or STDS and to re-sample them by the common granularity.

For example, we assume that space time dataset A has a temporal granularity of one month. Space time dataset B has a granularity of 7 days. Their temporal extent starts at 1 January 2001 and ends 1 January 2002. We assume that the time-stamped map layers of A and B have the same interval size as the granularity. There are no temporal gaps between the time-stamped map layers. The common temporal granularity between A and B is one day. To perform spatio-temporal operations between the space time datasets A and B, we need to re-sample them to a common granularity of one day. This re-sampling operation will always be performed temporally. The re-sampling operation will result in the intermediate space time datasets A^* and B^*. This operation is performed on the fly in memory, based on the temporal metadata and will not affect the original space time datasets. Hence, between 28 and 31 in memory map objects of A^* with a daily interval size will point to the same physical map layer, dependent on the actual month. For space time dataset B^* seven in memory map objects will point to the same physically map layer. This step assures that temporal topological relations between space time dataset A^* and B^* are reduced to equal, after, before, precede and follow. Therefore the STTOP and the STTCOP are simplified to spatial, selection and comparison operators.

This approach simplifies the handling of time with temporal or spatio-temporal operators. However, resulting space time dataset may have many more time-stamped map layers than the original space time datasets with plenty of redundant spatial and temporal information. It can not handle space time datasets with invalid temporal topology. However, the GRASS GIS Temporal Framework provides convenient functionality to compute valid temporal topology and to perform temporal aggregation of space time raster datasets.

4. Solving Real World Problems

4.1. Landsat NDVI Computation

The spatial operator allows us to apply the normalised difference vegetation index (NDVI) formula to a series of satellite images. An important requirement is that the satellite images are spatio-temporal distinctive from each other. This is the case for Landsat scenes, all bands of a single scenes have equal spatio-temporal extents that are distinct from any other Landsat scenes. We assume that the near infrared channel raster layers of several Landsat scenes are stored in STRDS *Landsat8_NIR*. The red channel raster layers are stored in STRDS *Landsat8_RED*. The NDVI computation for a time series of several Landsat scenes can now be formulated as a simple mathematical expression:

$$NDVI = (Landsat8\_NIR - Landsat8\_RED)/(Landsat8\_NIR + Landsat8\_RED)$$

The resulting algebraic expression applied with the command line tool t.rast.algebra to run in parallel on 8 dedicated CPU cores looks as follows:

```
t.rast.algebra
   basename=ndvi nprocs=8
   expr="NDVI=(Landsat8_NIR - Landsat8_RED) /
                       (Landsat8_NIR + Landsat8_RED)"
```

4.2. Sentinel2A NDVI Computation

The simple NDVI formula for Landsat time series can not be applied to Sentinel2A scenes, since different Sentinel2A scenes have different spatial extent but sometimes equal time stamps. They are not spatio-temporal distinctive. The expression must be extended with the spatio-topological relation equivalent to solve this issue, so that only scenes with equal spatio-temporal extents are used for the computation. We compute the normalised difference vegetation index (NDVI) for our dataset that consists of 100 Sentinel2A scenes from Germany using bands 4 and 8. We selected Sentinel2A scenes that were produced between 13 February 2017 and 6 July 2017. In addition we applied a filter to select only scenes that have an areal size greater then 5000 km^2 and cloud cover lower than 2%. The resulting dataset contained 200 time-stamped raster map layers organised in two STRDS (*S2A_B08, S2A_B04*) with a total size of 43 GB. We applied temporal and spatial topological relations to differentiate between time-stamped map layers with equal time stamps but different spatial extents.

The command to compute the NDVI on 8 dedicated CPU cores is:

```
t.rast.algebra
   basename=ndvi -s nprocs=8
   expr="NDVI=(S2A_B08{-,equal|equivalent,1}S2A_B04)
                    {/,equal|equivalent,1} \
             (S2A_B08{+,equal|equivalent,1}S2A_B04)"
```

We made use of the -s flag in t.rast.algebra to assure that the computational region is derived from the spatio-temporal related raster map layers that are involved in a spatial operation.

The GRASS GIS REST processing engine actinia was used to compute the same algebraic expression with a deployment of 12 docker container. The dedicated program actinia-algebra[8] was implemented, to distribute the map algebra expressions generated by *t.rast.algebra* to all 12 container and to collect the result in a new STRDS. In Figure 6 is the process time shown to compute the NDVI from 100 Sentinel2A scenes using up to 8 CPU's on a many core setup and up to 12 parallel processes on a docker container deployment.

---

[8]   https://github.com/mundialis/actinia_core/blob/master/scripts/actinia-algebra.

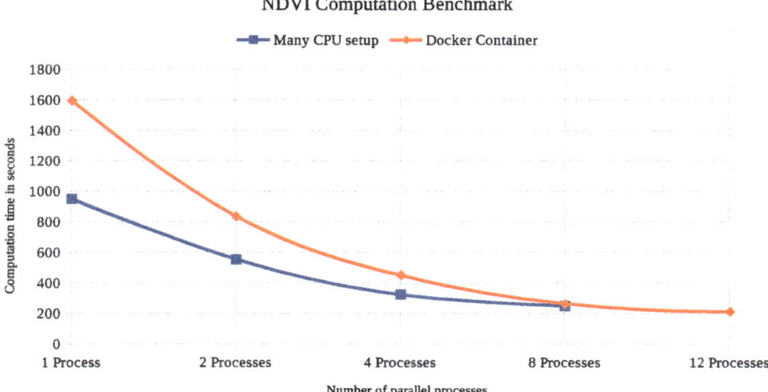

Figure 6. Computation of the normalised difference vegetation index (NDVI) based on 100 Sentinel2A scenes. The Sentinel2A dataset has a volume of 43 GB stored as two space-time raster datasets (STRDS). The many core setup benchmark was run on a virtualised Intel XEON System with 32 GB RAM, 8 Cores and 400 GB SSD. The docker container deployment was run on a 16 Core AMD Ryzen 1950 Threadripper with 32 GB RAM and 1 TB SSD. GRASS GIS 7.7 and actinia development version from March 2019 were used to create this benchmark.

4.3. Hydrothermal Coefficient Computation

The hydrothermal coefficient is a simple measure for agricultural drought and is commonly used in eastern Europe for monitoring meteorological conditions. The index is based on daily temperature and precipitation values and is sensitive to dry conditions specifically in the temperate climate zone. It is calculated as the relation of the accumulated precipitation values to the temperature sum above a baseline temperature threshold value of 10 for annual periods. The mathematical formulae is:

$$HTC = \frac{\sum P_{(T>10°C)}}{\sum T_{(T>10°C)} \cdot 0.1}$$

The index can be formulated by using the implicit aggregation feature of the spatio-temporal algebra for different temporal granularities in combination with a conditional statement for the threshold temperature. Therefore an STRDS with zeros and annual granularity has to be created as mask for which the daily meteorological STRDS are aggregated.

The algebraic expression for the hydrothermal coefficient can be formulated as:

```
t.rast.algebra
   basename=htc nprocs=8
   expr="HTC = (D {+,contains,l} if(T >= 10.0, P, 0.0)) /
               (D {+,contains,l} if(T >= 10.0, T / 10.0, 0.0))"
```

and is applied with the command line tool *t.rast.algebra* to run in parallel on 8 dedicated CPU cores, whereby three different STRDS are utilized:

- T := STRDS of daily temperatures,
- P := STRDS of daily precipitation
- D := STRDS of annual mask with 0 as pixel value

The result is shown in Figure 7. The index values range between 0 and 2, whereas zero values indicating intensive drought conditions and values above 1 represent a more humid year.

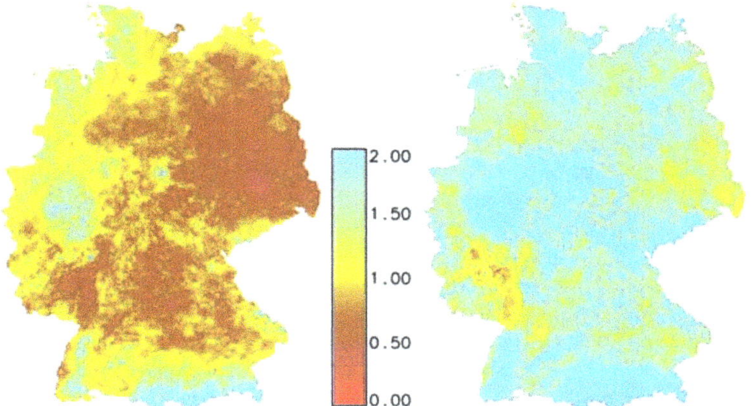

Figure 7. Comparison of the hydro-thermal coefficients (HTC) of Germany from the years 2003 and 2007. The HTC was computed using *t.rast.algebra* with the expression described above. The daily temperature and daily precipitation for the years 2003 and 2007 was provided by the German Weather Service DWD.

5. Discussion and Conclusions

We designed and implemented a new map algebra based on spatio-temporal topological relations. The syntax of the topology based map algebra was derived from the GRASS GIS map algebra module *r.mapcalc*. We extended the map algebra syntax by introducing a new spatio-temporal topological operator (STTOP) and a spatio-temporal comparison operator (STTCOP). Our topology based map algebra can therefore be easier understood by users, that are familiar with the syntax of existing map algebra implementations.

Our topology based map algebra allows selection operations as well as spatio-temporal operations on space-time datasets (STDS) based on spatio-temporal topological relations of the time-stamped map layers that are registered in the STDS. The selection part of our topology based map algebra works with all spatio-temporal datatypes that are available in GRASS GIS: STVDS, STRDS and STR3DS. Spatio-temporal operations can only be applied to STRDS and STR3DS. However, that means that our topology based map algebra works with 3 and 4 dimensions. The GRASS GIS space-time raster and 3D raster datasets can be interpreted as sparse data cubes with arbitrary time dimension. Hence, our topology based map algebra is designed to processes earth observation data cubes.

Spatial topological relations are based on the spatial extents of time-stamped map layers. Temporal topological relations are based on the temporal extents of time-stamped map layers. Spatio-temporal operations and topological relations must be specified as STTOP and STTCOP. In addition, useful spatial and temporal functions are available to modify STDS and to specify single maps and STDS of different types in a single expression. This is the first map algebra approach that allows the mixing of different datatypes in a single expression.

The STTOP and STTCOP are very powerful but its applications needs careful planning and can be very complex in nested expressions. The user must be aware of the temporal and spatial dimensions and the temporal topology of the STDS that are used in an expression. The usage of temporal and spatial topological relations must be carefully thought out to compute the required results.

The introduction of topological relations between time-stamped map layers in spatio-temporal operations leads to one-to-many relations between map layers. To address this issue we implemented the concept of implicit aggregation in case one-to-many relationships occur in the expression evaluation.

Both operators STTOP and STTCOP can be simplified in case the temporal topological relations between map layers are only equal, precedes and follows. Then simplified STTOP spatial operators and STTCOP comparison operators can be specified. This leads to less complex expressions and simplifies the application of our topology based map algebra.

We introduced the granularity mode in our topology based map algebra to simplify the handling of datasets with complex temporal topologies. In this mode all STDS that are specified in the algebraic expression will be temporally re-sampled to the greatest common temporal granularity. All time-stamped map layers will have the same temporal topology. Only equals, precedes and follows temporal topological relations are possible. Therefore, in granularity mode, all STTOP's and STTCOP's can be simplified to use only spatial and comparison operators in an expression.

Our topology based map algebra is not limited to a fixed computational region. The time-stamped map layers can have different spatial extents and spatial granularities. GRASS GIS will re-sample all raster and 3D raster maps on the fly based on nearest neighbour inter- or extrapolation. The algebra supports the definition of spatial relationships, so that only spatially related map layers are used for spatial operations. Our algebra supports the creation of computational regions based on the spatial extents of spatio-temporal-topologically related time-stamped map layers that are involved in a single spatial operation. This allows the application of algebraic expressions to time series of globally scattered satellite images without the need to create a computational region that includes the disjoint union of all spatial-extents of the time-stamped map layers that are used in the expressions. We demonstrated this in our NDVI computation of Sentinel2A satellite images example.

The implementation of our algebra is based on the GRASS GIS Temporal Framework. This framework provides many functionalities that were directly used in our algebraic approach. One functionality we would like to discuss is the ability to compute virtual space-time datasets (vSTDS) that can be used in other spatio-temporal operations in case of nested expressions. Each sub-expression is transformed in a vSTDS that consists of virtual map layers, that are defined by their spatio-temporal extents and an *r.mapcalc* operation. This functionality allowed us to implement lazy execution of the algebraic expressions. Hence, the full analysis of a topology based map algebraic expression eventually results in a list of independent *r.mapcalc* expressions and time stamping operations that can be executed massively parallel. This allows us to distribute the computation of any algebraic expression over many cores on a single computer, or over several computer nodes in a cluster or cloud environment. The many core approach was successfully demonstrated in the Sentinel2A NDVI computation example that includes a benchmark of processing time dependent on the number of used CPU cores.

The actinia processing engine was used to demonstrate the massively parallel processing of a single algebraic expression in a distributed container deployment. Hence, our topology based algebra can be deployed as a web service to perform complex algebraic computations on massive datasets. The actinia geo-processing engine allows to deploy GRASS GIS and other geo-tool based algorithms as REST service where earth observation data is physically stored. It runs on Telekom-, Google- and Amazon-Cloud platforms that provide direct access to earth observation data from satellites as Sentinel2A, Sentinel2B and Landsat. GRASS GIS and actinia are software components of the openEO initiative that provides standardized connections to and between earth observation providers. Two authors of this work are actively involved in the openEO initiative which opens the possibility to deploy our topology based map algebra as a standardized openEO processing service.

Author Contributions: S.G. performed the conceptualization, researched the methodology and wrote the original draft. T.L. and S.G. implemented the software, performed the formal analysis and validated the algebra and computational results. T.L. created the visualization of the hydro-thermal coefficient of Germany, reviewed and edited the original draft. E.P. performed supervision, validated the topological algebra concept, worked on the methodology, reviewed and edited the original draft.

Funding: Large parts of the topological algebra were implement by Thomas Leppelt in the 2013 OSGeo Google Summer of Code project *Temporal GIS Algebra for raster and vector data in GRASS*. Mentoring of this project was performed by Sören Gebbert. It is listed in the google melange archive at https://www.google-melange.com/archive/gsoc/2013/orgs/osgeo/projects/mastho.html.

Acknowledgments: We would like to thank mundialis GmbH & Co. KG that provided the hardware and software to compute the Sentinel2A based NDVI. We made use of the Sentinel2A and Landsat metadata search engine *EO-me* (https://eome.mundialis.de) provided by mundialis GmbH & Co. KG to select the required satellite images.

Conflicts of Interest: The authors declare no conflict of interest.

Appendix A. STTOP and STTCOP

Appendix A.1. STTOP Detailed Explanation

We demonstrate the application of temporal topological relations that results in implicit aggregation with the following expression

$$C = (A\{+, contains, l\}B)\{/, equals, l\}(A\{\#, contains, l\}B)$$

that includes several spatio-temporal topological operators. The temporal relation *contains* leads to implicit aggregation. In Figure A1 is the evaluation of the sub-expression, the application of the implicit aggregation and the assembling of the final result shown. The corresponding STRDS are identified using capital letters, the associated raster map layers have lower letters and a numerical index.

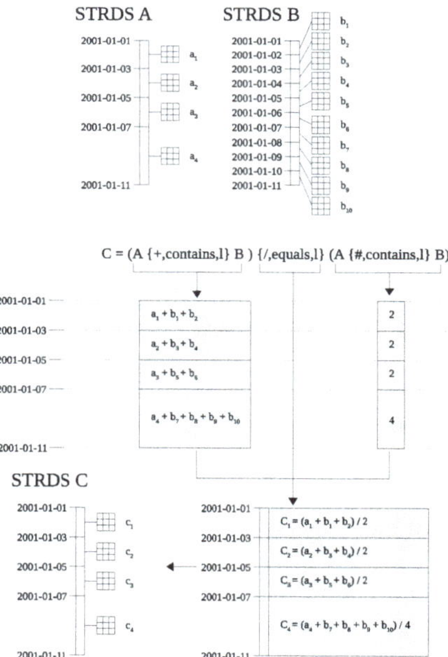

Figure A1. Visualization of nested expression evaluation with implicit aggregation based on one to many relationships between map layers of the expression: $C = (A\{+, contains, l\}B)\{/, equals, l\}(A\{\#, contains, l\}B)$. The STRDS A manages the raster map layers a_1, a_2, a_3 and a_4. The STRDS B manages the raster map layers b_1, b_2, b_3, b_4, b_5, b_6, b_7, b_8, b_9 and b_{10}. Both STRDS have interval time. The time intervals of raster map layers in STRDS A contain the time intervals of raster map layers in STRDS B. The result of this expression is the new STRDS C that contains 4 new raster map layers c_1, c_2, c_3 and c_4. The spatial operations between the spatio-temporal topological related raster map layers is expressed using the algebraic syntax specified in [27].

Appendix A.2. STTCOP Detailed Explanation

We designed three examples to illustrate the STTCOP and its application. The first example shown in Figure A2 describes the analysis of a conditional expression that involves date checks, implicit boolean and spatial aggregation. All intermediate results of the expression evaluation are visualised and its time-stamped components are shown. The STTCOP is shown in normal form $\{\&\&, contains, \&, 1\}$ as well as simplified forms $<=$ and $>=$. The STRDS A manages the raster map layers a_1, a_2, a_3 and a_4. The STRDS B manages the raster map layers b_1, b_2, b_3, b_4, b_5, b_6, b_7, b_8, b_9 and b_{10}. Both STRDS have interval time. The time intervals of raster map layers in STRDS A contain the time intervals of raster map layers in STRDS B.

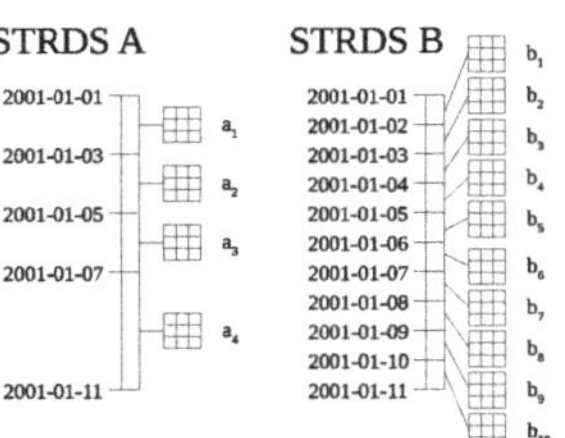

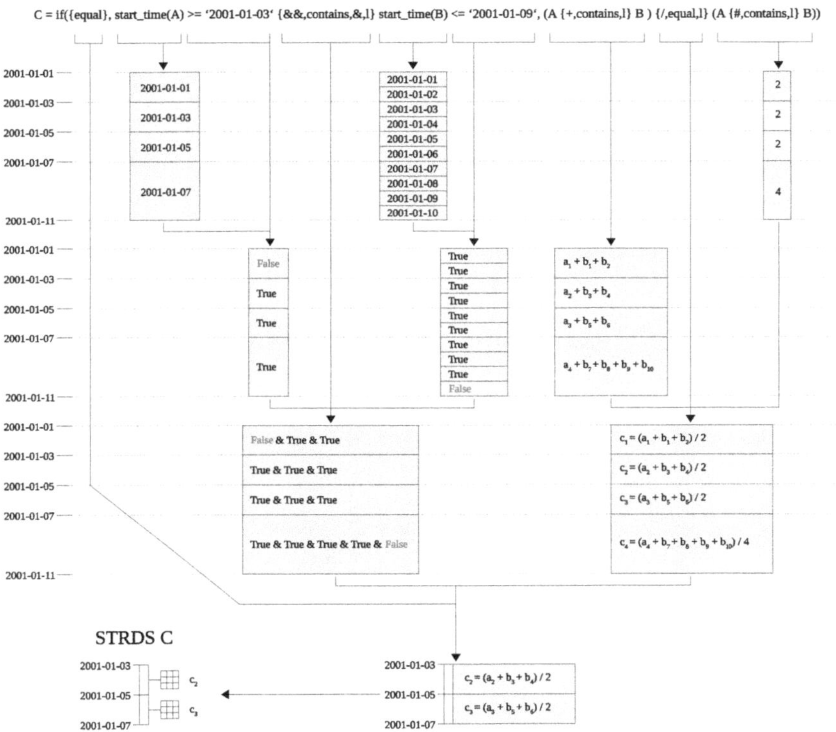

Figure A2. Visualisation of nested conditional expression evaluation with implicit aggregation of boolean values and spatial operations. The conditional statement selects only raster map layers from A that have a start time greater or equal to the date 3 January 2001 and raster map layers from B that have a start time smaller or equal to the date 9 January 2001. The boolean operation based on a temporal contains relation between the two sub-conditional statements is an AND operation with implicit AND aggregation. The conditional statement is connected with a temporal topological *equals* relation to the *then* expression. The result is the new STRDS C that contains 2 new raster map layers c_1 and c_2.

The second example shown in Figure A3 describes the analysis of a temporal selection operation with *if*, *then*, *else* statements and expressions. No spatial operation is performed. This example illustrates how time-stamped boolean values from conditional statements are applied to the *then* and *else* expressions.

Spatial conditional operations are handled differently then boolean conditional operations. The third example in Figure A4 shows the assembling of spatial conditional operations into the spatial operations of the *then* and *else* statements.

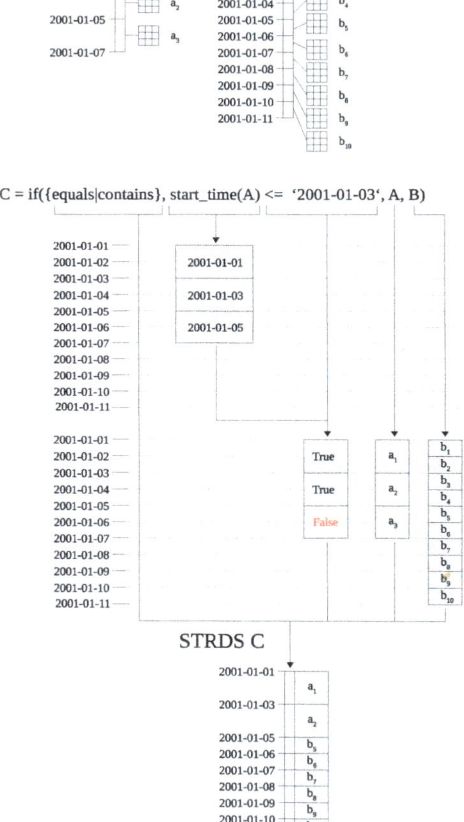

Figure A3. Visualisation of a simple nested conditional expression evaluation without spatial operations. The STRDS involved in the conditional expression are the same as in Figure A2. The conditional statement selects all raster map layers from STRDS *A* that have a start time lower or equal to the date 3 January 2001 and that have equal or contains topological relations. Otherwise raster map layers from STRDS *B* are selected that have no topological relations to the boolean values of the *if* statement or false boolean value if topological relations exist.

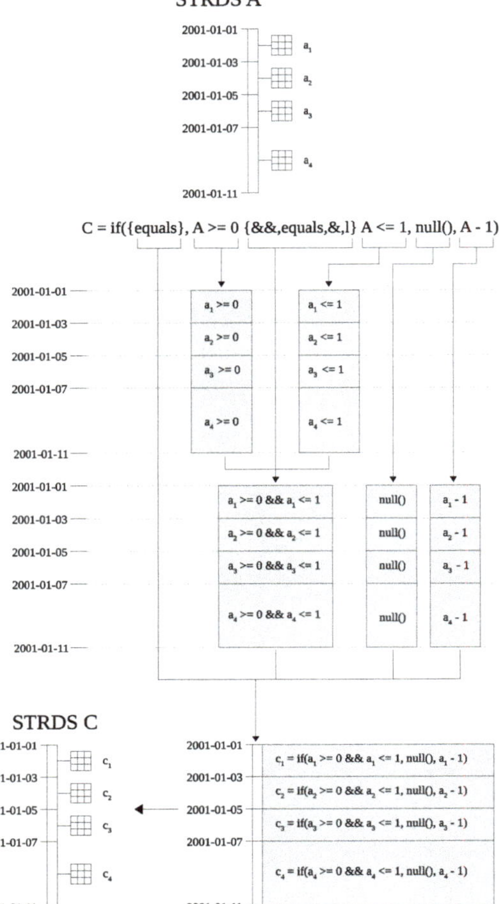

Figure A4. Visualisation of the assembling of conditional spatial operations in our topology based map algebra. The STRDS *A* contains the raster map layers a_1, a_2, a_3 and a_4. The conditional statement constrains that individual cells in raster map layers of *A* to be greater or equal to 0 and lower or equal to 1. These conditions are applied to the spatial operations in the *then* and *else* statements that have topological *equal* relations to the conditional statement. The resulting STRDS *C* contains 4 raster map layers that applied the spatial conditions and operations.

Appendix B. Spatio-Temporal Functions

Appendix B.1. Temporal Functions

Temporal functions in our topology based map algebra are introduced to perform temporal task based on time stamps of map layers. Available functions are listed in Table A1.

Table A1. Temporal functions in the topology based map algebra. All functions require a STDS as argument. The return type of the $td()$ function is a time-stamped list of scalars. All other temporal functions return a list of time-stamped map layers. The granularity size for buffering and shifting can be specified using absolute time like 1 month or 10 days or relative time that is represented by a scalar. All functions can be used on the left and right side of the STTOP.

Function	Description
$td(A)$	Return the interval size of each time-stamped map layer in STDS A as number of days and fraction of days
$buff\_t(A, size)$	Temporal buffer each map layer of STDS A with granule *size*
$tshift(A, size)$	Temporal shift each map layer of STDS A with granule *size*
$tsnap(A)$	Snap time instances and intervals of each map layer in STDS A

Appendix B.2. Spatial Functions

All spatial raster and 3D raster functions are displayed in Table A2.

Table A2. Spatial functions in the topology based map algebra. The argument for each function is a STRDS or STR3DS. Spatial functions are applied to all pixels of each map layer in a STDS.

Function	Description
$abs(A)$	return absolute values of all map layers registered in STDS A
$float(A)$	convert all map layers registered in STDS A to floating point
$int(A)$	convert all map layers registered in STDS A to integer [truncates]
$log(A)$	return natural logs of all map layers registered in STDS A
$sqrt(A)$	return square roots of all map layers registered in STDS A
$tan(A)$	return tangent of all map layers registered in STDS A (in degrees)
$round(A)$	round all map layers registered in STDS A to nearest integer
$sin(A)$	return sines of all map layers registered in STDS A (in degrees)
$sqrt(A)$	return square roots of all map layers registered in STDS A
$isnull(A)$	check if each pixel of each map layers registered in A is NULL
$isntnull(A)$	check if each pixel of each map layers registered in STDS A is not NULL
$null()$	set null values for each pixel for all map layers registered in STDS A

Appendix B.3. Special Functions

Special functions are used to identify STDS and map layers in expressions with mixed STDS types. Available functions are listed in Table A3.

Table A3. Special functions in the topology based map algebra.

Function	Description
$STRDS(A)$	A is a STRDS
$STR3DS(A)$	A is a STR3DS
$STVDS(A)$	A is a STVDS
$map(a)$	a is a single raster map layer without time stamp
$tmap(a)$	a is a single raster map layer with time stamp

Appendix B.4. Date and Time Functions

Supported date and time functions are listed in Table A4. All functions require a STDS as argument. Each function returns a *time-stamped list* of time components that can be used for temporal topological comparison operations. Each time-stamped component has the same time stamp as the raster map layer of the STDS it represents.

Table A4. Date and time functions in the topology based map algebra. Each function has a STDS as argument. Each function returns a list of time-stamped components of type datetime and scalar.

Function	The Type of the Time-Stamped Component
start_time(A)	Start time as HH::MM:SS
start_date(A)	Start date as yyyy-mm-DD
start_datetime(A)	Start datetime as yyyy-mm-DD HH:MM:SS
end_time(A)	End time as HH:MM:SS
end_date(A)	End date as yyyy-mm-DD
end_datetime(A)	End datetime as yyyy-mm-DD HH:MM
start_doy(A)	Day of year (doy) from the start time [1–366]
start_dow(A)	Day of week (dow) from the start time [1–7], the start of the week is Monday == 1
start_year(A)	The year of the start time [0–9999]
start_month(A)	The month of the start time [1–12]
start_week(A)	Week of year of the start time [1–54]
start_day(A)	Day of month from the start time [1–31]
start_hour(A)	The hour of the start time [0–23]
start_minute(A)	The minute of the start time [0–59]
start_second(A)	The second of the start time [0–59]
end_doy(A)	Day of year (doy) from the end time [1–366]
end_dow(A)	Day of week (dow) from the end time [1–7], the start of the week is Monday == 1
end_year(A)	The year of the end time [0–9999]
end_month(A)	The month of the end time [1–12]
end_week(A)	Week of year of the end time [1–54]
end_day(A)	Day of month from the start time [1–31]
end_hour(A)	The hour of the end time [0–23]
end_minute(A)	The minute of the end time [0–59]
end_second(A)	The second of the end time [0–59]

Appendix C. Examples

Appendix C.1. Temporal Selection Examples

An expression that selects all map layers from a STRDS A that are located during intervals of a STRDS B can be formulated as follows:

$$C \ = \ A \ \{:, during, l\} B$$

To select all map layers from STRDS A that have no equal time stamps to map layers of STRDS B but contain map layers from STRDS C the following expressions can be formulated:

$$D \ = \ A \ ! : B\{:, contains, l\} C$$

Appendix C.2. Spatio-Temporal Operation Examples

We developed several spatio-temporal operation examples to visualise the different aspects of our topology map algebra that are shown Figures A5–A9.

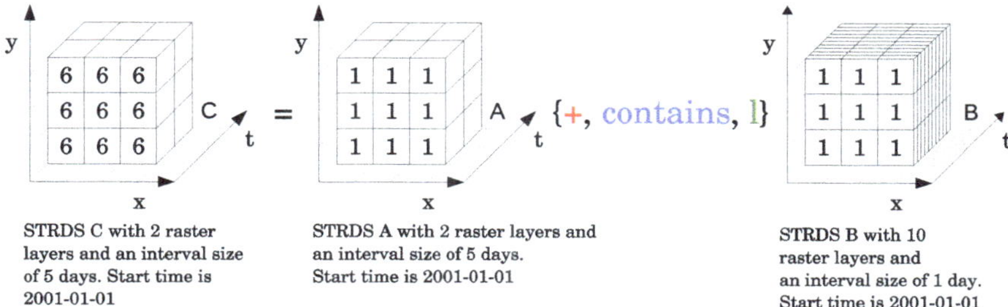

Figure A5. The the expression $C = A \{+, contains, l\}$ B is an example of one to many relationship from the left side to the right side of the spatio-temporal topological operator. It demonstrates the implicit aggregation of two time series of map layers using a contains relation between STRDS A and B.

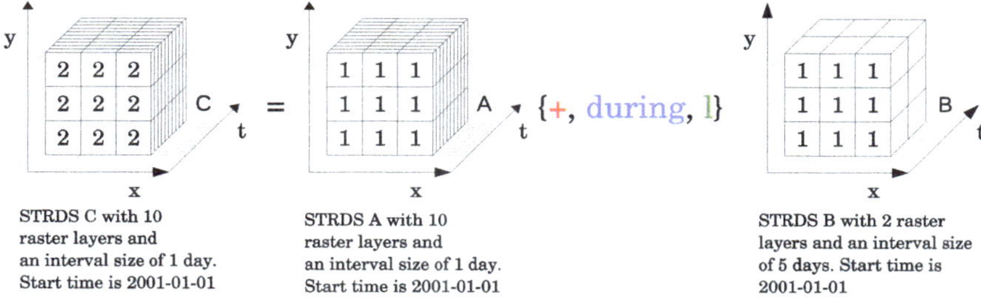

Figure A6. The the expression $C = A \{+, during, l\}$ B is an example of one to many relationship from the right side to the left side of the spatio-temporal topological operator. It demonstrates the dis-aggregation of two time series of map layers using a during relation between STRDS A and B.

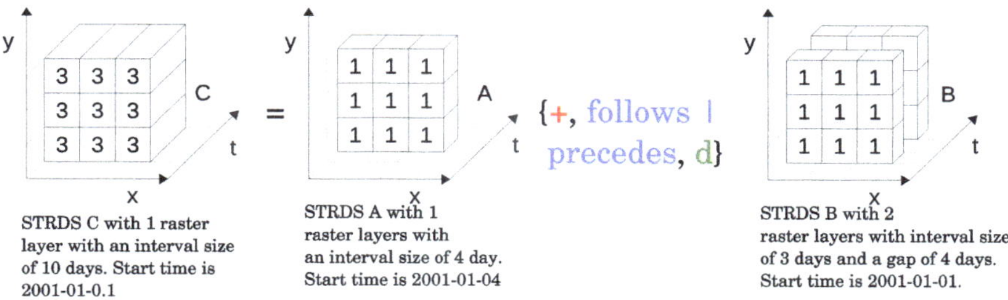

Figure A7. Implicit aggregation using topological neighbourhood relations and the application of the *disjoint union* temporal operator is demonstrated with the expression $C = A \{+, follows|preceedes, d\}$ B.

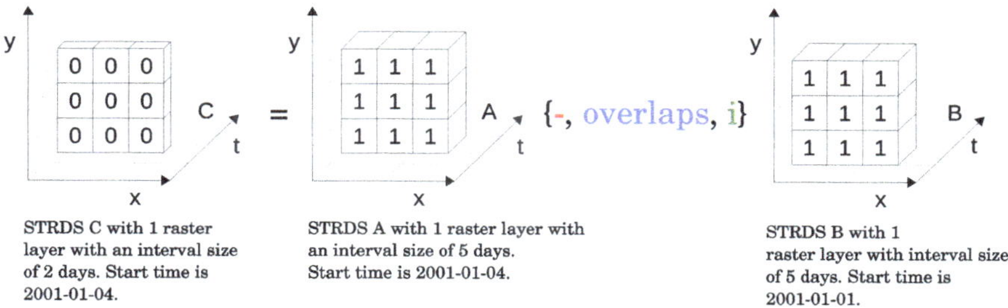

Figure A8. The temporal topological relation *overlaps* in conjunction with the *intersect* temporal operator are demonstrated with the expression $C = A \{-, overlaps, i\}$ B.

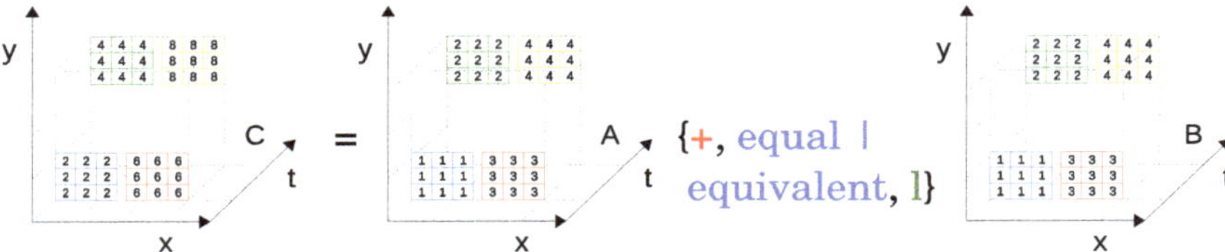

STRDS C with 4 raster Layers. Raster layers with values 2 and 6 have equal time instances of 2001-01-01 but different spatial extents. Raster layers with values 4 and 8 have equal time instances of 2001-01-08 but different spatial extents.

STRDS A with 4 raster Layers. Raster layers with values 1 and 3 have equal time instances of 2001-01-01 but different spatial extents. Raster layers with values 2 and 4 have equal time instances of 2001-01-08 but different spatial extents.

STRDS B with 4 raster Layers. Raster layers with values 1 and 3 have equal time instances of 2001-01-01 but different spatial extents. Raster layers with values 2 and 4 have equal time instances of 2001-01-08 but different spatial extents.

Figure A9. The expression $C = A \{+, equals|equivalent, l\} B$ is an example of the application of temporal and spatial topological relations to determine raster map layers that have equal time stamps but different spatial extents.

Appendix C.3. Spatio-Temporal Function Examples

- The following expression temporally buffers the STRDS A and B by one month and spatially summarise all time-stamped map layers of A and B that have *equals*, *overlaps* or *overlapped* temporal topological relations. The time stamps of the resulting map layers are the temporal union of the temporal topological related map layers.

$$C = buff\_t(A, \ '1\ month') \ \{+, equals|overlaps|overlapped, u\} \ buff\_t(B, \ '1\ month')$$

- The expression to selects all map layers from a STR3DS A that are located during intervals of a STVDS B can be formulated as follows:

$$C = STR3DS(A) \ \{:, during, l\} \ STVDS(B)$$

- The $map()$ function allows us to compute the soil-adjusted vegetation index $SAVI$ in case the canopy background adjustment factor was computed for each pixel and stored in raster layer L. The expression is as follows:

$$SAVI = ((NIR - RED) * (map(L) + 1)) \ / \ (NIR + RED + map(L))$$

Appendix C.4. Neighbourhood Operation Examples

- Compute the average of a moving window with the size of 5 time intervals of STRDS A:

$$B = (A[-2] + A[-1] + A + A[1] + A[2])/5.0$$

- Compute the gradient between the temporal neighbour map layers based on the interval size of STRDS A:

$$B = (A[-1] + A)/td(A)$$

References

1. ESA. Global Monitoring for Environment and Security. 2016. Available online: https://sentinel.esa.int/web/sentinel/home (accessed on 8 May 2012).
2. USGS. Landsat Missions. 2017. Available online: https://landsat.usgs.gov/ (accessed on 8 May 2012).
3. GSFC/NASA. Moderate Resolution Imaging Spectroradiometer (MODIS). 2016. Available online: http://modis.gsfc.nasa.gov (accessed on 8 May 2012).
4. Haylock, M.R.; Hofstra, N.; Tank, A.M.G.K.; Klok, E.J.; Jones, P.D.; New, M. A European daily high-resolution gridded data set of surface temperature and precipitation for 1950–2006. *J. Geophys. Res.* **2008**, *113*. [CrossRef]

5. Gebbert, S.; Pebesma, E. A temporal GIS for field based environmental modeling. *Environ. Modell. Softw.* **2014**, *53*, 1–12. [CrossRef]

6. Gebbert, S.; Pebesma, E. The GRASS GIS temporal framework. *Int. J. Geogr. Inf. Sci.* **2017**, *31*, 1273–1292, [CrossRef]

7. Tomlin, C.D. *Geographic Information Systems and Cartographic Modelling*; Number 910.011 T659g; Prentice-Hall: Upper Saddle River, NJ, USA, 1990.

8. Jeremy, M.; Roland, V.; Dana, T.C. Cubic Map Algebra Functions for Spatio-Temporal Analysis. *Cartogr. Geogr. Inf. Sci.* **2005**, *32*, 17–32. [CrossRef]

9. Mennis, J. Multidimensional Map Algebra: Design and Implementation of a Spatio-Temporal GIS Processing Language. *Trans. GIS* **2010**, *14*, 1–21. [CrossRef]

10. Gorelick, N.; Hancher, M.; Dixon, M.; Ilyushchenko, S.; Thau, D.; Moore, R. Google Earth Engine: Planetary-scale geospatial analysis for everyone. *Remote Sens. Environ.* **2017**. [CrossRef]

11. Neteler, M.; Mitasova, H. *Open Source GIS: A GRASS GIS Approach*, 3th ed.; Springer: New York, NY, USA, 2008. [CrossRef]

12. Mitasova, H.; Mitas, L.; Brown, W.; Gerdes, D.; Kosinovsky, I.; Baker, T. Modelling spatially and temporally distributed phenomena: New methods and tools for GRASS GIS. *Int. J. Geogr. Inf. Sci.* **1995**, *9*, 433–446. [CrossRef]

13. Mitasova, H.; Hardin, E.; Starek, M.; Harmon, R.S.; Overton, M. Landscape dynamics from LiDAR data time series. In *Geomorphometry 2011*; Hengl, T., Evans, I.S., Wilson, J.P., Gould, M., Eds.; geomorphometry.org: Redlands, CA, USA, 2011; pp. 3–6.

14. Neteler, M. Time series processing of MODIS satellite data for landscape epidemiological applications. *Int. J. Geoinf.* **2005**, *1*, 133–138.

15. Neteler, M. Estimating Daily Land Surface Temperatures in Mountainous Environments by Reconstructed MODIS LST Data. *Remote Sens.* **2010**, *2*, 333–351. [CrossRef]

16. Zorer, R.; Rocchini, D.; Delucchi, L.; Zottele, F.; Meggio, F.; Neteler, M. Use of multi-annual MODIS Land Surface Temperature data for the characterization of the heat requirements for grapevine varieties. In Proceedings of the 2011 6th International Workshop on the Analysis of Multi-temporal Remote Sensing Images (Multi-Temp), Trento, Italy, 12–14 July 2011; pp. 225–228. [CrossRef]

17. Neteler, M.; Bowman, M.H.; Landa, M.; Metz, M. GRASS GIS: A multi-purpose open source GIS. *Environ. Modell. Softw.* **2012**, *31*, 124–130. [CrossRef]

18. Zambelli, P.; Gebbert, S.; Ciolli, M. Pygrass: An Object Oriented Python Application Programming Interface (API) for Geographic Resources Analysis Support System (GRASS) Geographic Information System (GIS). *ISPRS Int. J. Geo-Inf.* **2013**, *2*, 201–219. [CrossRef]

19. Neteler, M.; Gebbert, S.; Tawalika, C.; Bettge, A.; Benelcadi, H.; Löw, F.; Adams, T.; Paulsen, H. Actinia: Cloud based geoprocessing. In Proceedings of the 2019 Conference on Big Data from Space (BiDS'2019), Munich, Germany, 19–21 February 2019; pp. 41–44. [CrossRef] [CrossRef]

20. Schramm, M.; Pebesma, E.; Wagner, W.; Verbesselt, J.; Dries, J.; Briese, C.; Jacob, A.; Mohr, M.; Neteler, M.; Mistelbauer, T.; et al. openEO—A Standardised Connection to and between Earth Observation Service Providers. In Proceedings of the 2019 Conference on Big Data from Space (BiDS'2019), Munich, Germany, 19–21 February 2019; pp. 229–232. [CrossRef]

21. Dyreson, C.; Grandi, F.; Käfer, W.; Kline, N.; Lorentzos, N.; Mitsopoulos, Y.; Montanari, A.; Nonen, D.; Peressi, E.; Pernici, B.; et al. A consensus glossary of temporal database concepts. *SIGMOD Rec.* **1994**, *23*, 52–64. [CrossRef]

22. Gray, J.; Chaudhuri, S.; Bosworth, A.; Layman, A.; Reichart, D.; Venkatrao, M.; Pellow, F.; Pirahesh, H. Data Cube: A Relational Aggregation Operator Generalizing Group-By, Cross-Tab, and Sub-Totals. *Data Min. Knowl. Discov.* **1997**, *1*, 29–53. [CrossRef]

23. Claramunt, C.; Jiang, B. Hierarchical Reasoning in Time and Space. In Proceedings of the 9th International Symposium on Spatial Data Handling, Beijing, 10–12 August 2000.

24. Bettini, C.; Dyreson, C.E.; Evans, W.S.; Snodgrass, R.T.; Wang, X.S. A glossary of time granularity concepts. In *Temporal Databases: Research and Practice*; Etzion, O., Jajodia, S., Sripada, S., Eds.; Springer: Berlin/Heidelberg, Germany, 1998; Volume 1399, pp. 406–413,

25. Allen, J.F. Maintaining knowledge about temporal intervals. *Commun. ACM* **1983**, *26*, 832–843. [CrossRef]

26. Egenhofer, M.; Al-Taha, K. Reasoning about gradual changes of topological relationships. In *Theories and Methods of Spatio-Temporal Reasoning in Geographic Space*; Frank, A., Campari, I., Formentini, U., Eds.; Springer: Berlin/Heidelberg, Germany, 1992; Volume 639, pp. 196–219. [CrossRef]

27. Shapiro, M.; Westervelt, J. *r. mapcalc: An Algebra for GIS and Image Processing*; Technical Report; Construction Engineering Research Lab (ARMY): Champaign, IL, USA, 1994.

Article

On-Demand Processing of Data Cubes from Satellite Image Collections with the gdalcubes Library

Marius Appel * and Edzer Pebesma

Institute for Geoinformatics, University of Münster, Heisenbergstraße 2, 48149 Münster, Germany
* Correspondence: marius.appel@uni-muenster.de

Received: 29 May 2019; Accepted: 26 June 2019; Published: 28 June 2019

Abstract: Earth observation data cubes are increasingly used as a data structure to make large collections of satellite images easily accessible to scientists. They hide complexities in the data such that data users can concentrate on the analysis rather than on data management. However, the construction of data cubes is not trivial and involves decisions that must be taken with regard to any particular analyses. This paper proposes on-demand data cubes, which are constructed on the fly when data users process the data. We introduce the open-source C++ library and R package gdalcubes for the construction and processing of on-demand data cubes from satellite image collections, and show how it supports interactive method development workflows where data users can initially try methods on small subsamples before running analyses on high resolution and/or large areas. Two study cases, one on processing Sentinel-2 time series and the other on combining vegetation, land surface temperature, and precipitation data, demonstrate and evaluate this implementation. While results suggest that on-demand data cubes implemented in gdalcubes support interactivity and allow for combining multiple data products, the speed-up effect also strongly depends on how original data products are organized. The potential for cloud deployment is discussed.

Keywords: earth observations; satellite imagery; R; data cubes; Sentinel-2

---

1. Introduction

Recent open data policies from governments and space agencies have made large collections of Earth observation data freely accessible to everyone. Scientists nowadays have data to analyze environmental phenomena on a global scale. For example, the fleet of Sentinel satellites from the European Copernicus program [1] is continuously measuring variables on the Earth's surface and in the atmosphere, producing terabytes of data every day. At the same time, the structure of satellite imagery is inherently complex [2,3]. Images spatially overlap, may have different spatial resolutions for different spectral bands, produce an irregular time series e.g., depending on latitude and swath, and naturally use different map projections for images from different parts of the world. This becomes even more complicated when data from multiple sensors and satellites must be combined as pixels rarely align in space and time, and the data formats in which images are distributed also vary.

Earth observation (EO) data cubes [4,5] offer a simple and intuitive interface to access satellite-based EO data by hiding complexities for data users, who can then concentrate on developing new methods instead of organizing the data. Due to its simplicity as a regular multidimensional array [6], data cubes facilitate applications based on many images such as time series and even multi-sensor analyses. At the same time, they simplify computational scalability because many problems can be parallelized over smaller sub-cubes (chunks). For instance, time series analyses often process individual pixel time series independently and a data cube representation hence makes parallelization straightforward.

Fortunately, there is a wide available array of technology that works with Earth observation imagery and data cubes. The Geospatial Data Abstraction Library (GDAL) [7] is an open source software library that is used on a large scale by the Earth observation community because it can read all data formats needed, and has high-performance routines for image warping (regridding an image to a grid in another coordinate reference system) and subsampling (reading an image at a lower resolution). However, it has no understanding of the time series of images, nor of temporal resampling or aggregation.

Database systems such has Rasdaman [8] and SciDB [9,10] have been used to store satellite image time series as multidimensional arrays. These systems follow traditional databases in the sense that they can organize the data storage, provide higher level query languages for create, read, update, and delete operations, as well as managing concurrent data access. The query languages of both mentioned systems also come with some basic data cube-oriented operations like aggregations over dimensions. However, array databases are rather infrastructure-oriented. They require a substantial effort in preparing and setting up the infrastructure, and databases typically require data ingestion, meaning that they maintain a full copy of the original data.

The Open Data Cube project (ODC) [11] provides open-source tools to set up infrastructures providing access to satellite imagery as data cubes. The implementation is written in Python and supports simple image indexing without the need of an additional copy. Numerous instances like the Australian data cube [4] or the Swiss data cube [5] are already running or are under development and demonstrate the impact of the technology with the vision to "support . . . the United Nations Sustainable Development Goals (UN-SDG) and the Paris and Sendai Agreements" [11].

Google Earth Engine (GEE) [3] even provides access to global satellite imagery including the complete Sentinel, Landsat, and MODIS collections. GEE is a cloud platform that brings the computing power of the Google cloud directly to data users by providing an easy-to-use web interface for processing data in JavaScript. The success of GEE can be explained not only by the availability of data, computing power, and an accessible user interface but also by the interactivity it provides for incremental method development. Scripts are only evaluated for the pixels that are actually visible on the interactive map, meaning that computation times are highly reduced by sub-sampling the data. GEE does not store image data as a data cube but provides cube-like operations, such as reduction over space and time.

Both, ODC and GEE provide Python clients but lack interfaces to other languages used in data science such as R or Julia. Additionally, the Pangeo project [12] is built around the Python ecosystem, including the packages xarray [13] and dask [14]. For data users working with R [15], two packages aiming at processing potentially large amounts of raster data are raster [16] and stars [17]. While raster represents datasets as two- or three-dimensional only and hence requires some custom handling of multispectral image time series, the stars package implements raster and vector data cubes with an arbitrary number of dimensions, and follows the approach of GEE by computing results only for pixels that are actually plotted, whereas raster always works with full resolution data. However, the stars package at the moment cannot create raster data cubes from spatially tiled imagery, where images come e.g., from different zones of the Universal Transverse Mercator (UTM) system.

Most of the presented tools to process data cubes including Rasdaman, SciDB, xarray, and stars assume that the data already come as a data cube. However, satellite Earth observation datasets are rather a collection of images (Section 2) and generic, cross language tools to construct data cubes from image collections are currently missing. In this paper, we propose on-demand data cubes as an interface on how data users can process EO imagery, supporting interactive analyses where data cubes are constructed on-the-fly and properties of the cube including the spatiotemporal resolution, spatial and temporal extent, resampling or aggregation strategy, and target spatial reference system can be user-defined. We present the gdalcubes C++ library and corresponding R package as a generic implementation of the construction of on-demand data cubes.

The remainder of the paper is organized as follows. Section 2 introduces the concept of on-demand data cubes for satellite image collections and presents our implementation as the gdalcubes software library. Two study cases on Sentinel-2 time series processing and constructing multi-sensor data cubes from precipitation, vegetation, and land surface temperature data evaluate the approach in Section 3, after which Sections 4 and 5 discuss the results and conclude the paper.

2. Representing Satellite Imagery as On-Demand Data Cubes with gdalcubes

2.1. Data Cubes vs. Image Collections

Earth observation data cubes are commonly defined as multidimensional arrays [6] with dimensions for space and time. We concentrate on the representation of multi-spectral satellite image time series as a data cube and here we therefore narrow it down to the following definition of a regular, dense *raster data cube*.

Definition 1. *A regular, dense raster data cube is a four-dimensional array with dimensions x (longitude or easting), y (latitude or northing), time, and bands with the following properties:*

(i)   *Spatial dimensions refer to a single spatial reference system (SRS);*
(ii)   *Cells of a data cube have a constant spatial size (with regard to the cube's SRS);*
(iii) *The spatial reference is defined by a simple offset and the cell size per axis, i.e., the cube axes are aligned with the SRS axes;*
(iv) *Cells of a data cube have a constant temporal duration, defined by an integer number and a date or time unit (years, months, days, hours, minutes, or seconds);*
(v)   *The temporal reference is defined by a simple start date/time and the temporal duration of cells;*
(vi) *For every combination of dimensions, a cell has a single, scalar (real) attribute value.*

This specific data cube type has a number of limitations and other definitions are more general (see e.g., [18–20]). However we will show how our implementation in the gdalcubes library allows the construction of such cubes from different data sources in Section 3, and help solve a wide range of problems.

As discussed in Section 1, satellite imagery is inherently complex and irregular. For example, a single Sentinel-2 image has different pixel sizes for different spectral bands. Multiple Sentinel-2 images may spatially overlap, and use different map projections (UTM zones). Furthermore, although the regular revisit time for Sentinel-2 data is five days (including both satellites), the temporal differences between images from adjacent orbits might be less than five days, leading to an irregular time series as soon as analyses cover larger spatial regions. Space agencies and cloud computing providers including new platforms such as the Copernicus Data and Information Access Services (DIASes), currently do not provide a data cube access to the data. Except for some platforms discussed in Section 1, including Google Earth Engine, the starting point for data users is often just the files, whether in the cloud or on a local computer. To efficiently build on-demand data cubes from irregularly aligned imagery, we define a data structure for image collections, representing how satellite-based Earth observation data products are distributed to the users.

Definition 2. *An image collection is a set of n images, where images contain m variables or spectral bands. Band data from one image share a common spatial footprint, acquisition date/time, and spatial reference system but may have different pixel sizes. Technically, the data of bands may come from one or more files, depending on the organization of a particular data product.*

Obviously, images in a collection should come from the same data product, i.e., measurement values must be comparable. Figure 1 illustrates how image collections are implemented in gdalcubes.

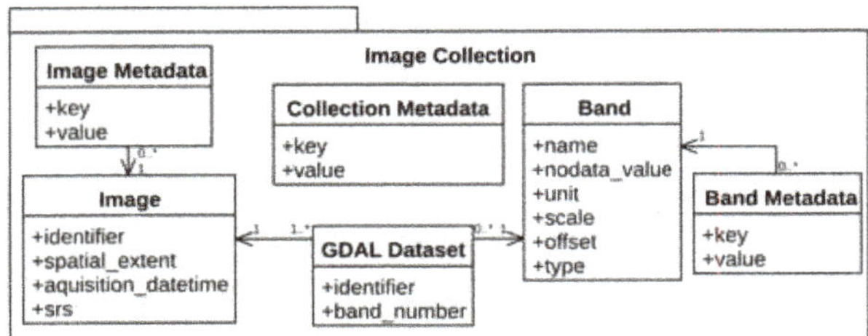

Figure 1. Data structure for image collections in gdalcubes. Geospatial Data Abstraction Library (GDAL) datasets refer to actual image data, which can be local or remote files, objects in cloud storage, sub-datasets in a more complex file format, or any other resources that GDAL can read.

2.2. Constructing User-Defined Data Cubes from Image Collections

Constructing data cubes involves some decisions that may include loss of information. These include the selection of the spatial reference system, the resolution in space and time, the area and time range of interest, and a resampling algorithm. Decisions may or may not be appropriate for particular analyses and we therefore delay the construction of the data cube until data must actually be read in the analysis. The idea is similar to how Google Earth Engine works: Users write their analysis and independently select parameters, like the area of interest and the resolution. We define a target cube with a data cube view, an object that defines the cube "geometry", and how it is created, including the target cube's:

- Spatial reference system;
- Spatiotemporal extent;
- Spatial size and temporal duration of cells (resolution);
- Spatial image resampling method, and;
- Temporal aggregation method.

The spatial resampling algorithm is applied when reprojecting, cropping, and/or resizing pixels of one image. The temporal aggregation method specifies how pixel values from multiple images that are covered by the same cell in the target data cube are combined. For example, if a data cube pixel has a temporal duration of one month, values from multiple images need to be combined, e.g., by averaging the five-daily values covered by a particular month. Similar to [21], who formalize a topological map algebra for analyzing irregular spatiotemporal datasets including satellite image collections, this allows to adapt the temporal granularity to the specific needs, and to make this explicit.

To lower memory requirements and to read and process data in parallel for larger cubes, we divide a target data cube into smaller chunks, whose spatiotemporal size can be specified by users and can be tuned to improve the performance of particular analyses. A chunk always contains data from all bands. Below, we summarize the algorithm to read a data cube chunk, given an image collection and a data cube view. The algorithm returns an in-memory four-dimensional dense array.

1. Allocate and initialize an in-memory chunk buffer for the resulting chunk data (a four-dimensional bands, t, y, x array);
2. Find all images of the collection that intersect with the spatiotemporal extent of the chunk;
3. For all images found:

    3.1. Crop, reproject, and resample according to the spatiotemporal extent of the chunk and the data cube view and store the result as an in-memory three-dimensional (bands, y, x) array;
    3.2. Copy the result to the chunk buffer at the correct temporal slice. If the chunk buffer already contains values at the target position, update a pixel-wise aggregator (e.g., mean, median, min., max.) to combine pixel values from multiple images which are written to the same cell in the data cube.

4. Finalize the pixel-wise aggregator if needed (e.g., divide pixel values by n for mean aggregation).

In the case of median aggregation, non-missing values from all contributing images are collected in an additional dynamically sized per-pixel buffer before the median can be calculated in the final step.

2.3. Data Cube Operations

Since data cubes as defined in this paper are simple multidimensional arrays, it is easy to express higher-level operators that take one (or more) data cubes as input and produce one data cube as a result. Examples include reduction over dimensions, applying arithmetic expressions on pixels, or focal window operations like image convolution. Table 1 lists some operations that are already implemented in gdalcubes.

Table 1. Implemented data cube operations in the current version of the gdalcubes library.

Operator	Description
raster_cube	Create a raster data cube from an image collection and a data cube view
reduce_time	Apply a reducer function independently over all pixel time series
reduce_space	Apply a reducer function independently over all spatial slices
apply_pixel	Apply an arithmetic expression on band values over all pixels
filter_pixel	Filter pixels with a logical predicate on one or more band values
join_bands	Stack the bands of two identically shaped cubes in a single cube
window_time	Apply a reducer function or kernel filter over moving windows for all pixel time series
write_ncdf	Export a data cube as a netCDF file
chunk_apply	Apply a user-defined function over chunks of a data cube

These operations can be chained, essentially constructing a directed acyclic graph of operations. The graph allows reordering operations in order to optimize computations and minimize data reads. Furthermore, chunks of data cubes can be processed in parallel.

2.4. The gdalcubes Library

The open-source C++ library and R package gdalcubes implement the concept of on-demand raster data cubes described above. The library includes data structures for image collections, raster data cubes, data cube views, and includes some high-level data cube operations (see Table 1). It uses the Geospatial Data Abstraction Library (GDAL) [7] to read and warp images, the netCDF C library [22] to export data cubes as files, SQLite [23] to store image collection indexes on disk, and libcurl [24] to perform HTTP requests. Additionaly it includes the external libraries tinyexpr [25] to parse and evaluate C expressions at runtime, date [26] for a modern C++ datetime approach, a tiny-process-library [27] to start external processes, and json [28] to parse and convert C++ objects from/to json. In the following, we focus on the description of the R package which simply wraps classes and functions from the underlying C++ library but does not add important features. The R package is available from the Comprehensive R Archive Network (CRAN)[1].

Figure 2 illustrates the basic workflow of how the package is used. At first, available images must be indexed to build an image collection. The image collection stores the spatial extent, the spatial reference system, the acquisition time of images, how bands relate to individual datasets or files, and where the image data can be found. The resulting image collection is a simple SQLite single file database with tables for images, bands, datasets, and metadata according to Figure 1.

---

[1] https://cran.r-project.org/package=gdalcubes

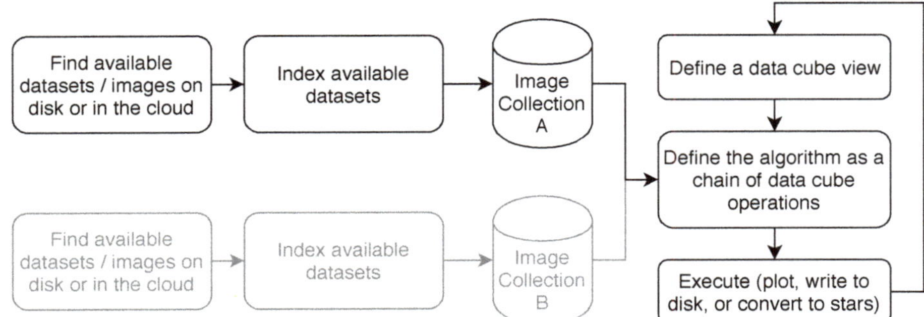

Figure 2. Typical analysis workflow for users of the R package gdalcubes.

Since we use GDAL to read image data, datasets can point to anything that GDAL can read, including local or remote files, object storage from cloud providers, sub-datasets in hierarchical file formats, compressed files, or even databases. GDAL dataset identifiers simply tell GDAL where to find and how to read the image data. The data structure with a one-to-many relationship between images and GDAL datasets and a one-to-many relationship between GDAL datasets and bands brings maximum flexibility in how the input collections can be organized. Images can be composed from a single file containing all bands (e.g., MODIS), from many files where one file contains data for one band (e.g., Landsat 8), or even from many files where files store some of the bands (e.g., grouped by spatial resolution). Again, files here are not limited to local files but refer to anything that GDAL can read. We chose SQLite for its portability and simplicity, relieving users from the need to run an additional database. To support fast spatiotemporal range selection and filtering, the image table contains indexes on the spatial extent and the acquisition date/time.

However, due to the variety of available EO products and its diverse formats and naming conventions, it is not trivial to extract all the information automatically. We abstract from specific products by defining collection formats for specific EO products. The package comes with a set of predefined formats including some Sentinel, Landsat, and MODIS data products. Further formats can be either user-defined or downloaded from a dedicated GitHub repository[2], where new formats will be continuously added. Internally, the collection formats are JSON files following a rather simple format that includes a description of the collection's bands and a few regular expressions on how to extract the needed fields, e.g., from a granule's file name.

After one or more image collections have been created, the typical workflow (Figure 2) is to define a data cube view that includes the area and time of interest, the target spatial reference system, and the spatiotemporal resolution, then define operations on the data cube, and finally plot or write the result to disk. These steps are typically repeated, where users refine the data cube view or the operations carried out on retrieved cubes. This fits well to incremental method development because users can try their methods on coarse resolution and/or a spatiotemporal extent first, before scaling the analysis to large regions and/or high resolution.

The workflow can also be identified in Figure 3, showing a minimal example R script to derive a preview image from a collection of Sentinel-2 Level 2A images by applying a median reducer over the visible bands at a 300 m spatial resolution. We first create an image collection with `create_image_collection()`, indexing available files on the local disk, then define a data cube view with `cube_view()`, and create the cube with `raster_cube()`. Calling this operation will however

---

2   https://github.com/appelmar/gdalcubes_formats

neither start any expensive computations nor read any pixel data from disk. Instead, the function immediately returns a proxy object that can be passed to data cube operations. We subset available bands of the data cube by calling `select_bands()` and apply a median reducer over time with `reduce_time()`. These functions also return proxy objects, containing the complete chain of operations and the dimensions of the resulting cube. Expressions passed as strings to data cube operations directly translate to C++ functions. In this case, the median reduction is fully implemented in C++ and does not need to call any R functions on the data. The `plot()` call finally executes the chain of operations and starts actual computations and data reads. The advantage of such a lazy evaluation is that no intermediate results must be written to disk but can be directly streamed to the next operation so that the order of operations can be optimized. In an example with 102 images from three adjacent grid tiles (summing to approximately 90 gigabytes), stored as original ZIP archives as downloaded from the Copernicus Open Access Hub [29] (see also Section 3, where we use the same dataset in the second study case), computations take around 40 s on a personal laptop with a quad-core CPU, 16 GB main memory, and a solid state disk drive. The resulting image is shown in Figure 4. The complete script has less than 20 lines of code and if users want to apply the same operation at a higher resolution, possibly for a different spatial extent and time range, only parameters that define the data cube view must be changed.

```r
library(gdalcubes)
library(magrittr)
gdalcubes_set_threads(8)

# 1. create an image collection from files on disk
files = list.files("/data/sentinel2_12a_archive", ".zip", full.names = TRUE)
S2.col = create_image_collection(files, format =  "Sentinel2_L2A")

# 2. create a data cube view for a coarse resolution overview
v = cube_view(srs="EPSG:3857", extent=S2.col, dx=300, dt="P5D",
aggregation="median", resampling = "bilinear")

# 3. create a true color overview image
raster_cube(S2.col, v) %>%
select_bands(c("B02", "B03", "B04")) %>%
reduce_time("median(B02)", "median(B03)", "median(B04)") %>%
plot(rgb=3:1,zlim=c(0,1200))
```

Figure 3. Example R script to derive a mosaic preview of Sentinel-2 images by calculating the median of visible bands over pixel time series.

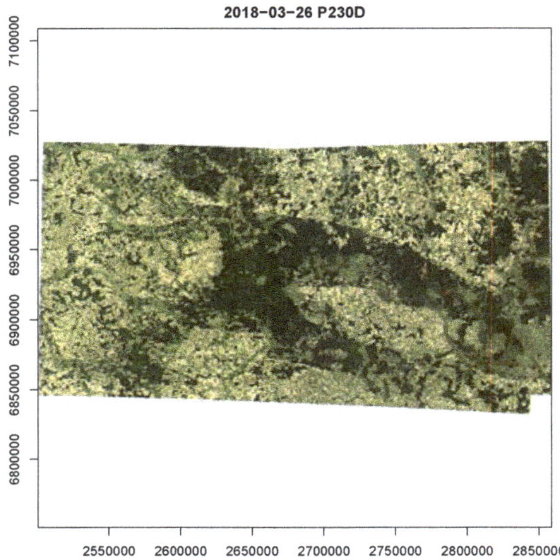

Figure 4. Output of R script in Figure 3, plotting median reflectances of visible Sentinel-2 bands over time.

Figure 5 shows a Google Earth Engine (GEE) script for applying a median (time) reduction over the same study region. The results are very similar but not identical on pixel level because of a few different images being used and because GEE reduced the entire image collections whereas gdalcubes creates data cubes with regular temporal resolution, which involves aggregating values from multiple images before applying the reducer.

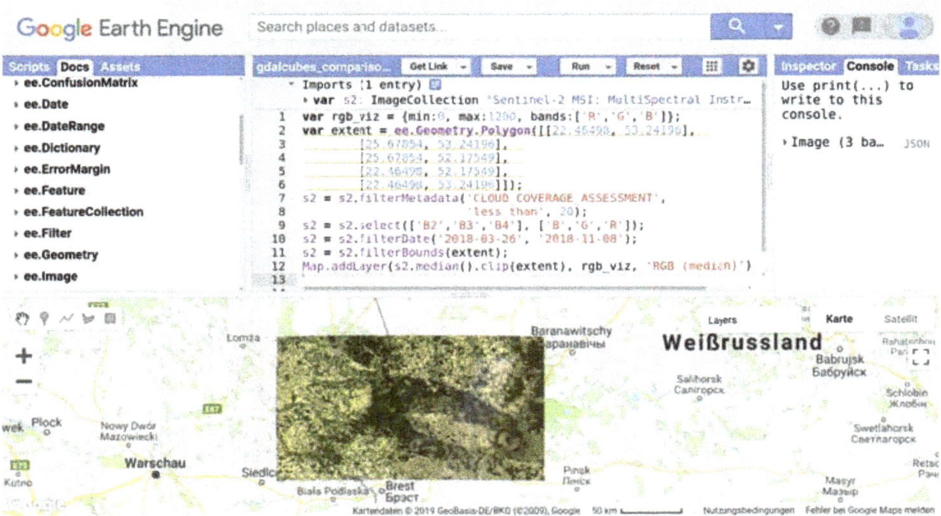

Figure 5. Screenshot of using Google Earth Engine [3] to apply a median RGB reduction of Sentinel-2 images for the same study area and time as used in Figures 3 and 4. Background imagery and map data © 2019 GeoBasis-DE/BKG (© 2019), Google.

3. Study Cases

We now demonstrate and evaluate the R package implementation in two study cases. The first study case focuses on demonstrating how gdalcubes can be used to combine different data products. The second study case processes Sentinel-2 time series and evaluates the scalability of computation times as a function of the resolution of the data cube view and the number of CPUs used. All computation time measurements have been performed on a Dell PowerEdge R815 Server with 4 AMD Opteron 6376 CPUs, summing to 64 CPU cores in total and 256 GB of main memory.

3.1. Constructing a Multi-Sensor Data Cube from Precipitation, Vegetation Data, and Land Surface Temperature Data

In this case study, we build a multi-sensor data cube including 16-daily vegetation index data from the Moderate-resolution Imaging Spectroradiometer (MODIS) product MOD13A2[3], daily land surface temperature data from the MODIS product MOD11A1[4], and daily precipitation data from the Global Precipitation Measurement mission (GPM) product IMERG[5] (using the daily accumulated, final-run product). Table 2 summarizes some important properties of the datasets used in this study case. Combining data from such sensors including meteorological and optical measurements is an important step in analyzing statistical dependencies between environmental phenomena. In this case, the combined resulting data cube can, for example, be used to study the resistance of vegetation against heat or drought periods.

Table 2. Summary of the data products as used in the first study case. Definitions: GPM, Global Precipitation Measurement mission; NDVI, normalized difference vegetation index; liquid_accum, liquid daily accumulated precipitation; LST_DAY, daytime land surface temperature; SRS, spatial reference system; MODIS, Moderate Resolution Imaging Spectroradiometer.

	MOD13A2	**GPM**	**MOD11A1**
Selected Variables	NDVI	liquid_accum	LST_DAY
Spatial Resolution	1 km × 1 km	0.1° × 0.1°	1 km × 1 km
Area of Interest	global (land only)	global (60° N–60° S full)	Europe (land only)
Temporal Resolution	16 days	daily	daily
Time Range	2014-01-01–2019-01-01	2014-01-01–2019-01-01	2014-01-01–2019-01-01
File Format	HDF4	GeoTIFF (zip compressed)	HDF4
SRS	MODIS sinusoidal	Lat/Lon grid	MODIS sinusoidal

The script to build a combined data cube is shown in Figure 6. We first create a common data cube view, covering Europe from the beginning of 2014 to the end of 2018 at a 10 km spatial and daily temporal resolution. Then, we create three separate raster data cubes and apply some individual operations, e.g., to compute 30-day precipitation means from daily measurements. We then combine the cubes using two calls to the `join_bands()` function, which collects the bands from two identically shaped data cubes. Since the MOD13A2 product covers land areas only, we ignore any pixels in the combined cube without vegetation data by calling `filter_predicate()`. Expressions passed to the `apply_pixel` and `filter_predicate` functions are translated to C++ functions, with `iif` denoting a simple one line if-else statement. Finally, we export the cube as a netCDF file. Figure 7 shows a resulting temporal subset of a cube derived at a 10 km spatial resolution. Computation times to execute the script varied between 40 and 240 min on a 50 km and 1 km spatial resolution respectively, meaning that by reducing the number of pixels in the target data cube by a factor of 2500, we could reduce

---

[3] https://lpdaac.usgs.gov/products/mod13a2v006/
[4] https://lpdaac.usgs.gov/products/mod11a1v006/
[5] https://pmm.nasa.gov/data-access/downloads/gpm

computation times by a factor of 6. In this case, data users would additionally need to reduce the area and/or time range of interest to try out methods and get interactive results within a few minutes.

```
v.europe = cube_view(srs= "EPSG:3035", extent=list(left=2500000, right = 6000000,
          top = 5500000, bottom = 1500000, t0 = "2014-01-01", t1 = "2018-12-31"),
          dx=10000, dt="P1D")

MOD13A2.col = image_collection("MOD13A2_global_2014_2018.db")
GPM.col     = image_collection("GPM.db")
MOD11A1.col = image_collection("MOD11A1_2014_2018.db")

GPM.cube =
    raster_cube(GPM.col, v.europe) %>%
    select_bands("liquid_accum") %>%
    apply_pixel("liquid_accum / 10", names = "PREC") %>%
    window_time(expr = "mean(PREC)", window = c(30,0))

MOD13A2.cube =
    raster_cube(MOD13A2.col, v.europe) %>%
    select_bands("NDVI") %>%
    apply_pixel("NDVI / 1e4", names="NDVI")

MOD11A1.cube =
    raster_cube(MOD11A1.col, v.europe) %>%
    select_bands("LST_DAY") %>%
    apply_pixel("LST_DAY  * 0.02", names="LST") %>%
    window_time(expr = "mean(LST_30D)", window = c(30,0))

join_bands(MOD13A2.cube, GPM.cube) %>%
    join_bands(MOD11A1.cube) %>%
    filter_predicate("iif(isnan(NDVI), 0, 1)") %>%
    write_ncdf("combined.nc")
```

Figure 6. R script to combine data cubes from three different data products. The construction of the image collection is omitted here.

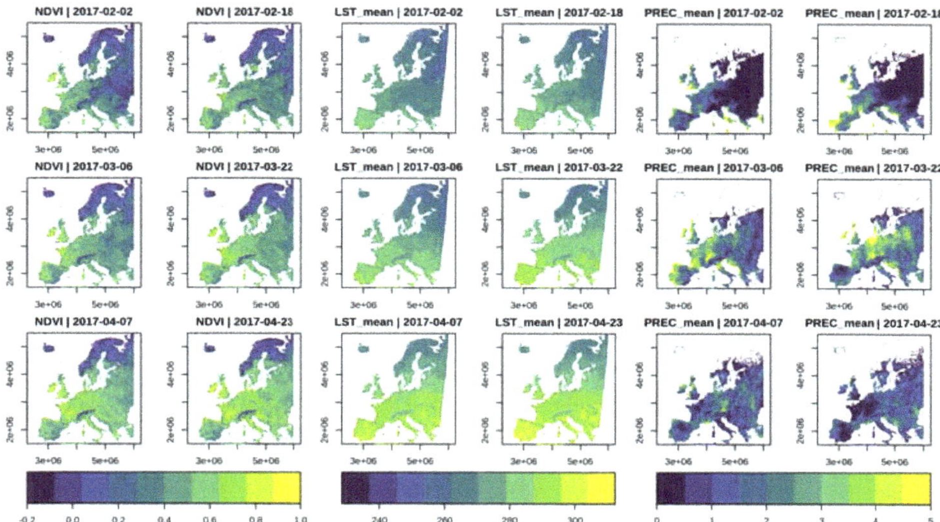

Figure 7. Temporal subset of the combined data cube with NDVI measurements (**left**), average daytime land surface temperature (K) during the last 30 days (**center**), and average daily precipitation (mm) during the last 30 days (**right**).

3.2. Processing Sentinel-2 Time Series

In a second study case, we applied a time series analysis on a collection of 102 Sentinel-2 images. The dataset covers the border region of Poland and Belarus, covering a total area of approximately 25,000 km^2. The images have been recorded between March and November 2018 and come from three different grid tiles and two different UTM zones. Images sum up to approximately 90 gigabytes and are stored as original compressed ZIP archives, downloaded from the Copernicus Open Access Hub [29]. Figure 8 shows the chain of data cube operations to detect permanent water bodies. We first compute the normalized difference water index (NDWI) based on green and near infrared reflectance, then simply classify all pixels with NDWI larger than or equal to zero as water (value 1), other pixels as no water (value 0), and then derive the mean over all pixel time series, representing the proportion of time instances where a pixel has been classified as water. In the last step, we set all pixels with value less than or equal 0.1 to NA and export the resulting image as a netCDF file. Figure 9 illustrates the study area and the results of the water detection on a low and high resolution in a map.

```
raster_cube(S2.col, v) %>%
    select_bands(c("B03", "B08")) %>%
    apply_pixel("(B03-B08)/(B03+B08)", names = "NDWI") %>%
    apply_pixel("iif(NDWI >= 0, 1, 0)", names = "water") %>%
    reduce_time("mean(water)") %>%
    filter_predicate("water_mean > 0.1") %>%
    write_ncdf(tempfile())
```

Figure 8. R script to detect permanent water bodies from a Sentinel-2 data cube.

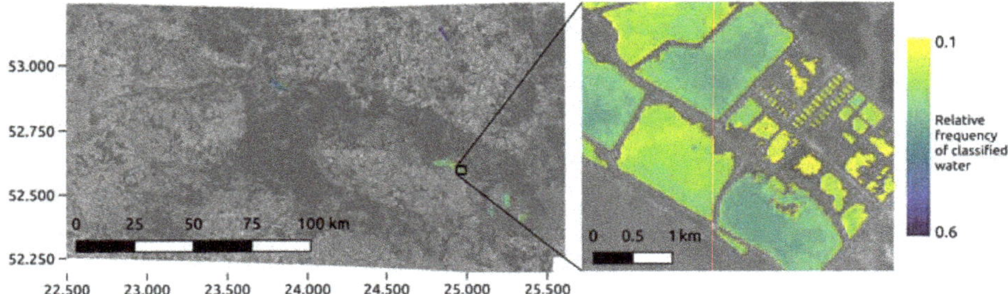

Figure 9. Result map from water detection in the second study case. The right part illustrates the results at a high spatial resolution.

To evaluate how the implementation scales with the spatial resolution of the target data cube, we vary the spatial resolution in the data cube view and measure computation times for executing the code in Figure 8. At a fixed spatial resolution with pixels covering an area of 100 m × 100 m, we vary the number of used CPUs.

Figure 10 (left) shows how the speedup changed if the number of pixels was reduced by a certain factor. Reducing the number of pixels by a factor of 10,000, i.e., working with 1 km × 1 km pixel size compared to 10 m × 10 m, reduced the computation times by a factor of approximately 100. Although this may seem like a rather low effect, it means that we can process the whole dataset within less than a minute on low resolution as opposed to approximately one hour at full resolution. Computation times reduced consistently with an increasing number of CPUs (Figure 10). For example, we have been able to process 7.61 times more pixels per second when using eight threads compared to using a single thread.

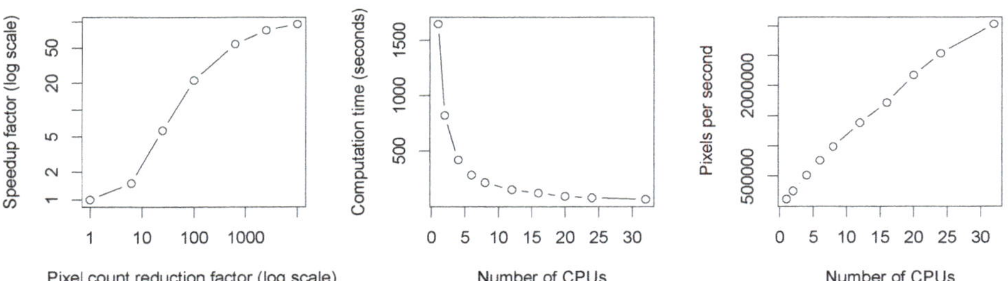

Figure 10. Computational results for the second study case. The left plot shows the achieved speedup factors depending on the reduction of pixels in the target data cube. For example, reducing the number of pixels by a factor of 100 resulted in a speedup of around 20, compared to computation times with a 10 m by 10 m spatial resolution. The center and right plots show computation times and pixel throughput respectively as a function of the number of used CPUs.

4. Discussion

Today, data cubes are increasingly used as the basis for further analysis of large Earth observation image time series. For the creation of data cubes from image collections, resampling and/or aggregation in space and time is needed, in addition to image warping. As discussed below, the presented approach does this on-the-fly and interfaces existing open source software to process Earth observation data cubes.

4.1. Interactive Analyses of Large EO Datasets

The case studies have demonstrated that on-demand raster data cubes, where users define the shape of the target cube, allow the reduction of computation times and thus improve interactivity in analyses of large satellite image collections. This approach is similar to Google Earth Engine in that it only reads the pixels that are actually needed, as late as possible (lazy evaluation). However, the magnitude of reduction in computation times depends on particular data products. In the example of Sentinel-2 time series processing, gdalcubes makes use of provided image overviews or pyramids when working on a coarse resolution cube view. In contrast, MODIS products do not include such overviews and hence the full data must be read first, which reduces the gain in interactivity. In these cases, users may need to reduce the area and temporal range of interest to yield acceptable computation times. Under certain situations it might pay off to build overview images manually using the GDAL implementation before using gdalcubes. This also becomes important in cloud environments, where overviews may even reduce costs associated with data access or transfer.

There is currently quite some discussion about whether so-called analysis-ready data, which are essentially data cubes, should be processed for large scale imagery (e.g., [5] and CEOS-ARD[6]) in order to make these data usable for a larger community. As this creation is a very expensive operation, we argue in line with [3] that it is hard to create data cubes with parameters that satisfy every researcher, and that the on-the fly creation of data cubes retains maximum flexibility in this respect. More research on quantifying the loss of statistical accuracy or power due to resampling and working at lower-than-maximum resolutions is still needed.

4.2. Scalable and Distributed Processing in the Cloud

The examples shown here were executed on a local machine. Several days were needed to download the data, whereas the time for processing in the case study was much lower. While this is acceptable for medium-sized datasets, it becomes impossible for large scale, high resolution analyses. The rational trend is to move computations to cloud platforms where the data is already available. These include Amazon Web Services, the Google Cloud, and specialized EO data centers such as the Copernicus DIASes. Since the gdalcubes implementation uses GDAL to read imagery, it can directly access object storage from major cloud providers. Image collections then simply point to globally unique object storage identifiers and hence image collection indexes can be shared. Furthermore, though not yet available in the R package, the C++ implementation of gdalcubes comes with a prototypical server application, providing a simple REST-like API to process specific chunks of a cube. Running several of these gdalcubes worker instances in containerized cloud environments would allow process distribution over many compute instances.

An interesting open question is how the image collection index performs with much larger datasets. In the case studies with up to 34,000 images in a collection (global vegetation index data from MODIS for 5 years), we could not see any performance decreases so far. The image collection typically consumes a few kilobytes per image and images can be added incrementally. However, since the underlying table structure in the SQLite database only uses one-dimensional indexes on the spatial extent, acquisition time, and identifier of images, this might not scale well e.g., for the full Sentinel-2 archive. Implementations with more advanced indexes as in the S2 Geometry Library[7] might be needed in these cases.

---

[6]   http://ceos.org/ard
[7]   http://s2geometry.io

4.3. Interfaces to Other Software and Languages

The presented examples demonstrate how the gdalcubes library can be used in R. All data cube operations, the construction of data cubes from files, and the export as netCDF files are however implemented in C++. The R package is a thin software layer that makes the C++ library easily usable for R users. Since other languages such as Python or Julia also allow one to interface the C++ code, writing interfaces in these languages is feasible and relatively straightforward.

The data model to represent data cubes in memory is rather simple. A chunk is nothing more than a one-dimensional contiguous double precision C++ vector with additional attributes storing the dimensionality. As a result, it is also possible to interface and extend gdalcubes with linear algebra, image processing (e.g., Orfeo ToolBox [30]), or other external libraries including the NumPy C application programming interface [31].

The only other software systems we know of that can create regular data cubes from image collections are GRASS GIS [21], Open Data Cube [11], and the (non-open source) Google Earth Engine [3]. The open source library gdalcubes introduced here is a nice addition to these as it is relatively easy to integrate in scripting languages such as R, Julia, or Python, and can work in conjunction with software that can process data cubes such as GRASS GIS [32], R packages raster [16] and stars [17], and Python packages numpy [33] and xarray [13]. At the moment, a more user-friendly package to interface gdalcubes with the Python ecosystem is missing.

4.4. Limitations

The presented work focused on representing satellite imagery as raster data cubes and the implementation always uses a four-dimensional array with two spatial, one temporal, and one band dimension (see Definition 1). Hence, it is not directly applicable for higher dimensional data such as climate model output with vertical space or in cases where it is useful to represent time as two dimensions (e.g., year and day of year, or time of forecast and time to forecast). Furthermore, it currently represents raster data cubes only. Fortunately, the existing R package stars [17] implements generic multidimensional arrays, including support for rectilinear and curvilinear rasters and some first attempts to bring together functionalities from both packages are currently in progress.

The case studies demonstrated that the speed-up effect of on-demand data cubes on interactivity for lower resolution analyses strongly depends on particular datasets. One very important factor is whether the data comes with image pyramids/overviews, as well as the data format. In this regard, modern approaches such as the cloud-optimized GeoTIFF format with additional overviews seem very promising.

Similar to Google Earth Engine [3], the gdalcubes library is not well suited to problems that are hardly scalable and perform global analyses where results depend on distant pairs of pixels. There are also some parameters like the selection of chunk size, which are not easy to automatically optimize.

5. Conclusions

This paper proposes an approach to the on-demand creation of raster data cubes and presents an open source implementation in the gdalcubes library. It presents a generic solution to convert and combine irregular satellite imagery to regular raster data cubes, thereby supporting interactive incremental method development. This makes it easier for data users to exploit the potential of Earth observation data cubes such as combining data from several sensors and satellites. The organization of particular data products has a strong effect on speedups for computations on sub-sampled data. As the library has been written in C++, interfaces to scripting languages like Python and Julia could be developed easily; a gdalcubes R interface has been published on the Comprehensive R Archive Network (CRAN).

Author Contributions: Conceptualization, M.A.; Funding acquisition, M.A. and E.P.; Investigation, M.A.; Methodology, M.A.; Project administration, E.P.; Resources, M.A.; Software, M.A.; Supervision, E.P.; Visualization, M.A.; Writing—initial draft, M.A.; Writing—review & editing, E.P.

Funding: This research was funded by Deutsche Forschungsgemeinschaft under DFG project *S3-GEP: Scalable Spatiotemporal Statistics for Global Environmental Phenomena*, number 396611854.

Acknowledgments: The authors would like to thank two anonymous reviewers for their helpful comments and suggestions.

Conflicts of Interest: The authors declare no conflict of interest.

References

1. Copernicus—The European Earth Observation Programme. Available online: https://ec.europa.eu/growth/sectors/space/copernicus_en (accessed on 14 June 2019).
2. Lillesand, T.; Kiefer, R.W.; Chipman, J. *Remote Sensing and Image Interpretation*; John Wiley & Sons: New York, NY, USA, 2015.
3. Gorelick, N.; Hancher, M.; Dixon, M.; Ilyushchenko, S.; Thau, D.; Moore, R. Google Earth Engine: Planetary-scale geospatial analysis for everyone. *Remote Sens. Environ.* **2017**, *202*, 18–27. [CrossRef]
4. Lewis, A.; Oliver, S.; Lymburner, L.; Evans, B.; Wyborn, L.; Mueller, N.; Raevksi, G.; Hooke, J.; Woodcock, R.; Sixsmith, J.; et al. The Australian Geoscience Data Cube—Foundations and lessons learned. *Remote Sens. Environ.* **2017**, *202*, 276–292. [CrossRef]
5. Giuliani, G.; Chatenoux, B.; Bono, A.D.; Rodila, D.; Richard, J.P.; Allenbach, K.; Dao, H.; Peduzzi, P. Building an Earth Observations Data Cube: Lessons learned from the Swiss Data Cube (SDC) on generating Analysis Ready Data (ARD). *Big Earth Data* **2017**, *1*, 100–117. [CrossRef]
6. Lu, M.; Appel, M.; Pebesma, E. Multidimensional Arrays for Analysing Geoscientific Data. *ISPRS Int. J. Geo-Inf.* **2018**, *7*, 313. [CrossRef]
7. Warmerdam, F. The geospatial data abstraction library. In *Open Source Approaches in Spatial Data Handling*; Springer: Berlin/Heidelberg, Germany, 2008; pp. 87–104.
8. Baumann, P.; Dehmel, A.; Furtado, P.; Ritsch, R.; Widmann, N. The Multidimensional Database System RasDaMan. In Proceedings of the 1998 ACM SIGMOD International Conference on Management of Data, SIGMOD '98, Seattle, WA, USA, 1–4 June 1998; ACM: New York, NY, USA, 1998; pp. 575–577.
9. Stonebraker, M.; Brown, P.; Zhang, D.; Becla, J. SciDB: A Database Management System for Applications with Complex Analytics. *Comput. Sci. Eng.* **2013**, *15*, 54–62. [CrossRef]
10. Appel, M.; Lahn, F.; Buytaert, W.; Pebesma, E. Open and scalable analytics of large Earth observation datasets: From scenes to multidimensional arrays using SciDB and GDAL. *ISPRS J. Photogramm. Remote Sens.* **2018**, *138*, 47–56. [CrossRef]
11. Open Data Cube. Available online: https://www.opendatacube.org (accessed on 23 May 2019).
12. Pangeo—A Community Platform for Big Data Geoscience. Available online: https://pangeo.io (accessed on 23 May 2019).
13. Hoyer, S.; Hamman, J. xarray: ND labeled Arrays and Datasets in Python. *J. Open Res. Softw.* **2017**, *5*, 10. [CrossRef]
14. Rocklin, M. Dask: Parallel computation with blocked algorithms and task scheduling. In Proceedings of the 14th Python in Science Conference, Austin, TX, USA, 6–12 July 2015; pp. 126-132.
15. R Development Core Team. *R: A Language and Environment for Statistical Computing*; R Foundation for Statistical Computing: Vienna, Austria, 2008; ISBN 3-900051-07-0.
16. Hijmans, R.J. *raster: Geographic Data Analysis and Modeling*; R Package Version 2.9-5; 2019. Available online: https://CRAN.R-project.org/package=raster (accessed on 27 June 2019).
17. Pebesma, E. *stars: Spatiotemporal Arrays, Raster and Vector Data Cubes*; R Package Version 0.3-1; 2019. Available online: https://CRAN.R-project.org/package=stars (accessed on 27 June 2019).
18. Baumann, P.; Rossi, A.P.; Bell, B.; Clements, O.; Evans, B.; Hoenig, H.; Hogan, P.; Kakaletris, G.; Koltsida, P.; Mantovani, S.; et al. Fostering Cross-Disciplinary Earth Science Through Datacube Analytics. In *Earth Observation Open Science and Innovation*; Mathieu, P.P., Aubrecht, C., Eds.; Springer International Publishing: Cham, Switzerland, 2018; pp. 91–119.

19. Nativi, S.; Mazzetti, P.; Craglia, M. A view-based model of data-cube to support big earth data systems interoperability. *Big Earth Data* **2017**, *1*, 75–99. [CrossRef]

20. Strobl, P.; Baumann, P.; Lewis, A.; Szantoi, Z.; Killough, B.; Purss, M.; Craglia, M.; Nativi, S.; Held, A.; Dhu, T. The Six Faces of The Datacube. In Proceedings of the 2017 Conference on Big Data from Space (BIDS' 2017), Toulouse, France, 28–30 November 2017; pp. 28–30.

21. Gebbert, S.; Leppelt, T.; Pebesma, E. A Topology Based Spatio-Temporal Map Algebra for Big Data Analysis. *Data* **2019**, *4*, 86. [CrossRef]

22. Rew, R.; Davis, G. NetCDF: An interface for scientific data access. *IEEE Comput. Graph. Appl.* **1990**, *10*, 76–82. [CrossRef]

23. SQLite. Available online: https://www.sqlite.org (accessed on 24 May 2019).

24. Stenberg, D.; Fandrich, D.; Tse, Y. libcurl: The Multiprotocol File Transfer Library. Available online: http://curl.haxx.se/libcurl (accessed on 24 May 2019).

25. Tinyexpr. Available online: https://github.com/codeplea/tinyexpr (accessed on 24 May 2019).

26. Date. Available online: https://howardhinnant.github.io/date/date.html (accessed on 24 May 2019).

27. Tiny-process-library. Available online: https://gitlab.com/eidheim/tiny-process-library (accessed on 24 May 2019).

28. JSON for Modern C++. Available online: https://github.com/nlohmann/json (accessed on 24 May 2019).

29. Copernicus Open Access Hub. Available online: https://scihub.copernicus.eu (accessed on 24 May 2019).

30. Inglada, J.; Christophe, E. The Orfeo Toolbox remote sensing image processing software. In Proceedings of the 2009 IEEE International Geoscience and Remote Sensing Symposium, Cape Town, South Africa, 12–17 July 2009; Volume 4, pp. IV:733–IV:736.

31. NumPy C-API. Available online: https://docs.scipy.org/doc/numpy/reference/c-api.html (accessed on 14 June 2019).

32. Neteler, M.; Bowman, M.; Landa, M.; Metz, M. GRASS GIS: A multi-purpose Open Source GIS. *Environ. Model. Softw.* **2012**, *31*, 124–130. [CrossRef]

33. Walt, S.V.D.; Colbert, S.C.; Varoquaux, G. The NumPy Array: A Structure for Efficient Numerical Computation. *Comput. Sci. Eng.* **2011**, *13*, 22–30. [CrossRef]

Article

A Portal Offering Standard Visualization and Analysis on top of an Open Data Cube for Sub-National Regions: The Catalan Data Cube Example

Joan Maso [1],*, Alaitz Zabala [2], Ivette Serral [1] and Xavier Pons [2]

[1] CREAF, Fac. Ciencies, Campus UAB, 08193 Bellaterra, Barcelona, Spain
[2] Departament de Geografia, Universitat Autònoma de Barcelona, Edifici B, 08193 Bellaterra, Barcelona, Spain
* Correspondence: joan.maso@uab.cat; Tel.: +34-93-581-1771

Received: 14 June 2019; Accepted: 8 July 2019; Published: 10 July 2019

Abstract: The amount of data that Sentinel fleet is generating over a territory such as Catalonia makes it virtually impossible to manually download and organize as files. The Open Data Cube (ODC) offers a solution for storing big data products in an efficient way with a modest hardware and avoiding cloud expenses. The approach will still be useful up to the next decade. Yet, ODC requires a level of expertise that most people who could benefit from the information do not have. This paper presents a web map browser that gives access to the data and goes beyond a simple visualization by combining the OGC WMS standard with modern web browser capabilities to incorporate time series analytics. This paper shows how we have applied this tool to analyze the spatial distribution of the availability of Sentinel 2 data over Catalonia and revealing differences in the number of useful scenes depending on the geographical area that ranges from one or two images per month to more than one image per week. The paper also demonstrates the usefulness of the same approach in giving access to remote sensing information to a set of protected areas around Europe participating in the H2020 ECOPotential project.

Keywords: Open Data Cube; Earth Observations; interoperability; visualization; Sentinel; Analysis Ready Data

---

1. Introduction

The territory of Catalonia has been analyzed by remote sensing from different thematic angles such as forest fire patterns and effects [1], land use and land cover change [2], agriculture statistics and abandonment [3,4], forest dynamics [5], or even air temperature [6]. These works required a considerable amount of time spent in data preparation and organization for requesting and downloading, as well as in correcting it geometrically and radiometrically [7]. To avoid the repetition of the Landsat imagery processing for each study, the Department of Environment of the Catalan Government and Centre de Recerca Ecològica i Aplicacions Forestals (CREAF) created the SatCat data portal that organizes the historical Landsat archive (from years 1972 to 2017) over Catalonia in a single portal and that provides visualization and download functionalities based on OGC Web Map Service (WMS) and Web Coverage Service (WCS) international standards [8]. Still, CREAF inverts a considerable amount of processing work on maintaining the portal up-to-date and to incorporate the increasing flow of the new Landsat and Sentinel 2 sensors. The two main reasons are the number of processing steps that the raw satellite data requires to make it useful and the difficulties on organizing a big series of imagery scenes in a way that are easy to manage.

To solve the first issue, United States Geological Survey (USGS) and European Space Agency (ESA) have made a considerable effort facilitating the access to optical satellite imagery, processed up to a level that is ready for immediate use on land studies. Recently, Analysis Ready Data (ARD) is distributed in a processing level that is geometrically rectified and free of the effects of the atmosphere, making it ideal for immediate use for vegetation and land use studies [9]. Since March 2018, ESA has been distributing the Sentinel 2A and 2B data at 2A processing level (Bottom of Atmosphere Reflectance), which can be directly downloaded from the Copernicus Open Access Hub (https://scihub.copernicus.eu/) also covering the Catalan area. The downloading process is not only available in the portal, but it is also facilitated by an API that makes the automation of the regular updates programmatically possible. At the time of writing this paper, USGS is only providing Landsat ARD for the USA territory (https://www.usgs.gov/land-resources/\T1\textcompwordmarknli/\T1\textcompwordmarklandsat/\T1\textcompwordmarkus-landsat-\T1\textcompwordmarkanalysis-ready-data) even if it is possible to request on-demand processing of the Landsat series (4-5 TM, 7 ETM and 8 OLI/TIRS) for other regions of the world, including Catalonia.

By applying successive technical improvements in their sensors, satellites are responding to the user demands in terms of spatial and temporal resolution. Obviously, this increment in resolution is accompanied by an increase in the number of files and/or in file sizes. In the past, it was possible to deal with file names and data volume, but is currently becoming increasingly impossible, even when the small region is needed to study. Data distribution is based on bulky scenes with a long list of file names difficult to classify and maintain, which must be downloaded and processed one by one with Geographic Information System (GIS) and Remote Sensing (RS) software. Creating long time series of a small area might take long processing time in repetitive clipping steps and format transformations. In this situation, extracting a time profile of a single position in the space requires visiting a subset of the long list of files. There is a need for having the remote sensing data organized in a way that it is possible to formulate spatio-temporal queries easily. Current distribution in the form of big scenes is not optimized to respond most of such queries, and the fact that some datasets are growing with a continuous flow of more recent scenes until the end of the operational life of the platform adds another level of complexity. One way of making temporal analysis practicable is to organize or index the data in a way that data extraction of a subset of a continuously growing dataset in four dimensions (two spatial, one temporal, and one radiometric) is fast.

The Earth Observations Data Cubes (EODC) is an emerging paradigm transforming how users interact with large spatio-temporal Earth Observations (EO) data [10]. They provide a better organization of data by indexing it, faster processing speeds by ingesting scenes as well as improved query languages to easily retrieve a convenient subset of data. The Open Data Cube (ODC) [11,12] is an implementation of the EODC concept that takes advantage of the existence of ARD to promote and generalize the creation of national data cubes. The Swiss Data Cube (SDC) is the second operational implementation, after Australia [13], and delivers nationwide information using remotely-sensed time-series Earth Observations [14]. As will be described later in detail, by permitting for a local installation of the software and the data, the ODC allows for full control of the data sources and the processes and algorithms applied to them. This is particularly convenient to guarantee total reproducibility of the results obtained when analyzing the data, but it will require an investment in hardware as well as an initial effort on populating the system with the right products. In contrast, solutions such as Google Earth Engine (GEE) provides a cloud environment were multi-petabyte archive of georeferenced datasets that includes Earth observation satellites and airborne sensors (e.g., USGS Landsat, NASA MODIS, USDA NASS CDL, and ESA Sentinel), weather and climate datasets, as well as digital elevation models can be immediately used [15]. Local or global analysis is possible by adopting GEE programming API that provides build in functions, or by programming lower level code. At the time these lines were written, GEE did not include Sentinel 2 Level 2A imagery [16] (the product being the focus of the presented work). GEE offers unprecedented computer capacities to the uses, removes the need for the user to design parallel code and takes care of the parallel processing automatically. This means that users can only express large computations by using GEE primitives and some operations that cannot be parallelized in GEE simply cannot be performed effectively in this environment. GEE also imposes some limits and other defenses that are necessary to ensure that users do not monopolize the system that can prevent some user applications [17].

The ODC provides a good starting point for data organization and data query. Still, using an installation of the ODC requires experience in python programming language and in a set of libraries that the ODC builds on top. The ODC offers a Python Application Programming Interface (API) to query and extract data indexed or ingested by the ODC. This library relays on the Python XArray library that is a data structure and a set of functions to deal with multidimensional arrays that are described with metadata and semantics. XArray internally uses NumPy that is a consolidated pure multidimensional array library. To manipulate the data in the ODC the user needs to be knowledgeable on the ODC API, the XArray and NumPy libraries. In addition, if the user wants to get a graphical representation of the data, extra knowledge on Python graphical libraries is also required. Some examples of practical applications where the ODC could be applied are detecting changes in landscape, tracking the evolution of agricultural fields, or the stability of protected areas. Unfortunately, the technical level of expertise for applications similar to the ones mentioned before relegates the use of the ODC to a subset of RS experts skilled in Python programming.

The first part of the paper demonstrates that the same approach used for national data cubes is also useful for smaller areas such as a province or a region of a country and shows how the ODC can be scaled down to the Catalonia region and further down to the level of an European protected area. This paper also demonstrates that the combination of a modest computer infrastructure with semiautomatic processes allows a single individual to manage a data cube. The second part of this paper proposes some additions to the ODC that make possible the exploration of time series of data for users that have no background in programming languages or scripting widening the potential audience. The solution is based on a combination of OGC WMS and WCS services with an integrated HTML5 web client that performs animations and temporal series statistics. The third part of the paper illustrates how a non-programmer can start exploring the data using the HTML5 enabled MiraMon Web Map Browser, and for example, compute the number of non-cloudy measurements for each pixel of the region, and finally discusses the scalability of the solution in space and time.

2. Scaling Down the Data Cube to a Level of a Province, Region or Protected Area

Catalonia is an autonomous community in North-East of Spain, which extends across 32,000 km^2. For about 10 years we have manually build the SatCat Landsat archive that exposes geometrically corrected imagery coming from all Landsat series over the complete Catalan territory spanning from the first Landsat 3 image in 1972 to the current Landsat 8 optical product. ESA was solely responsible for Landsat products in Europe for some time, while USGS is mainly responsible for operating and distributing Landsat imagery. Both organizations have provided data to the SatCat. On top of the archive, we setup a standard WMS service encoded in C and developed using libraries coming from the MiraMon software. To be able to retrieve the data fast, the MiraMon WMS server requires preparing it as internal tiles and at a diverse set of resolutions with an automatic interpolation and cutting tool. In addition, the service also adopts the OGC WCS 1.0 standard for subsetting and downloading. The result is an easy to explore dataset that once presented to the public became widely used both for visual exploration and downloading. To introduce a new scene in the time series, a manual repetitive workflow and protocol was established. The temporal and spatial resolution makes the update of series still manually manageable today. Catalonia is divided in 2 Landsat paths one for the east side (GironaBarcelona) and one for the west side (LleidaTarragona) that are kept separated in two layers to preserve the original scenes. Due to the differences in the number of bands and resolutions of the different instruments, each sensor type is also presented as a different layer. The system benefits from the TIME and extra dimensions concept in the WMS standard to define a product that has a time series composed by several original bands. Due to the intrinsic limitations of the WMS interface that provides pictorial representations of the imagery, the browser is not able to change the visualization of the data or to perform any kind of analysis. To facilitate the time series interpretation an animation mechanism was developed, consisting of requesting individual WMS time frames that are presented to the user as a continuous film. In the next section of the paper, we will present an uncommon way of using the WMS standard used in our new map browser that overcomes these limitations.

With the advent of Sentinel 2A and 2B, we have realized that it is no longer possible to process all incoming data manually with the same human resources. As part of the process of rethinking the workflow of building a Sentinel 2 data series to make it more automatic, we have considered the data cube approach as a way to better organize the data but also as a way to maintain the product up to date with the most recent scenes. We realized that the methodology described by the ODC to setup national data cubes can be used on a region of any size so we decided to build a Catalan Data Cube (CDC) starting by the new data flows coming from the ESA Copernicus program.

As for hardware, we were constrained to low budget alternatives. Instead of going for the last state of the art technology, we opted to use a dual Xeon main-board ASUS Z8NA-D6 with 2 Intel X5675 processors with a total of 12 cores (24 virtualized cores) with 48 GB DDR3 1333 Mhz RAM and four 8 TB disks for about 1500 EUR. After setting up a Windows server operating system, we proceeded to setup up the miniconda environment, the PostgreSQL database, the cubeenv software (the ODC), and the Spyder IDE. The process is fully documented in the ODC website[1].

We wanted to start by importing a Sentinel 2 Level 2A product. The process of downloading the data from the ESA Sentinel Data Hub can be made automatic by combining the wget tool with the instructions for batch scripting provided by ESA[2]. There are three steps in the process: First, a wget request to the ESA Sentinel Data Hub to find the resources to be downloaded from the product type S2MSI2A in an interval of time and space. As a result we get an atom file with a maximum of 100 hits. Then we needed to download them one by one with a short script (a Python code) to interpret the atom file (that is in XML format) and to create the list of the URLs corresponding to the granule of each hit and download each individual granule. Each granule is provided as a ZIP file following the SAFE format [18] that needs to be uncompressed.

The ODC works with a PostgreSQL database to store the metadata about the elements in the data cube but it will keep the image data as separated in the hard drive, only retaining a link to them. In order to import resources in the ODC, specific YAML files containing metadata about the resources are necessary to populate the database. The ODC GitHub provides abundant information and examples on how to create the necessary YAML files and how to index imagery in the data cube for the most common products. Two types of YAML files need to be created: A product description for the S2MSI2A product as a whole (mainly specifying details about the sensor and the bands acquired by it) and files that describe the peculiarities of each granule (mainly in terms of spatial and temporal extent). ESA has only recently distributed the S2MSI2A product (since March 2018) so no example of the description of it has been found. We created it ourselves following a YAML file for scenes generated ad-hoc with the Sen2Cor software that produces a very similar SAFE file structure. We also used a Python script that transforms a SAFE folder result from a Sen2Cor execution into a YAML file describing the granule peculiarities that we could adapt to the product generated by ESA. Once we had these YAML files, we were able to index the granules into the data cube. By indexing this kind of data the ODC still relies on the original JPEG2000 formats. JPEG2000 format is designed for high compression and fast extraction of a subset but it still requires some time to decompress each scene, making it not appropriate for queries requiring big time series that results in too slow responses for the ODC API. Fortunately the ODC offers a solution called ingestion. The ingestion process creates another product in the data cube that automatically converts the JPEG2000 files into a tile structure composed by NetCDF files in a desired projection. The ingestion process offers the possibility to mosaic all granules of the same day in a single time slice reducing fragmentation of the Sentinel 2 granules.

---

[1]   https://datacube-core.readthedocs.io/en/latest/ops/conda.html
[2]   https://scihub.copernicus.eu/userguide/BatchScripting

In the current implementation of this functionality (called 'solar day'), the process does not take any consideration on which is the best pixel and covers one image with the next, which is something that could be improved in future versions of the ODC software. The ingestion process can be repeated to add new dates in a preexisting time series as soon as they become available and integrated in the data cube into a fully automatic batch process.

Once this process is completed, the data is perfectly organized in the data cube and can be accessed in an agile manner. The cubeenv Python API allows for writing Python scripts that extract a part of a time series as tiles and represent them on screen by using matplotlib. In practice, the data cube offers a spatio-temporal query mechanism that does not require any knowledge about file names, folder structure, or file format. The result is an array of data structures, one for each available time slice. The number of slices in the time series and the dates associated with them is determined by the API. The Python API is a very useful feature to test the ingested data and ensure that all data is correctly available. Some examples on how to do so are provided as Jupyter notebooks in the ODC website. Using the Python library, data can be exploited by analytical algorithms generating new products, which can eventually be reintroduced in the data cube. That might produce great results but the need for Python skills makes it too complicated for most of the GIS community or the Catalan administration and inaccessible to the general public.

To import the Sentinel 2 product for Catalonia in the CDC for the available period from March 2018 to March 2019, involved the download of 1562 granules (204,386 files) resulting in a volume of 1.18 TB (1.301.116.256.901 bytes) of the original scenes in JPEG2000 format that required a downloading and indexing time of about 258 h. Once ingested, it resulted in 132 scenes in a volume of 816 GB (876.598.015.657 bytes) and distributed in 6093 NetCDF files requiring an ingestion time of about 23.4 h.

3. Adding Easy and Interoperable Visualization to the Data Cube

Another Python script was prepared that retrieves the data cube tiles and convert them into the MiraMon server format at a variety of resolutions. The script determines the number of time slices that the data cube can provide at that point in time and it is able to detect if a certain time slice was already prepared in a previous iteration or needs to be prepared now. For each time slice, if the preparation is successful, the script is also able to edit the server configuration file (that is an INI file) to add it to the WMS server and the web map client configuration file (that is a JSON file) to add a time step making the whole update process fully automatic.

Actually, the ODC provides support for WMS services by using the https://github.com/opendatacube/datacube-ows code. Nevertheless, in this development we wanted to take advantage of a new functionality that we have recently developed in the MiraMon server and map browser for the ECOPotential project: MiraMon implementation of the WMS OGC Web Services serves imagery in JPEG or PNG for standard WMS clients as well as in raw format where the arrays of numeric values (original data) of the cells is transmitted directly to the client (we call it IMG format). The use of a raw numerical format opens a myriad of new possibilities in the client side. Now the map browser JavaScript code is able to request IMG format asynchronously (with AJAX) and get the return as a JavaScript binary array (new characteristic in HTML5). HTML5 also adds support for a graphics library that works on an area of the screen that is called canvas. In the canvas, we can get access to each pixel of the screen area and modify it. Binary arrays are dynamically converted into arrays of RGBA values that will then covering the whole canvas area. The conversion can be as complex as we want, ranging from applying a grey scale color map, creating an RGB combination from 3 bands (3 WMS binary array requests), and up to a complex pixel-to-pixel operation involving several bands and thus many WMS binary array requests [19]. Our CDC WMS client can be accessed at http://datacube.uab.cat/cdc/ (see Figure 1). The description of most of the functionalities implemented in the map browser is out of the scope of this paper that will only focus on the time series analysis.

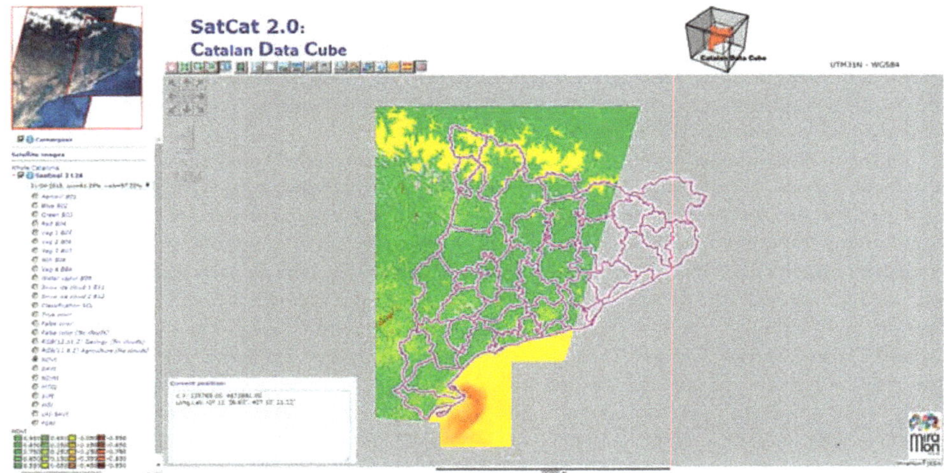

Figure 1. The Catalan Data Cube WMS browser.

4. Exploratory Analysis of Time Series on the Catalonia Data Cube

In implementing the map browser of the Catalan Data Cube we have gone even further in the use of binary arrays by exploring the potential of using an array of arrays in supporting time series visualization and analytics. A time series is a multidimensional binary array in JavaScript allowing for the extraction of data in any direction of the x,y,t space including spatial slices, but also temporal profiles and 2D images where one dimension is time.

4.1. Summarizing Scene Area and Cloud Coverage of Each Scene

The first thing we did was to generate a list of temporal scenes that could be more meaningful for the user. While downloading data from the Sentinel hub, we are only accepting granules that have less than 80% of cloud coverage. If we combine this with the fact that Catalonia is only covered partially with a single granule and that both Sentinel 2A and 2B are integrated in a single product; it is really difficult to anticipate the real coverage of each daily time slice, if the user is not proficient in the distribution of the satellite paths. For this reason, the list of scene names is a composition of the date, the percentage of the area or interest (a polygon that includes the Catalan territory) covered by scenes with less clouds than CDC limit (80% or less) and the percentage of area covered by scenes that shows the ground (not covered by clouds). In this way, we can differentiate a scene that only covers a small fraction of Catalonia, but with the available part mostly "visible" (not covered by clouds), so it can still be useful to study the small areas as a continuous extent (see Figure 2).

4.2. Animation of the Temporal Evolution and Temporal Profiles

The main window of the map browser allows showing one scene at a time (see Figure 1). To perform time series analysis, we have included a new window (that can be opened by pressing the video icon:) where the set of time slices of the area defined in the main window will be requested, visualized, and analyzed. For each time slice, the components are downloaded, saved in memory and rendered in a HTML independent division (HTML DIV) that contains a canvas. Creating the animation effect is as simple as hiding all divisions and making visible one of them in sequence (see Figure 3e). Modern browsers are fast in doing this kind of hiding operations allowing for speeds higher than 10 frames per second, thus resulting in a quite convincing video effect.

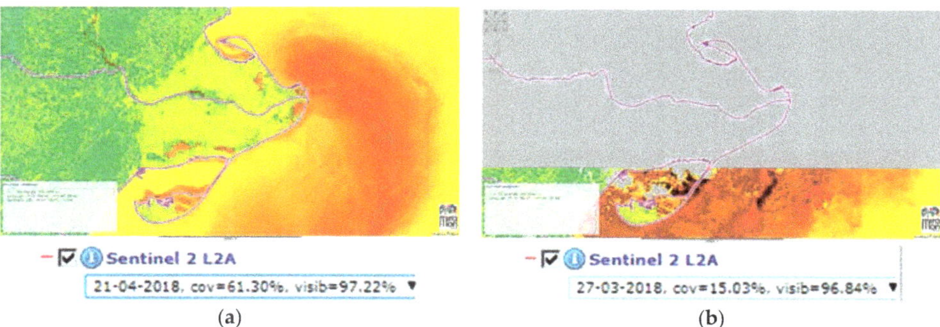

Figure 2. Two scenes offered in the Catalan Data Cube map browser and how they are shown in the table of contents (legend): (**a**) scene offers less coverage than (**b**) scene (as can be seen in "cov" percentages), but both scenes have high visibility ("vis" percentage, i.e., the ground is not obscured by clouds) in the covered areas.

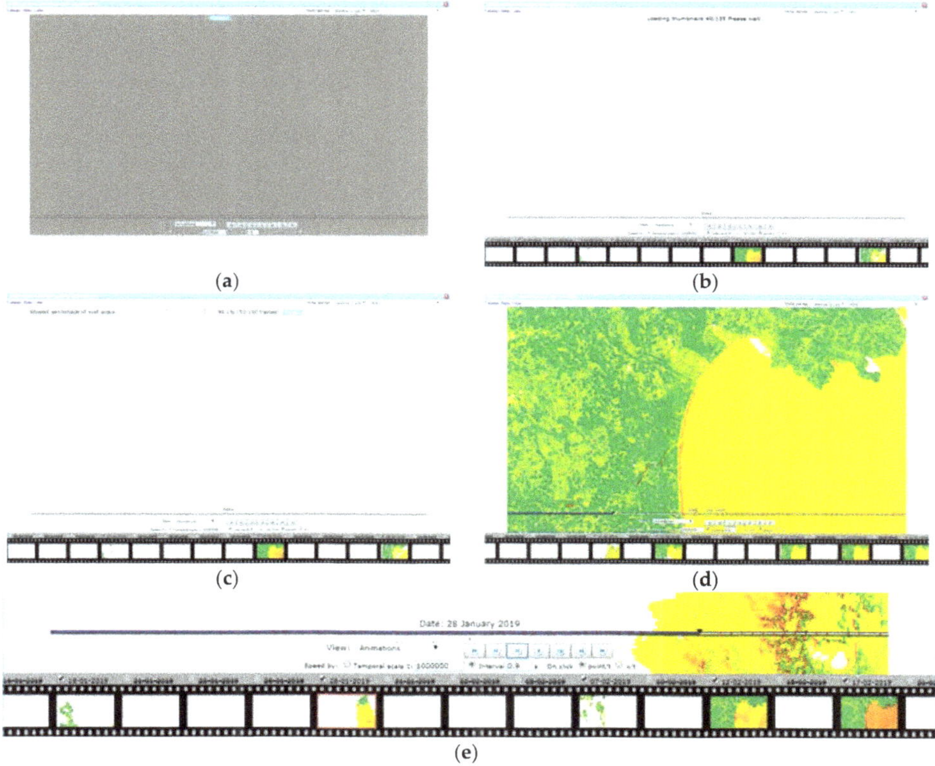

Figure 3. Time series visualization of the Catalan Data Cube WMS browser: (**a**) layer selection and thumbnails download start when pressing the *Load* button; (**b**) thumbnails download process; (**c**) selection of the full resolution images to download using the slider of percentage of void space; (**d**) resulting full resolution animation; and (**e**) detail of the temporal controls.

In the case of the CDC, downloading and saving the time series binary arrays is a stress test for both the web browser and the map server. Indeed, a time series from a year will involve about 140 scenes that will result in up to 140 WMS GetMap requests of individual time slices, if we are visualizing a band, or 280 WMS GetMap request if we visualize a dynamically generated NDVI. In a state of the art screen the map requires 1500×800 pixels or more; and considering that each pixel is represented by 2 bytes (short integer binary array), this will require transmitting a total of 310 MB of uncompressed information for a single band and 620 MB for an NDVI. The situation is partially mitigated by the use of RLE compression during the transmission, but still a considerable amount of data is transmitted from the server to the client. Another factor introduced to mitigate the situation was to consider that, due to the number of scenes having a partial coverage, requesting a time series of a local area will necessarily result in some scenes completely blank. The map browser turns this into an advantage by initially requesting only small quick looks of the scenes (see Figure 3a,b). This is going to increase the user experience by representing a pre-visualization of the time series in the form of a film, and it is also useful for calculating an estimation of the percentage of nodata values of the area in a certain time-slice. If the quick look is containing nodata values only, it is automatically discarded. Moreover, by operating a slider, the user can then decide to download only the scenes that have a minimum percentage of information in them, reducing the need for requesting unnecessary bad images at full screen resolution (see Figure 3c,d).

After the user has decided about the accepted percentage of coverage (and presses the *Load* button), all the WMS GetMap requests at screen resolution start. Once the images have been downloaded and the multidimensional binary array has been completed, an animation begins showing the sequence of scenes (as the "Animation" option, by default, is selected in the dropdown list on the temporal analysis control, as shown in Figure 3e).

By default, the *On click* option is set to point/t and thus when the user clicks on any part of the image, a temporal profile diagram of this point is shown. The profile of the point is presented with the time evolution of the spatial mean value of the pixels from each scene (grey solid line), as well as the mean ± standard deviation as a reference (grey dashed lines). The user can continue clicking in the screen to add temporal profiles of other pixels to the graphic (see Figure 4). These profiles can be copied to the clipboard as tab separated values that can be later pasted in a spreadsheet for further analysis.

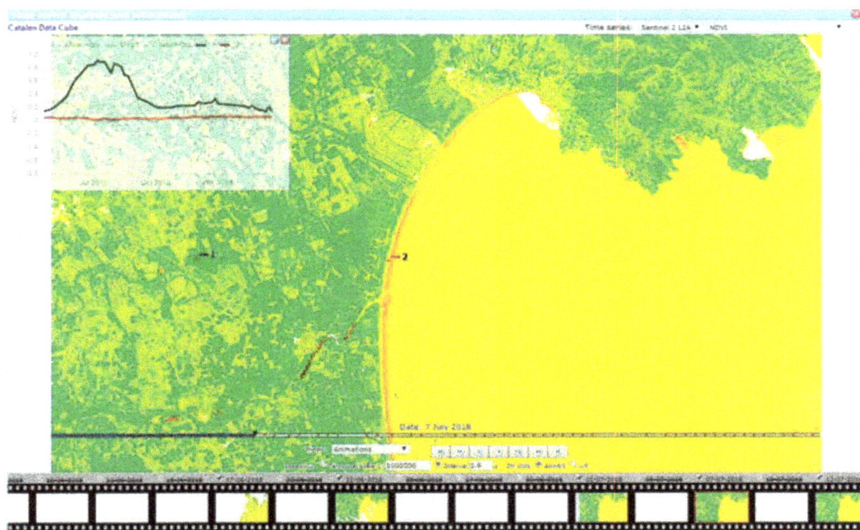

Figure 4. Dynamic NDVI layer animation in the Catalan Data Cube including a temporal profile for a crop (showing phenological dynamics, in black) and sand (almost a constant signal, in red) points (centered in Roses Gulf area).

4.3. Temporal Statistics

Having a multidimensional binary array in memory opens the door to expand the temporal analysis possibilities. To demonstrate some of them, we have implemented initial statistics, available through the dropdown list *View* in the temporal analysis control area. For quantitative values, we can calculate the mean, the mode and the standard deviation of each pixel of the set of scenes resulting in a new image (see Figure 5a). For categorical values, it is possible to calculate an image that represents the number of scenes, which contain a particular class or the modal value of the time series (see Figure 5b).

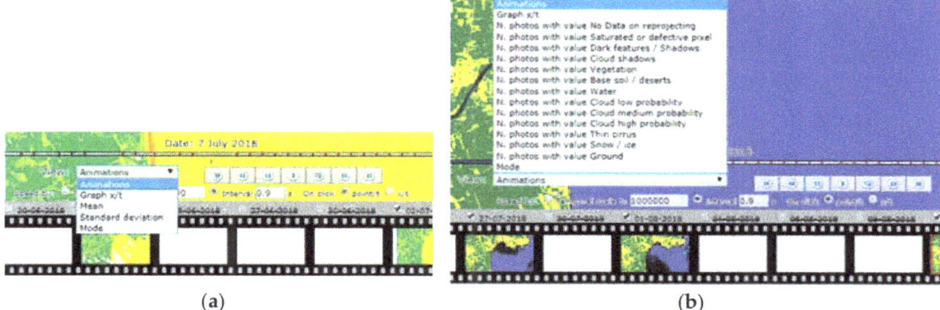

(a)   (b)

Figure 5. Options of temporal statistics of the Catalan Data Cube WMS browser: (**a**) for quantitative values (such as NDVI); (**b**) for categorical values (such as Scene Classification map, SCL).

As an example, mean and standard deviation for NDVI are shown in Figure 6. The mean NDVI is useful to see the overall behaviour of a certain area, and thus being able to identify some covers. Moreover, an overall image representing the amount of variation of a variable can be created by applying the standard deviation expression to each pixel of the time series. The result is an image that has low values in pixels that remain constant over time and high values in pixels that have more variability over time. In an NDVI time series the standard deviation is represented in a grey scale, where white colours represent agricultural fields that have big dynamics in NDVI, grey values are stable forest areas, and dark values are invariant human built environments (such as cities, roads, and paths) that have a much more constant NDVI value over time.

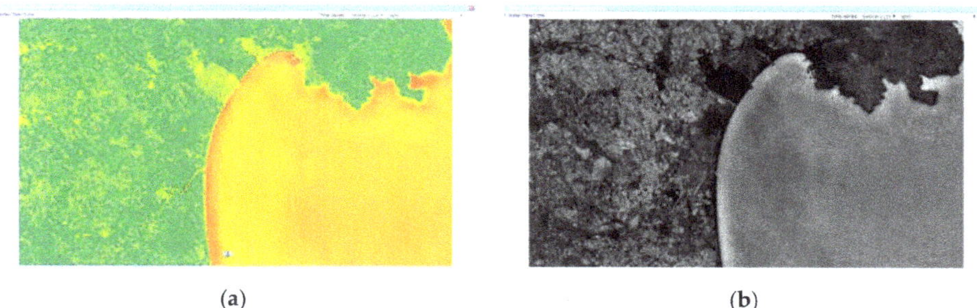

(a)   (b)

Figure 6. Temporal statistics of the Catalan Data Cube WMS browser for the NDVI variable along the first year of Sentinel 2 acquisitions (27/03/2018 to 24/03/2019): (**a**) mean; (**b**) standard deviation.

Another interesting addition is the possibility of generating an x/t graph. On this view, a 2D image is formed by selecting a horizontal line in the animation (Y coordinate that will remain constant) and representing an image where each row is the selected Y line in one scene. In the resulting 2D graph, time progresses down and "vertical changes" in color reflect changes over time (see Figure 7). White patches are nodata values present in some of the images in the selected horizontal line. In the future, other time series derived statistics will be incorporated.

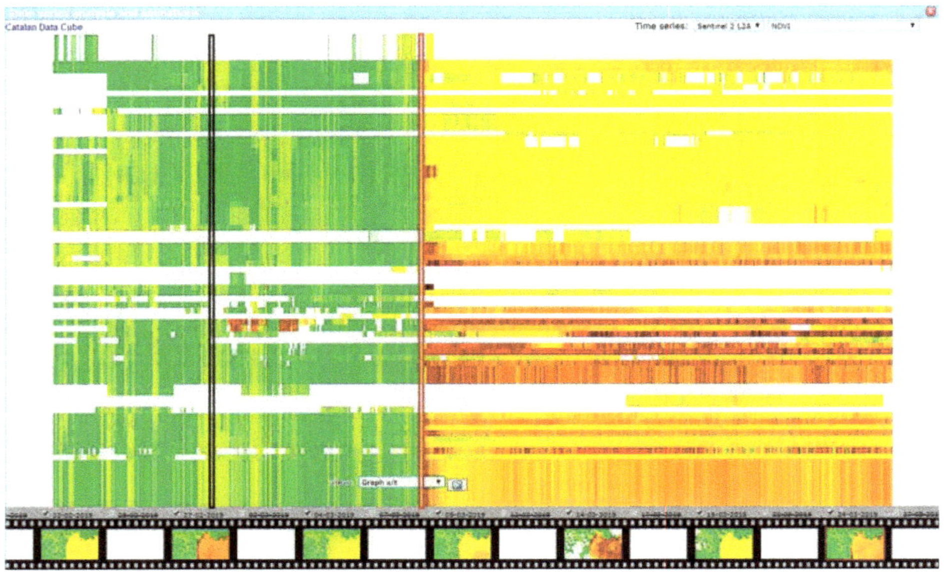

Figure 7. x/t graphic in the Catalan Data Cube. Black and red rectangles show the temporal profile (in vertical starting from above with the first date and ending below with the last date) for the same crop (black) and sand (red) points used in Figure 4.

4.4. Coverage Evolution in Catalonia along First Year of Sentinel-2 Acquisitions

Sentinel 2 level 2A product generated and distributed by ESA incorporates a categorical band (Scene Classification map, SCL) that classifies the pixels of each image in 12 classes (using L2A_SceneClass algorithm[3]). This is particularly useful to estimate if a pixel actually represents the ground or is not very meaningful because it is a missing value, a saturated pixel, or it is covered by clouds, shadows, etc. In the map browser, it is possible to generate a dynamic band that reclassifies all categories that represent a valid value for "ground" (i.e., vegetation, non-vegetated—i.e., bare ground—, water and snow or ice) as a single category. This virtual band will be dynamically computed for all time scenes when needed. Then, in the temporal analysis window, we will be able to see the temporal evolution of this category as well as to generate an image representing the number of scenes that contain the "ground" category (see Figure 5b). Due to the swath of the Sentinel 2A and B, many scenes partially cover the Catalan territory, a situation that is modulated by the cloud presence, which is more frequent in some regions. The image gives us the actual level of Sentinel 2 imagery recurrence for each part of Catalonia (see Figure 6).

The result visually reflects that the east part of Catalonia and the North-West part only had about 20 samples in the first year of acquisitions (less than one sample every 15 days). The South-West part presented an average of 40 samples. The central part of Catalonia had more fortune with at least 60

---

3 Extracted from https://sentinel.esa.int/documents/247904/685211/\T1\textcompwordmarkS2+L2A+Product+\T1\textcompwordmarkDefinition+\T1\textcompwordmarkDocument/\T1\textcompwordmark2c0f6d5f-60b5-48de-bc0d-e0f45ca06304.

samples, but it is the lower left part of the center area the one that received the maximum number of samples with up to 86 samples (about a sample every 5 days). This is an interesting result that suggests that future efforts in temporal analysis of Sentinel 2 data over Catalonia should focus in the lower left part of the central area, where the temporal resolution is higher and we should expect a better accuracy. This is particularly important for phenological studies that are looking for changes in the vegetation status that can suddenly happen in only 2–3 days (for example the start of the growing season). Sample availability is highly correlated with the common area between orbits that dramatically decreases the revisiting period in certain areas according to its distribution. This is the main reason for the high values in the central part of the image in Figure 8, as this area is the overlapping between the swaths of the 051 and 008 orbits (see Figure 9a,b). Moreover, the sample availability is also related to cloud coverage, thus we can assume some relation to annual rainfall. Figure 9c shows a map of the mean annual rainfall in Catalonia where we can see a certain spatial correlation with the image availability: In certain areas, it is lower (marked by a blue polygon) than in others (marked by an orange rectangle) (see Figure 9a) as they correspond to areas with higher rainfall (in blue in Figure 9c) and lower rainfall (in reddish in Figure 9c), respectively.

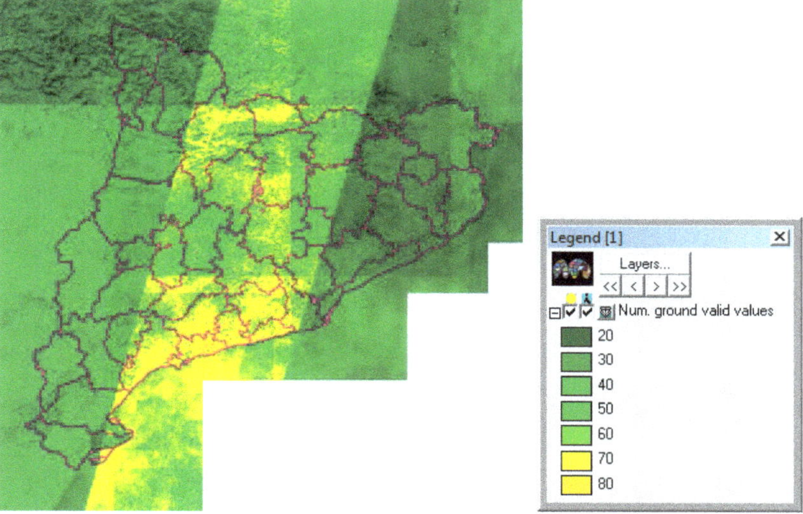

Figure 8. Number of scenes with visible ground for each pixel over Catalonia along first year of Sentinel-2 acquisitions (27/03/2018 to 24/03/2019) in the Catalan Data Cube.

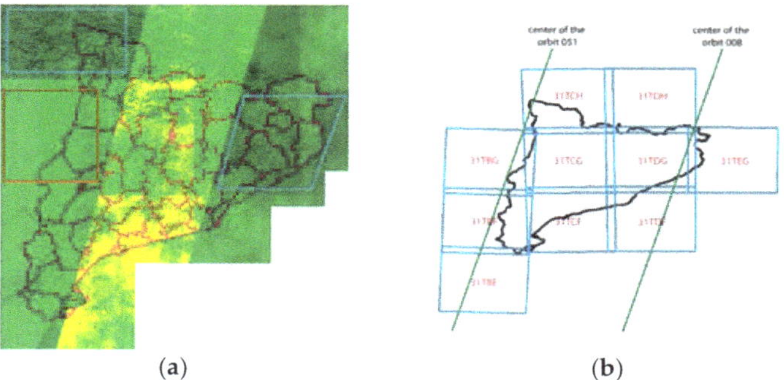

(a)   (b)

Figure 9. *Cont.*

(c)

Figure 9. Image availability over Catalonia: (**a**) number of scenes with "Ground" class; (**b**) orbits distribution over Catalonia; and (**c**) mean annual rainfall[4].

5. Discussion & Perspectives

5.1. Can We Apply the Methodology to Other Regions?

As we have explained before, the approach described in this paper was designed with the aim to make the Sentinel 2 imagery collection automatic for Catalonia and to build the CDC. Nevertheless, the same methodology is *a priori* applicable to any other region. To demonstrate this, we have deployed another data cube for the protected areas in the H2020 ECOPotential project (www.ecopotential-project. eu). In ECOPotential, we provide 25 protected areas with the available remote sensing data and derived products that can be used by protected area managers in their maintenance tasks and their decision making processes. To make data exploration easy, we also used an integrated map browser (maps.ecopotential-project.eu) that allows selecting the protected area, and then browse to the data prepared by the different ECOPotential partners. We called it "Protected Areas from Space".

In ECOPotential we have also experienced the difficulties in retrieving, storing and sorting the relevant information. Due to the heterogeneity of the sources and products, each addition needed to be understood in terms of format and data model, analyzed for the best representation in terms of colors and legends and integrated in the browser. Variety was the most challenging aspect of this big data problem. In the end, this resulted in the largest map browser our organization has ever prepared, with a total of 277 different layers distributed among the 25 protected areas (each layer can present more than one variable and, in many cases, several time frames). However, when trying to incorporate Sentinel 2 we faced the problem of *volume* of information due to the number of scenes and granules that needed to be processed. It was clear that the manual methodology applied for the rest of the products were completely impractical in this case.

After developing the CDC, we revisited the ECOPotential map browser and we adapted the methodology applied in the CDC to the extent of the protected areas. The result was another data cube that contains the data for approximately 2/3 of the protected areas (only the marine and bigger areas where left out) covering a total of 15 of the 25 protected areas. Figure 10 shows the distribution of the protected areas and the ones included in the data cube. We decided to index all granules for all protected areas in the data cube as a single product. Due to the extension of the area, the ingestion process needed to be separated in different products, one for each UTM fuses (see Table 1). When the WMS service is created, each protected area has its own layer name. In numbers this resulted in one year of Sentinel 2 Level 2A data; 2553 granules consisting in 199360 jp2 files requiring 4.16 TB and

---

4   Source: Servei Meteorològic de Catalunya: http://www.meteo.cat/climatologia/atles_climatic/

resulting in 1693 daily scenes (considering all protected areas). The Figure 11 shows a few examples of the use of Sentinel 2 images from the ECOPotencial Data Cube in the Protected Areas from Space map browser.

Table 1. UTM fuses covering all 15 selected protected areas in ECOPotential.

Ingestion CRS	Protected Area
UTM 28N	La Palma
UTM 29N	Peneda Gerês, Doñana
UTM 30N	Sierra Nevada
UTM 31N	Camargue
UTM 32N	Gran Paradiso, Swiss National Park
UTM 33N	Abisko, Bayerischer Wald, Murgia Alta Park, Northern Limestone (KalKalpen)
UTM 34N	High Tatra, Ohrid Prespa, Samaria
UTM 36N	Har Ha Negev

After demonstrating that the approach could be easily generalized to other regions of the world we decided to publish the Python scripts that we used in a GitHub repository (https://github.com/joanma747/CatalanDataCube) acknowledging that they are actually deeply inspired and based on previous routines exposed by the ODC developer team.

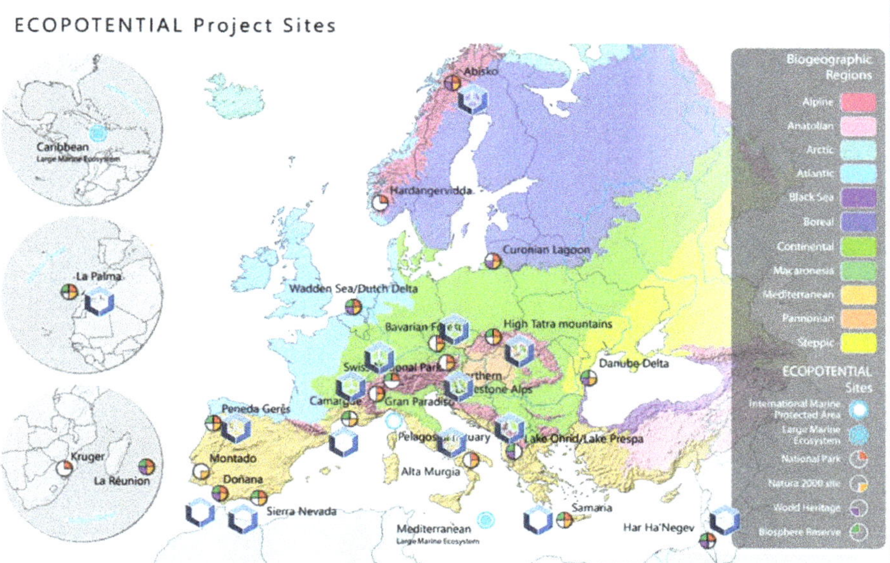

Figure 10. Protected areas in the ECOPotential project. The protected areas selected for the ECOPotential data cube are represented with the ODC logo next to them.

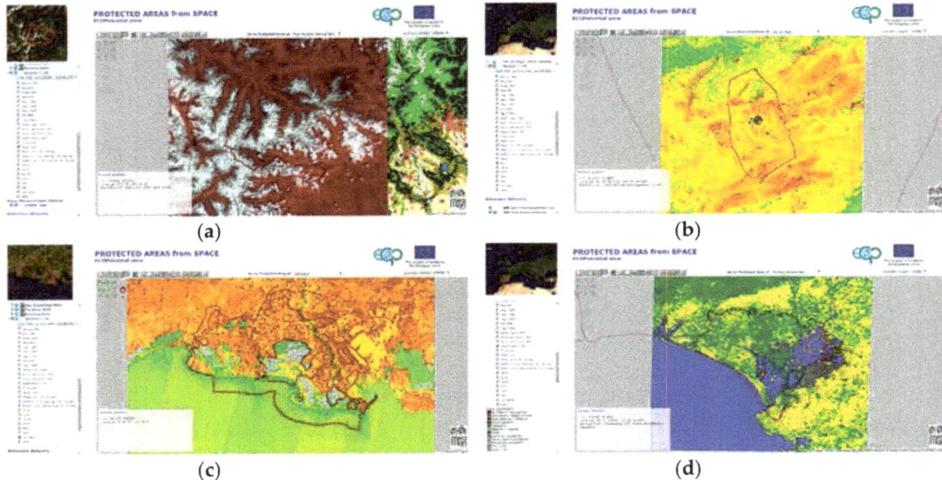

Figure 11. Use of Sentinel 2 images from the ECOPotential Data Cube in the Protected Areas from Space map browser: (**a**) false color RGB combination over the mountain ecosystem of Gran Paradiso National Park in Italy (in red, highest values of vegetated areas); (**b**) Soil Adjusted Vegetation Index (SAVI) dynamic calculation over the arid ecosystem of Har Ha Negev National Park in Israel (in green, highest values of SAVI; in brown-orange, lowest values); (**c**) Leaf Area Index (LAI) dynamic calculation over the coastal ecosystem of the Camargue National Park in France (in green highest values of LAI; in brown-orange, lowest values); and (**d**) Scene Classification map (SCL) provided by ESA over the coastal ecosystem of Doñana National Park.

5.2. Does the Presented Open Data Cube Approach Scale up?

The data cube approach adopted by the CDC, represents an important step towards in the simplification of the data preparation and data access of Earth Observation using inexpensive hardware. However, for the approach to be agile it requires to use local storage in the form of random access storages: Hard drives. Mass market hard drives are limited to 12 TB and dual Intel Xeon main boards support only 6 SATA drives. This type of hardware limits the capacity to a maximum of 72 TB. The use of extension boards adds a maximum of 12 SAS drives (depending on the hardware) thus extending the storage capacity to 144 + 72 = 216 TB. Assuming (a) a conservative scenario where the storage capacity is not going to increase significantly in the next years, (b) the amount of Sentinel 2 data will remain constant in the 2 TB used by the indexed and ingested data from March 2018 to March 2019 presented before (c) we might store up to five similar products (including other platforms such as Landsat and other Sentinel as well as some derived products) and (d) an annual increase of the data availability that will grow by a factor of about 1.7 every year (as shown by Reference [20] for the last 4 years), this approach can be supported by a single similar computer for about 10 years for a region of the same size than Catalonia and in a similar latitude. For a territory 10 times bigger, the approach can only be applied for 5 years.

Of course, there are other options for digital storage but they will require a much bigger investment. This might sound discouraging but it actually means that the ODC is the right solution for today data volume and it is enough for the big data research questions that we are facing now. In the future, other solutions will be needed, such as the possibility to move the whole data cube to the cloud.

The approach to the time series analytics in the client side presented here also has its limitations. In ten years time, a time series of about 130 time frames will become 1300 frames requiring 10 times more memory space for storing the binary arrays. Fortunately, in the 64 bit systems used today, there is almost no logical direction memory limit and the amount of physical memory is regularly increasing so we do not expect to reach any ceiling in the next decade. Note that the quantity of memory needed to handle a time series in the client does not depend on the size of the territory studied or on the scale used, because the client requests only an invariable number of screen pixels to the server.

Another factor to take into account is the amount of time required to transmit a time series from the server to the client. Currently, it takes about 3 m to complete the transference. Assuming that generation and transmission times remain constant, a 10 year long time series might require an unacceptable amount of time. Time series is now being requested as a sequence of independent WMS GetMap requests, one for each time frame. The same WMS standard can be used to request a multidimensional data cube specifying a time interval. By allowing the server to generate the complete multidimensional data, a new more compact multidimensional data structure might be applied that might take advantage of the redundancy in the time series resulting in a considerable decrease in the transmission time.

For the determination of statistical images, an extrapolation of what is happening today with a single year will result in unacceptable computational times of several minutes. The solution for this is a combination of better memory structures with a code optimization. JavaScript code is interpreted by the browser during the runtime, thus becoming very sensible to inefficiencies in certain costly operations that once detected are easily avoidable. The suggested memory structures should favor the easy extraction of the time series for a single point in space keeping, the extraction of a single time frame reasonable efficient.

5.3. Can the Screen Based Analytics Be Translated into a Full Resolution Analytics?

The main difference between the original SatCat and the new web portal is the ability of the latter to work with the actual values derived from the satellite measurements instead of being limited to show pictorial representations. This allows for some data analytics and pixel-to-pixel and local neighbour processing in the client side. The processing is done at screen resolution and happens each time that the user pans or zooms, creating the illusion of a pre existing result. Some analytics (such as filtering) is propagated to the entire time series giving the impression that the process is spread to the time dimension. The browser is able to save the status of the session and when the same page is loaded again, the illusion continues. However, the user might examine the results and become convinced that the tested processing is satisfactory and that he wants to execute it at full resolution and save it as a new product. We are considering adding this functionality to the CDC and we are evaluating three options. The first option consists of adding a piece of software that transforms the JavaScript operations into Python routines, which the ODC could execute directly in the server side. This solution has the advantage that the new results could be immediately exposed as new products in the data cube, but has the disadvantage that the hardware of the CDC needs to execute the processing. We have already described that the whole idea was to demonstrate that a modest hardware could be used for the data cube so this solution will not scale up to many users. Another approach could be to implement a OGC WCS access into the data cube. In this case, users could use the WCS GetCoverage operation to download the information and execute the processing in another facility or in their own computers. This approach might require too much time and bandwidth to complete the transfer. A final solution could be to implement a WCPS in the CDC. WCPS [21] provides a collection of operations and a query language to remotely execute analytical processes. The WCPS standard is agnostic on where the processing is done. The WCPS implementation (such as Rasdaman) could use the data coming from the CDC and redirect and distribute the processing among the available processing facilities providing a more scalable solution.

6. Conclusions

The ODC has been advertised as a platform allowing nations to organize and process remote sensing products provided by the main Earth Observation satellites. The amount of data that some of these products generate requires automatic solutions for simplifying the data download and use. This paper demonstrates that the approach is equally useful for a sub-national region such as Catalonia or for smaller natural regions that require information for their management, such as the protected area network of the ECOPotential project. Even if ODC helps in organizing the data, it still requires knowledge on the ODC API and Python programming, limiting the accessibility of the data to expert users. The paper proposes the addition of another layer of software consisting of a web map browser that combines the interoperability of the state of the art OGC WMS standard with the new possibilities offered by the HTML5 to present data in a way that everybody can explore and understand without any need of programming skills. In addition, it demonstrates a promising strategy based on multidimensional binary arrays allowing for some time series analytics that the current web browser can execute and that will be extended in the future. Currently, animations, temporal profiles, x/t images, and images representing mean values and variations can be generated by the user in the map browser without any server intervention.

The paper proposes the use of modest off-the-shelf computer hardware and concludes that the current approach can reach some limits in 10 years time but still can offer a solution to analyze the state and the evolution of the planet valid up to the next decade. The paper also analyzes the real availability of Sentinel 2 data over Catalonia showing huge variations in the area due to the distribution of the Sentinel 2 paths and swath overlaps, but also the effect of clouds, concluding that, in 2018, some South and central regions of Catalonia get more than one useful image per week, while in others situated in the Pyrenees and Eastern areas, only an average of one or two images per month were useful.

This work is inspired by the original SatCat Landsat service for the Catalonia region (with coverage starting in the 1972) and extends and complements it with a new useful service that is maintained automatically, incorporates Sentinel 2 data, and provides time series analysis.

Author Contributions: Conceptualization, J.M.; Formal analysis, A.Z.; Methodology, J.M.; Supervision, X.P.; Visualization, I.S.; Writing—original draft, J.M. and A.Z.; Writing—review & editing, I.S.

Funding: This work has been partially funded by the Spanish MCIU Ministry through the NEWFORLAND research project (RTI2018-099397-B-C21/C22 MCIU/AEI/FEDER, UE). The projects ERA-Planet – GEOESSENTIAL and ECOPotential, leading to this application, have received funding from the European Union's Horizon 2020 research and innovation programme under grant agreements No. 689443 and 641762 respectively. Xavier Pons is recipient of an ICREA Academia Excellence in Research Grant (2016–2020).

References

1. Díaz-Delgado, R.; Pons, X. Spatial patterns of forest fires in Catalonia (NE of Spain) along the period 1975–1995: Analysis of vegetation recovery after fire. *For. Ecol. Manag.* **2001**, *147*, 67–74. [CrossRef]
2. Poyatos, R.; Latron, J.; Llorens, P. Land use and land cover change after agricultural abandonment: The case of a Mediterranean mountain area (Catalan Pre-Pyrenees). *Mt. Res. Dev.* **2003**, *23*, 362–368. [CrossRef]
3. Gonzalez-Alonso, F.; Cuevas, J.M.; Arbiol, R.; Baulies, X. Remote sensing and agricultural statistics: Crop area estimation in north-eastern Spain through diachronic Landsat TM and ground sample data. *Int. J. Remote Sens.* **1997**, *18*, 467–470. [CrossRef]
4. Juan José, V.-M.; Ninyerola, M.; Zabala, A.; Domingo-Marimon, C.; Gonzalez-Guerrero, O.; Pons, X. Environmental and socioeconomic factors of abandonment of rainfed and irrigated crops in northeast Spain. *Appl. Geogr.* **2018**, *90*, 155–174. [CrossRef]
5. Vidal-Macua, J.J.; Ninyerola, M.; Zabala, A.; Domingo-Marimon, C.; Pons, X. Factors affecting forest dynamics in the Iberian Peninsula from 1987 to 2012. The role of topography and drought. *For. Ecol. Manag.* **2017**, *406*, 290–306. [CrossRef]
6. Cristóbal, J.; Ninyerola, M.; Pons, X. Modeling air temperature through a combination of remote sensing and GIS data. *J. Geophys. Res. Atmos.* **2008**, *113*. [CrossRef]

7. Pons, X.; Pesquer, L.; Cristóbal, J.; González-Guerrero, O. Automatic and improved radiometric correction of Landsat imagery using reference values from MODIS surface reflectance images. *Int. J. Appl. Earth Obs. Geoinf.* **2014**, *33*, 243–254. [CrossRef]

8. Masó, J.; Pons, X. Adding Functionalities to WMS-WCS Clients: Animation and Download. In Proceedings of the XXII International Cartographic Conference: Mapping Approaches into a Changing World, A Coruña, Spain, 9–16 July 2005; ISBN 0-958-46093-0.

9. Egorov, A.V.; Roy, D.P.; Zhang, H.K.; Hansen, M.C.; Kommareddy, A. Demonstration of percent tree cover classification using Landsat analysis ready data (ARD) and sensitivity analysis with respect to Landsat ARD processing level. *Remote Sens.* **2018**, *10*, 209. [CrossRef]

10. Baumann, P.; Mazzetti, P.; Ungar, J.; Barbera, R.; Barboni, D.; Beccati, A.; Bigagli, L.; Boldrini, E.; Bruno, R.; Calanducci, A.; et al. Big Data Analytics for Earth Sciences: The EarthServer Approach. *Int. J. Digit. Earth* **2016**, *9*, 3–29. [CrossRef]

11. Killough, B. Overview of the Open Data Cube Initiative. In Proceedings of the IGARSS 2018—2018 IEEE International Geoscience and Remote Sensing Symposium, Valencia, Spain, 22–27 July 2018; pp. 8629–8632. [CrossRef]

12. Strobl, P.; Baumann, P.; Lewis, A.; Szantoi, Z.; Killough, B.; Purss, M.B.J.; Craglia, M.; Nativi, S.; Held, A.; Dhu, T. The Six Faces of the Data Cube. In Proceedings of the 2017 conference on Big Data from Space, Toulouse, France, 28–30 November 2017; pp. 32–35.

13. Lewis, A.; Oliver, S.; Lymburner, L.; Evans, B.; Wyborn, L.; Mueller, N.; Raevksi, G.; Hooke, J.; Woodcock, R.; Sixsmith, J.; et al. The Australian Geoscience Data Cube—Foundations and Lessons Learned. *Remote Sens. Environ.* **2017**, *202*, 276–292. [CrossRef]

14. Giuliani, G.; Chatenoux, B.; De Bono, A.; Rodila, D.; Richard, J.-P.; Allenbach, K.; Dao, H.; Peduzzi, P. Building an Earth Observations Data Cube: Lessons Learned from the Swiss Data Cube (SDC) on Generating Analysis Ready Data (ARD). *Big Earth Data* **2017**, *1*, 100–117. [CrossRef]

15. Teluguntla, P.; Thenkabail, P.; Oliphant, A.; Xiong, J.; Gumma, M.K.; Congalton, R.G.; Yadav, K.; Huete, A. A 30-m landsat-derived cropland extent product of Australia and China using random forest machine learning algorithm on Google Earth Engine cloud computing platform. *ISPRS J. Photogramm. Remote Sens.* **2018**, *144*, 325–340. [CrossRef]

16. Mahdianpari, M.; Salehi, B.; Mohammadimanesh, F.; Homayouni, S.; Gill, E. The first wetland inventory map of newfoundland at a spatial resolution of 10 m using sentinel-1 and sentinel-2 data on the google earth engine cloud computing platform. *Remote Sens.* **2019**, *11*, 43. [CrossRef]

17. Gorelick, N.; Hancher, M.; Dixon, M.; Ilyushchenko, S.; Thau, D.; Moore, R. Google Earth Engine: Planetary-scale geospatial analysis for everyone. *Remote Sens. Environ.* **2017**, *202*, 18–27. [CrossRef]

18. Gatti, A.; Galoppo, A. Sentinel-2 Products Specification Document, Issue 14.5. 2018. Available online: https://sentinel.esa.int/documents/247904/685211/Sentinel-2-Products-Specification-Document (accessed on 09 July 2019).

19. Masó, J.; Zabala, A.; Serral, I.; Pons, X. Remote Sensing Analytical Geospatial Operations Directfly in the Web Browser. *Int. Arch. Photogramm. Remote Sens. Spat. Inf. Sci.* **2018**, *XLII–4*, 403–410. Available online: https://www.int-arch-photogramm-remote-sens-spatial-inf-sci.net/XLII-4/403/2018/isprs-archives-XLII-4-403-2018.pdf (accessed on 09 July 2019). [CrossRef]

20. Soille, P.; Burger, A.; Marchi, D.; de Kempeneers, P.; Rodriguez, D.; Syrris, V.; Vasilev, V. A versatile data-intensive computing platform for information retrieval from big geospatial data. *Future Gener. Comput. Syst.* **2018**, *81*, 30–40. [CrossRef]

21. Baumann, P. The OGC web coverage processing service (WCPS) standard. *Geoinformatica* **2010**, *14*, 447–479. [CrossRef]

Article

Achieving the Full Vision of Earth Observation Data Cubes

Steve Kopp *, **Peter Becker ***, **Abhijit Doshi, Dawn J. Wright**, **Kaixi Zhang and Hong Xu**

Esri, 380 New York St, Redlands, CA 92373, USA
* Correspondence: skopp@esri.com (S.K.); pbecker@esri.com (P.B.)

Received: 1 June 2019; Accepted: 4 July 2019; Published: 6 July 2019

Abstract: Earth observation imagery have traditionally been expensive, difficult to find and access, and required specialized skills and software to transform imagery into actionable information. This has limited adoption by the broader science community. Changes in cost of imagery and changes in computing technology over the last decade have enabled a new approach for how to organize, analyze, and share Earth observation imagery, broadly referred to as a data cube. The vision and promise of image data cubes is to lower these hurdles and expand the user community by making analysis ready data readily accessible and providing modern approaches to more easily analyze and visualize the data, empowering a larger community of users to improve their knowledge of place and make better informed decisions. Image data cubes are large collections of temporal, multivariate datasets typically consisting of analysis ready multispectral Earth observation data. Several flavors and variations of data cubes have emerged. To simplify access for end users we developed a flexible approach supporting multiple data cube styles, referencing images in their existing structure and storage location, enabling fast access, visualization, and analysis from a wide variety of web and desktop applications. We provide here an overview of that approach and three case studies.

Keywords: data cube; image cube; image data cube; imagery; Landsat; Sentinel; earth observation; GIS; web services; web application; analysis; GIS

---

1. Introduction

The history of earth observation data, and evolution of information technology and the internet have created a transition of scene-based and project-based thinking to imagery as a seamless time series. Killough (2018) described a goal of the CEOS Open Data Cube as increased global impact of satellite data. To achieve this goal, we see four requirements, some already in process, all still evolving.

- Easier to access to data
- Easier to use data
- Imagery big data analytics in the cloud
- Improved usability through tailored imagery web applications

The first three decades of Landsat and the broader earth observation community were dominated by specialists working on individual projects often on individual satellite scenes of a single date or a few dates from a single senor. The data was expensive, required special software and knowledge to extract anything more than the most basic of information, and required significant computational power. Projects involving large geographies and many time steps were limited to research institutions and government science agencies.

In 1984 a single Landsat scene cost $4400 [1], which would be over $10,000 today. In 2009, just before the advent of geospatial cloud computing, the U.S. government made all Landsat available at no cost. Since 2009, there has been a 100-fold increase in use of Landsat data [2]. Until it became free, the idea of doing time series analytics on dozens or hundreds of Landsat scenes over an area was so cost prohibitive as to limit applied research. This policy change has resulted in a rapid increase in use and value to the community [3].

Beginning in the late 1990s, the early days of Earth observation imagery on the internet were dominated by simple visualization and finding data to download then perform local analysis. From the beginning, starting with Microsoft TerraServer in 1998, it was clear that seamless mosaics would be the expected user experience. Data search and download web sites of the era remained primarily scene based, with the notable exception of the USGS Seamless Server [4]. Recognizing significant duplicate work by its customers downloading and assembling tiled data, USGS assembled seamless collections of geographic feature data and digital elevation models, and provided a web interface for visualization and to interactively select an area of interest and download a seamless mosaic. This one was one of the earliest publicly accessible seamless open data implementations based upon Esri technology. Keyhole also recognized the need for seamless imagery and its technology was later purchased and launched as Google Earth in 2005. Esri began providing access to web services of Landsat imagery in 2010. These multispectral, temporal image services provided access to the Landsat GLS Level 1 data. In 2013 Esri began serving Landsat 8 from private hosting, and in 2015 transitioned to publishing web image web services of the AWS Open Data Landsat collection.

Over the last decade, the availability of petabytes of free earth observation data through AWS Open Data, Google Earth Engine, and others, combined with fast, low-cost cloud computing has created a paradigm shift in the earth observation imagery community. There are now numerous web sites [5] providing free access to cloud hosted Landsat, Sentinel, and other data, and an increasing number of sites providing analysis capabilities against this data.

In recent years, several approaches to storing and serving large Earth observation data have been developed [6,7]. One notable success was the Australia data cube [8] which significantly advanced the community vision of what could be possible if one's thinking was less constrained by storage, computation, and data costs. They highlighted the unexploited value of long image time series and developed new preprocessing workflows to support them. The project was significant enough that the Committee on Earth Observation Satellites (CEOS) data cube team launched the Open Data Cube initiative developing software and workflows to make it easier for others to create similar data cubes [9]. There are currently nine data cubes using the Open Data Cube technology in production or operation, with a goal to have 20 by 2022 [10].

'Data cube' is a generic term used to describe an array of multiple dimensions. Data cubes help to organize data, simplifying data management and often improving the performance of queries and analysis. In its simplest form it can be thought of as a 3D spreadsheet where three axes may represent sales, cities, and time. In a mapping context, the two primary dimensions of a data cube are typically the latitude and longitude position. Other common dimensions of geospatial data are time, depth (when working with geologic or oceanographic data), or altitude (when working with atmospheric data). Data cubes of earth observation imagery are typically three dimensions; latitude, longitude, and time, Figure 1.

A query of an image data cube at a specific time, will return an image map such as that seen on the top level of the cube in Figure 1. A query of an image data cube at a specific location will return the time series of values at that location, like a vertical probe dropped on top of the cube. The cube structure also simplifies data aggregation operations such as weekly, monthly, or annual analysis.

After years of talk of democratizing geospatial data [11] we are finally seeing notable progress [12]. 'Democratizing' is not simply about making data and analysis more available, it also requires improving accessibility in a way which leads to widespread adoption and awareness. This means it is necessary to understand potential end users and provide a relevant and approachable user experience. The availability of image data cubes, cloud hosted image analysis, and web technology means this can best be accomplished through easily configurable web applications tailored to the knowledge and goal of the user.

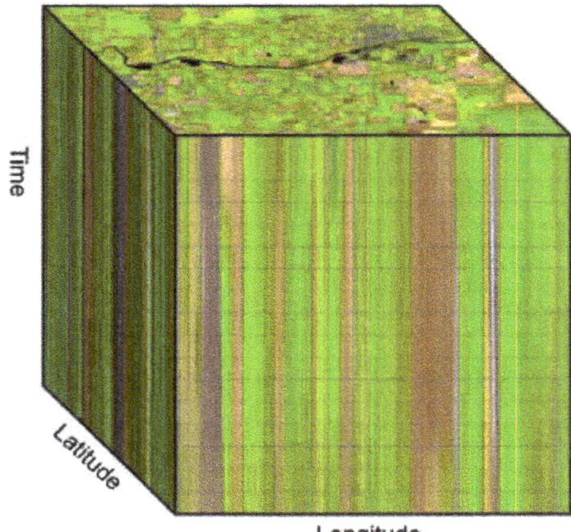

Figure 1. Earth observation data cube and its dimensional axes.

Google Earth increased global geospatial awareness by providing access to data in a simple user experience suitable to the skills and needs of its users. Prior to this, GIS and web maps were more of a niche few people were aware of, now everyone knows how to navigate a web map. The same is possible with earth observation data and geospatial data science through the use of image data cubes and tailored web applications.

This paper describes a collection of geospatial technology components which in combination establish a complete platform for Earth observation data exploitation, from data ingestion and data management to big data analytics to sharing and dissemination of modeling results. The first section of the paper describes the technology components, their relevance toward achieving the full vision of Earth observation data cubes as well as insight to lessons learned along the way. The second section of the paper will present three application case studies applying this approach to a variety of data cubes.

2. Method

To obtain the full value of image data cubes, development and integration is required in a number of areas: data preprocessing, storage optimization, data cube integration, analysis, and sharing.

2.1. Image Preprocessing—Building Analysis Ready Data

Processing imagery before loading into a data cube structure to create Analysis Ready Data (ARD) is a foundational aspect of modern earth observation data cubes. ARD preprocessing has been described as a fundamental requirement [13], disruptive technology [13], and cornerstone of the Open Data Cube initiative [10]. ARD preprocessing makes it easier for a wider audience of non-imagery experts to perform more correct analysis, thus expanding the potential number of people correctly applying earth observation imagery [13]. There are multiple approaches and programs for ARD processing [10,13].

Many image processing workflows to create ARD involve resampling the pixels, which decreases their quality. While all resampling creates some artifacts, choices can be made appropriate to the application. For most scientific applications, it is desired to minimize resampling such that only one resampling takes place and the type of resampling is most applicable to the type of analysis. Where spectral fidelity is important nearest neighbor resampling is recommended, and for applications where textural fidelity is important, bicubic or similar sampling is recommended. One of the decisions in a data cube preprocessing workflow is to what sampling should be applied and how soon after the imagery is acquired should the data be processed, or if there is a need to re-process all the data when new transformation parameters are determined.

Ideally, the data would remain as sensed by the sensors and the user performing the analysis would define the geometric transforms, projection, pixel size, pixel alignment, as well as radiometric corrections to be applied to the imagery for analysis specific to their needs. This would require providing access to the source data, which increases the complexity of any analysis, and adds an unnecessary size limit on the type of problem to be solve. Hence the data cube concept requires the selection and agreement on all these factors and the data is data is preprocessed to these specifications. The intent and assumption is the data is only resampled once and all subsequent processing will be done with the defined pixel alignment.

There are multiple approaches and programs for ARD processing [10,13] and CEOS has developed a set of guidelines for ARD processing known as CARD4L [14].

2.1.1. Radiometric Preprocessing

To improve analysis results, the radiometry of the sensor needs to be calibrated and corrected for a variety of sensor anomalies, as well as effects of atmosphere, for which multiple approaches are available [13]. Determination of multiple radiometric transforms is complex and relies upon auxiliary data that may not initially be available at the time of image acquisition. What may be of sufficient accuracy for one set of scientific analysis may not be sufficient for another. Tradeoffs are evaluated, and compromises reached to determine a level of processing that is sufficient for the majority of the users based on the best available auxiliary data. For most data cube applications, the data is processed to an agreed level of surface reflectance.

In addition to correcting pixel values to surface reflectance, additional calculations are often performed to flag pixels such as cloud, and cloud shadow, which can be problematic for some analysis techniques. Cloud masks can be computationally expensive to create, and compress well for efficient storage, and therefore are frequently part of the preprocessing workflow and stored as a binary image mask.

A balance needs to be found between processing and storage. If a parameter can be computed by a computationally simple local function from the existing datasets, then it is generally better to compute this when needed instead of during preprocessing which requires additional processing and storage.

For example, computing an NDVI from two bands is computationally insignificant and would result in a new product that requires significant storage and is therefore typically not part of the preprocessing workflow. NDVI and other band indices are more often implemented as a dynamic process performed when needed, as described in Section 2.4.1.

Geometrically referencing the pixels, computing surface reflectance, and creating cloud masks are computationally expensive and appropriate as part of the preprocessing workflow.

2.1.2. Geometric Preprocessing

From a geometric perspective, remote sensing instruments do not make measurements in predefined regular grids. Performing temporal analysis that involves multiple data sources involves resampling the data to a common coordinate system, pixel size, and pixel alignment. Unfortunately, the transforms from ground to sensor pixels are not usually accurately available as soon as the image is collected, and the transformation parameters may improve over time.

A disadvantage of storing anything but the unrectified imagery is that the data is resampled to a defined coordinate system, pixel size and alignment. A key aspect of Analysis Ready Data is to attempt to ensure that all data is sampled to the same grid. For both the Australian Data Cube and US Landsat ARD a single coordinate system was selected, and all data processed from the Level 0 data directly to the defined coordinate system. As the spatial extent of data cubes increases the selection of appropriate projection becomes more challenging. For multiple regions, a current solution is to split the world into multiple regional data cubes each of which has a suitable local coordinate system or to use UTM Zones.

The tradeoffs in choosing an appropriate coordinate system exists for all global datasets. For Landsat and Sentinel 2 level 1 and 2 products UTM zones are being used. Each Landsat scene is defined by a unique path/row and is assigned the most suitable UTM zone. This results in pixels being aligned for all scenes allocated to the same UTM zone, but resampling is required between scenes from different UTM zones. For the Sentinel 2 scenes, each scene is cut into granules that each have their appropriate projection and data that fall at the intersection of two UTM zones is processed and stored in both projections. As the granules go towards the north and south pole there is an increasing amount of overlap and duplication, Figure 2. Techniques have been developed [15] for addressing the UTM zone overlap of Sentinel 2 data.

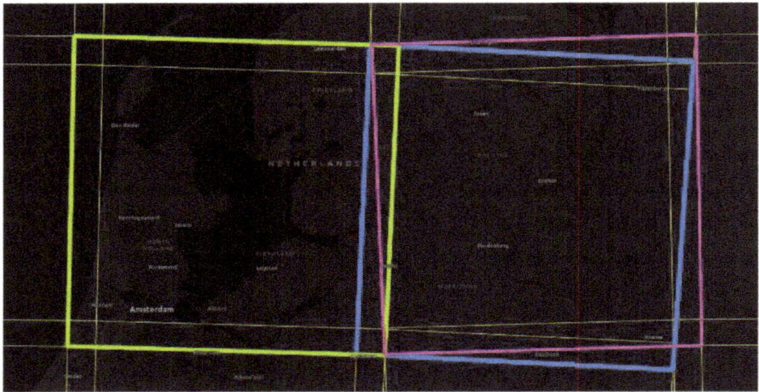

Figure 2. Sentinel 2 tile boundaries as thin yellow lines, with three highlighted tiles overlapping in central Netherlands. The green tile is predominantly in UTM zone 31, and the blue and magenta tiles are predominantly in UTM zone 32.

2.1.3. Tiling

The Australia data cube introduced a preprocessing step to cut all imagery into 1-degree tiles to improve parallel computation performance on their HPC hardware [8,16]. This step has continued into the Open Data Cube project and others. The advantage of tiling is that any pixel block or pixel location is addressed by a simple equation that defines the name and location of the file. However, because satellites do not collect data in 1-degree tiles, and the orbital track prevents collecting square areas, many more tiles are created, and many tiles are predominantly empty. To know if a pixel exists for a specific location it is not sufficient to know if a tile exists, but one needs to know the footprint of the scene or needs to read the data to determine if it is set to NoData. Hence tiling data for processing or storage reasons can noticeably increase the data access time during use.

The alternative is to maintain the data as individual scenes as acquired by the satellite, where each scene is referenced to a single file which has a corresponding footprint polygon that defines the extent of the data. For any pixel location, a simple query of the footprint polygon determines if a scene exists and has data. Such geometry queries are highly optimized and efficiently eliminate requests fetching files, blocks, or pixels which contain little or no information.

From a file structure perspective both the tiled datasets and the scene datasets are internally tiled, meaning that the pixels inside of a file are grouped into regularly size contiguous chunks so that all pixels in a defined area are stored close together speeding up data access. Such tiling schemes typically use internal tile sizes of 256 or 512 pixels along each axis.

2.2. Optimizing Data Storage

Once the data is preprocessed, how the data is stored will impact accessibility, application utility, performance, and cost. Our focus is to maximize accessibility with potentially thousands of simultaneous users, make the data useful for visualization and analysis, and to balance interactive visualization performance against infrastructure cost.

Earth observation data cubes are now growing to petabytes of data leading to potentially significant infrastructure costs. Although very fast random-access storage is available the cost for such data volumes can be prohibitively high. Currently, the best approach from a cost and reliability perspective is object storage in public commercial clouds such as Amazon AWS, Microsoft Azure, and Google Cloud. Regardless, if a data cube is terabytes in a local file server, or petabytes in a cloud system, performance of the data will be impacted by choices of file structures and compression. For the best user experience, faster is better, and there are tradeoffs to balance performance and cost.

2.2.1. Compression

One way to reduce storage cost and improve performance is to compress the imagery. File compression algorithms can be lossless (pixel values do not change), lossy (pixel values are changed to achieve greater compression) or controlled lossy (pixel values change to a maximum amount defined by the user). When imagery is used for analysis, lossless or controlled lossy compression is preferred. For visualization, where exact radiometric values are less crucial, lossy compression is typically used because of its higher compression ratio. The type of compression chosen impacts not only pixel values and size, but also storage cost, performance, and computation cost.

Deflate compression [17] is commonly used in recent data cubes as part of Cloud Optimized GeoTIFF (COG) [18] such as Landsat AWS and Open Data Cube. It is lossless, has relatively low compression, with the benefit of low computation cost/short time for compression and decompression.

JPEG 2000 wavelet compression used in the Sentinel AWS data cube can be used as lossless or lossy. As a lossy compression it provides greater compression than is achieved with deflate compression, however the computation required to compress and decompress the data is significantly more.

LERC (Limited Error Raster Compression) [19] is a recent compression that can be lossless or controlled lossy and provides very efficient compression and decompression of imagery data especially higher bit depth data, such as 12 bit and more. It is our recommended compression for multispectral earth observation data cubes. Unlike other lossy compression algorithms, LERC allows the user to define a maximum amount of change that can occur to a pixel value. By setting this to slightly smaller than the precision of the data the compression maintains the required precision while maximizing the compression and performance.

Advantages of LERC include that it is 3–5 times faster than deflate for both compression and decompression, depending upon bit depth, and allows the error tolerance to be defined. When used in a lossless mode LERC typically achieves about 15% higher compression ratio (smaller file) than deflate for Landsat and Sentinel type data.

To test the performance and size of different compression algorithms and formats we selected a collection of Landsat ARD files that were less than 10% cloud and with average 20% NoData (areas outside the image with no value). These where converted using GDAL version 2.4.1 to different formats and lossless compressions. The times to create and write the files and the resulting file sizes were recoded to compare data creation times, shown in Table 1. We also recorded time to read the full image at full resolution into memory.

Table 1. Test results comparing image compression and image format. The left side of the table shows values normalized to a ratio of the size and times for Tiled TIFF with no compression. The right side of the table shows values for the same test normalized to a ratio of the size and times for COG Deflate.

	As a Factor of Tiled TIFF No Compression			As a Factor of COG Deflate		
	Size	Time to Write	Time to Read Full	Size	Time to Write	Time to Read Full
Tiled TIFF None	1.00	1.00	1.00	1.65	0.12	0.93
Tiled TIFF Deflate	0.67	2.34	0.96	1.10	0.28	0.90
Tiled TIFF LZW	0.58	5.18	0.98	0.96	0.61	0.92
Tiled TIFF LERC	0.53	1.69	1.02	0.87	0.20	0.96
MRF LERC	0.53	2.09	0.91	0.87	0.25	0.85
COG Deflate	0.61	8.48	1.07	1.00	1.00	1.00
JPEG 2000	0.33	7.42	1.02	0.54	0.87	0.95

2.2.2. Image Format

The format of the file storage also impacts performance and therefore cost. Common choices include JPEG 2000, netCDF, tiled GeoTIFF, COG, and MRF [19,20]. Each has advantages and disadvantages. JP2 provides the highest compression but requires significantly more computation. NetCDF has been used in the past but is not recommended here because the data structure is not well suited for direct access from cloud storage. Tiled GeoTIFF provides a simple to access format, that can optionally include internal or external image pyramids, but is not specifically optimized for cloud usage. Multiple compression algorithms are supported including deflate, LZW, and recently LERC. COG (cloud optimized GeoTIFF) is very similar to Tiled GeoTIFF, but assumes pyramids exist for the image and stores the pyramids at the start of the file to improve performance when browsing a zoomed-out low resolution image. However, this change can significantly increase file creation time as seen in Table 1. File creation performance for COG can be faster than in these tests if all data restructuring is done in memory.

MRF is similar to tiled GeoTIFF, but information is structured into separate files for metadata, index, and pixel data which enables different compressions including LERC and ZenJPEG making it well suited for cloud storage of data cubes. ZenJPEG is an implementation to enable the correct storage of NoData values with JPEG for improved lossy compression at 8 or 12 bits per channel. The fact that the metadata and index is kept separately enables applications to easily fetch and cache it, thus reducing repeated requests for the information when accessing large numbers of files. All these formats are directly supported by GDAL [21] which is used by most applications for accessing geospatial imagery.

Our tests as reported in Table 1 found MRF with LERC compression is about 13% more compressed than COG Deflate, and MRF LERC is significantly faster to write. Reading MRF LERC is about 15% faster than COG Deflate.

2.2.3. Image Pyramids

For analysis applications at the full resolution, only the base resolution of the imagery needs to be stored. For many applications it is advantageous to enable access at reduced resolutions so that overviews of the data can be accessed quickly. This can be achieved by storing reduced resolution datasets, also referred to as pyramids, which increase storage requirements. For uncompressed data typical pyramids stored with a sampling factor of 2 would increase the storage requirements by 33%. With compressed data the additional storage requirement is a bit more due to the higher frequency of the lower resolutions. Wavelet based formats such as JP2 inherently contain reduced resolutions, but the storage is less efficient to access as described above. If storage cost is not a significant factor, storing pyramids with a factor of 2 between levels is most suitable. From an analysis perspective, the existence of reduced resolution datasets provides limited value.

Alternatives to reduce storage include skipping the first level, changing the sampling factor to 3 or using lossy compression for the reduced resolution datasets. An additional consideration is the sampling used to create the reduced resolution datasets. In most cases, simple averaging is performed between the levels, but naturally this is not appropriate for datasets that are nominal or categorical such as cloud masks.

2.2.4. Optimizing Temporal Access

Storing each captured satellite scene as a file or object per scene provides simplicity from a storage perspective but is not optimum for temporal access. To access a profile through time for a specific location requires a file to be opened for each temporal slice and band read. Retrieving a temporal profile for a stack of 10 scenes that each have 8 bands would require 80 files to be opened, which may be relatively quick. However, since one of the advantages of data cubes is to manage an extensive time series, a time series analysis on a Landsat data cube might need to open hundreds or thousands of files, not suitable for interactive data exploration using a web client drilling through a time series at a location.

Putting the data into a single file does not resolve the problem as the access pattern has an impact on efficiency. If the data is structured as slices then access for any specific slice is optimized, but in that pattern, temporal access still requires sparse requests. If the data are pixel aligned through time during the preprocessing, they can alternatively be structured into many small cubes (versus tiles). This results in faster temporal access since the data for a temporal search are stored closer together, but performance for accessing slices is reduced.

A simple approach with optimal performance when retrieving both image areas as a single time slice, and a time series at a location is to replicate the data and transpose it such that accessing pixels representing different times is equivalent to accessing pixels along a row of pixels in the non-transposed version, essentially rotating the axis of the data, Figure 3. Although this doubles the storage volume of the full resolution data, it also maintains the best possible simultaneous performance for both query types.

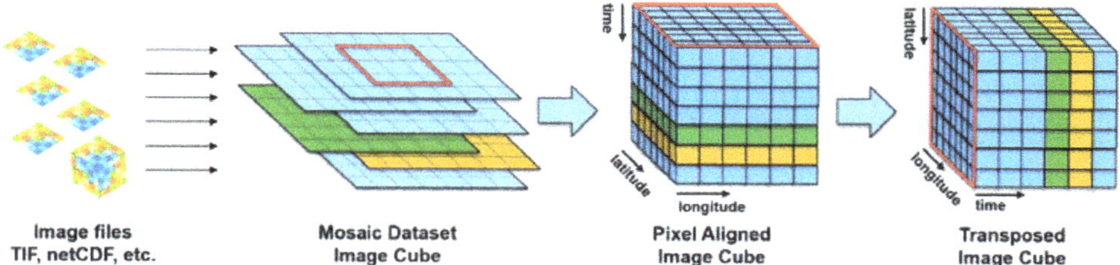

Figure 3. A collection of georeferenced images with time stamps can be organized for easier access as a mosaic dataset image cube. To simplify usage and improve performance the data can be processed into a pixel aligned image cube. Optimal performance for temporal queries can be achieved by transposing the data storage along the time dimension.

An efficient balance to optimizing temporal access to a data cube is to store the data as efficiently as possible as slices (traditional scenes by date), and when repeated temporal analysis of an area of interest is required, generating a transposed data structure that is optimum for the analysis.

There are many decisions to make in designing and preprocessing a data cube. These decisions result in optimizations and compromises based on the requirements of the application and resources of the data cube creator. The result will be many different data cubes of different styles, flavors, and capabilities. The following sections address how these different data cubes can be seamlessly integrated into a platform for visualization and analysis.

2.3. Building and Referencing Image Data Cubes

A mosaic dataset [22] is an ArcGIS data structure which holds metadata about images, a pointer to the image pixels, as well as processing workflows to be executed when the pixels are accessed. The mosaic dataset is how ArcGIS builds and manages data cubes, as well as the mechanism for integrating third party data cubes such as those described in Section 3 of this paper.

Mosaic datasets do not store pixels, but instead the pixel data remains in its original or most appropriate form and is referenced by the mosaic datasets. The mosaic dataset ingests and structures all the metadata and stores it as a relational database such as a file geodatabase or any enterprise database such as Postgres or SQL Server, as well as cloud optimized databases such as AWS RDS Aurora. This enables high scalability and allows the mosaic dataset to be updated while in use.

The mosaic dataset can also store processing rules to transform stored pixels into required products, such as a pansharpened image, or a variety of band indices such as vegetation, water, etc.

The processing is defined by function chains that can include both geometric and radiometric transformations of the data. In a simple case, a mosaic dataset may consists of a collection of preprocessed orthoimages in a single coordinate system, or could be a collection of preprocessed orthoimages in a variety of coordinate systems projections which are reprojected on the fly when the data is requested. The processing chains can be complex, for example a mosaic dataset may reference imagery directly acquired by a sensor and the processing chains can define both geometric transformations (such as orthorectification to a defined digital terrain model) and radiometric transforms (such as to convert sensor data into surface reflectance). Mosaic datasets can be created using a wide range of tools within ArcGIS including dialogs, graphical model workflows, Python scripts, and Jupyter notebooks.

To simplify the user experience with common image data sources, Esri works with satellite and camera vendors to create definitions of how to recognize and read particular sensor data, include appropriate metadata, and how to process it. For example, if a user runs an NDVI function on Landsat 5 data, it knows the appropriate spectral bands to use without requiring the user to specify which bands contain red and near infrared information. This sensor metadata and processing information is stored as a raster type definition [23]. ArcGIS includes a wide range of raster types supporting many data products from DigitalGlobe and Airbus as well as USGS. MTL for Landsat and Dimap for ESA Sentinel 1 and 2, and more. New raster types can be easily created by users in Python to ingest metadata and define processing for any structured pixel data. Custom Python raster types are how ArcGIS is taught to read metadata and pixels from Digital Earth Australia and Open Data Cube YAML files, and the other examples later in this paper.

2.4. Analysis

ArcGIS provides two types of analysis on imagery: dynamic or on-the-fly analysis performed at the current display extent and resolution, and raster analytics typically performed at full resolution analysis with persistent results.

2.4.1. Dynamic Image Analysis

Dynamic image analysis is the most common approach for interactive data exploration, rapid prototyping of analysis, or creating highly interactive web applications for exploring data cubes and performing computationally simple analysis. It is most suitable for local (per-cell) and focal (neighborhood) operations, and supports function chaining to allow multiple operations to be performed in a single step. This is done as a single operation with no intermediate files created. When the map is zoomed or panned the calculation is recomputed for the new area and result displayed, with no noticeable performance difference between this computation and simply drawing a band combination with no calculation.

There are over 150 raster functions provided, covering a wide range of image processing and analysis capabilities, a graphical editor for constructing function chains, as well as the ability to author custom raster functions with Python. For overlapping imagery, the system can provide a single output by blending the pixels, aggregating, or computing a variety of statistics per pixel location. To extend the provided functions the system can return and utilize a Python NumPy, XArray, or ZArray for integration with a wide range of scientific analysis libraries.

Processing workflows can be defined in advance as part of the mosaic dataset definition, and also controlled or modified as desired by a client application. For example, a web application can expose multiple decades of imagery, a variety of band index options, and a change detection tool, which allows an end user to select a band index of interest and dates of interest, e.g. to find areas with a significant change in vegetation, the server would compute NDVI for each of two dates, compute the change, and apply a threshold and colormap to the result. Due to the way the imagery is stored and served, such a calculation can return a result dynamically as the map is panned, zoomed, or a different or band index is chosen.

2.4.2. Big Data Image Analysis

Although dynamic image services provide an excellent experience for performing on-the-fly processing they are not optimum when processing a large extent at high resolution. To perform more rigorous analysis, such as segmentation and classification, or create persistent results we developed a distributed computation framework for local, focal, and global analysis, which automatically chunks and scales to the to the number of processes requested by the user. The system scales efficiently from a small number of parallel processes on a local machine to hundreds of processes in a distributed cluster, and can utilize GPUs when available. To perform analysis on data cubes, we developed efficient search, parallel read/write, and distributed computation capabilities, which can be deployed to an on-premise private cluster, or in commercial cloud such as Amazon or Azure. The system includes the same analysis capabilities as the dynamic image services described above, plus highly optimized versions of more complex analysis such as segmentation, classification, shortest path, terrain and hydrologic analysis. The system also integrates with third party analytic packages through Python, R, CNTK, TensorFlow, and more. These optimized algorithms working on massive distributed data in publicly accessible cloud storage allowing anyone to perform analysis which previously would be performed on a supercomputer.

Many image processing tasks are computationally simple and can become I/O bound in a parallel compute setting. To illustrate the effectiveness of parallel I/O capabilities of the system we ran an image processing workflow on the Landsat GLS 1990 collection. The data was 7422 multispectral scenes stored in Amazon S3. The analysis workflow was for each scene to generate a NoData mask, perform a top of atmosphere correction, calculate a modified soil adjusted vegetation index, slice this into categories, and write the output image. The computation was performed on a single node, 32 core AWS c3.8XLarge with 60gb of RAM. We ran 200 processes against the 32 cores and the task completed in 2 h and 48 min averaging approximately 44 scenes per minute.

For large computation-intensive workflows, it is helpful to distribute the compute load over multiple nodes in a cluster, each with multiple cores. New algorithms were architected and developed to scale efficiently in a distributed compute environment where the software framework automatically chunks and distributes work to each node and core. The input data for this test was a 397 gb, 1-meter resolution area of the National Agriculture Imagery Program (NAIP), approximately 100 billion pixels. The analysis workflow performed was a mean shift segmentation, classification, and write the output file. The computation was performed on a 10 node x 20 cores each Azure Standard_DS15_v2 with 140 gb of RAM. We ran 200 processes and the task completed in 1 hour and 13 min. A Python example of a similar workflow is shown in Figure 4.

```python
import arcpy
# Check out the ArcGIS Image Analyst extension license
arcpy.CheckOutExtension("ImageAnalyst")
# source dataset
inRaster = "https://landsatsample.esri.com/server/rest/services/landsat7_one_scene/ImageServer"
# 1. mean shift segmentation
seg_raster = arcpy.ia.SegmentMeanShift(inRaster, spectral_detail = 14.5,
spatial_detail = 10, min_segment_size = 20,  band_indexes = "4 3 2")
# 2. train random forest
ecd_file = "c:/output/moncton_sig_rf.ecd"
arcpy.ia.TrainRandomTreesClassifier(seg_raster, train_features =
"c:/test/train.gdb/train_features", out_definition =  ecd_file, in_additional_raster
= inRaster, maxNumTrees = 50, maxTreeDepth = 30, maxSampleClass = 1000, attributes =
"COLOR;MEAN;STD;COUNT;COMPACTNESS;RECTANGULARITY")
# 3. classify
classifiedraster = arcpy.ia.ClassifyRaster(seg_raster, ecd_file, inRaster)
# 4. save classification output
classifiedraster.save("c:/output/image_classified.tif")
```

Figure 4. Python script example for segmenting and classifying imagery from an image service.

Big data analysis workflows can be authored in ArcGIS Pro and run locally or in a remote cloud server. Because the underlying tools and APIs are the same, the only difference between running on a local desktop and in a remote cloud server is one parameter on the dialog. It is similarly easy to publish that workflow to the server so others can access it. Analysis workflows can be authored from dialogs, graphical modelers, Python, and Jupyter notebooks, Figure 5. Published services are available via REST and OGC protocols.

```python
In [1]:  import arcpy
         arcpy.CheckOutExtension('ImageAnalyst')

Out[1]:  'CheckedOut'

In [2]:  # Load the time-series landsat mosaic dataset as a mdim raster
         landsatRaster=arcpy.Raster("https://landsatsample.esri.com/server/rest/services/landsat7_time_series/ImageServer", True)

In [3]:  # calculate NDVI for each time point
         ndviRaster = arcpy.ia.Foreach(landsatRaster, 'NDVI', {'VisibleBandID':4,'InfraredBandID':3, 'ScientificOutput': True})

In [4]:  # calculate difference between NDVI at time T and NDVI at time T+1
         ndvi_change = arcpy.ia.Foreach(ndviRaster, 'Minus', {'Raster2':'@1'})

In [5]:  # persist the output
         ndvi_change.save(r"C:\output\ndvi_change.crf")
```

Figure 5. Jupyter notebook example in Python for computing the NDVI difference between two dates from an image service.

2.5. Sharing the Value of Image Data Cubes

To achieve the vision of expanding the use of imagery and increasing its impact it is necessary to grow the community of people using and understanding imagery. This can be achieved through web services and web applications.

2.5.1. Sharing Web Maps and Image Services

Web services define a protocol for two pieces of software to communicate through the internet, for example for data from one server to appear as a map in a client application. Imagery can be served with a variety of protocols [24,25] including OGC WMS, OGC WCS, REST, etc. Map services such as OGC WMS are the most common way to serve imagery and is essentially a prebuilt picture of the data.

It is most useful as a background image when used with other data in a web map but is of limited value for data exploration and analysis because the original data values are often not available. For example, if provided as jpeg which is lossy compressed, or the data is higher than 8-bit pixel values and needs be rescaled to fit into an 8-bit RGB PNG. However, because the display has been predefined, map services are the easiest to use for an imagery novice who does not understand bit depth and stretching values for display, or edge effects of multi-date imagery.

Image services support interactive image exploration and analysis, and are a cornerstone of image analysis on the web. They are commonly published as REST, SOAP, or OGC WCS. Image services provide web applications access to original pixel values, and all bands and dates the publisher chooses to share. Image services can support a wide range of dynamic image analysis and exploration described above in Section 2.4.1, as well as full resolution analysis described in Section 2.4.2, as well as extraction for download.

2.5.2. Sharing Analysis

To increase the impact of imagery it should be transformed from data to information through some analytic workflow. There are a variety of ways to share analysis tasks and workflows to enable non-experts to be able to extract useful and reliable information from imagery. Analysis workflows can be defined as part of a mosaic dataset and published as part of the image service, and when accessed by a client application the analytic workflow is executed dynamically. Analytic workflows can also be published as geoprocessing services, OGC WPS, Python scripts, and as Jupyter notebooks. Using analytic web services, applications can combine data and analysis from multiple locations to solve a unique new task possibly not envisioned by the service publisher. For example, combining a fire burn scar difference image analysis service from one publisher with a population overlay service from another publisher to compute the number of people impacted by a fire.

ArcGIS image analysis workflows are most often authored in ArcGIS Pro, which was designed from inception to be well integrated for publishing and consuming cloud-based web services. Workflows are designed using dynamic analysis for rapid prototyping, and then shared to the cloud for scaled up big data analytics at full resolution, which can be accessed by any web client application, including ArcGIS Pro.

2.5.3. Web Applications for Focused Analytic Solutions

Providing a relevant and approachable user experience to the breath of potential users and skill levels is best accomplished through configurable web applications tailored to the knowledge and goal of the end user. The web applications are powered by cloud hosted analysis services and image services published from image data cubes.

Web applications can be created that access all forms of geospatial web services. Applications consume predefined services and can also interact with the dynamic services defining the processing to be applied. Users can also compose new workflows through web applications. Such processing can be on-the-fly processing or can also include running analysis tools that interact with cloud servers performing large computation tasks. These web applications not only act as clients to servers, but can also retrieve the data values for the servers enabling the creation of highly interactive user environments that make use of the processing capabilities available in modern web browsers.

With the emergence of cloud technology, we have seen the continuous growth of Software as a Service [26]. Esri's SaaS solution, ArcGIS Online, is part of the overall ArcGIS platform, designed to work seamlessly with ArcGIS Pro and is included with all ArcGIS licenses. It provides access to petabytes of geospatial data and hosted analytic services, as well as providing infrastructure for hosting users' data and all types of geospatial services including image services and analysis services. It also includes capabilities to author and host custom web applications. Applications can be written using in Javascript, or using a drag and drop user interface such as Web AppBuilder. For imagery services there are a collection of prebuilt web control widgets including spectral profile, time series profile, scatterplot, change detection, and more. Configurable web application templates are also provided for common web application needs such as image change detection and interactively browsing image time series. Using a web application template a novice can progress from a web map to a fully hosted and publicly accessible web application in minutes.

It is now easy enough to build custom web applications integrating image services and image analytics that almost anyone can create an interactive application to share specific data, analysis, and stories. It is no longer necessary for a stakeholder or executive to navigate a generic geospatial software to understand a drought or zoning problem, their staff familiar with the data and analysis can create and share a simple web application to lead them through understanding the problem in a way that is relevant to their knowledge and needs. Representative of their ease and popularity, there are currently over 3 million web applications in ArcGIS Online, with approximately 1000 new Web AppBuilder applications created each day.

3. Results

3.1. Landsat and Sentinel in Amazon

In 2015 Amazon Web Services in collaboration with USGS and the assistance of Planet began loading Landsat 8 Level-1 data into an Amazon S3 bucket and provided open public access [27]. Using the raster type and mosaic dataset approach described in Section 2, Esri provides visualization and analysis of this data through a free, easy to use web client, Figure 6, as well as web service access for visualization and analysis in other applications.

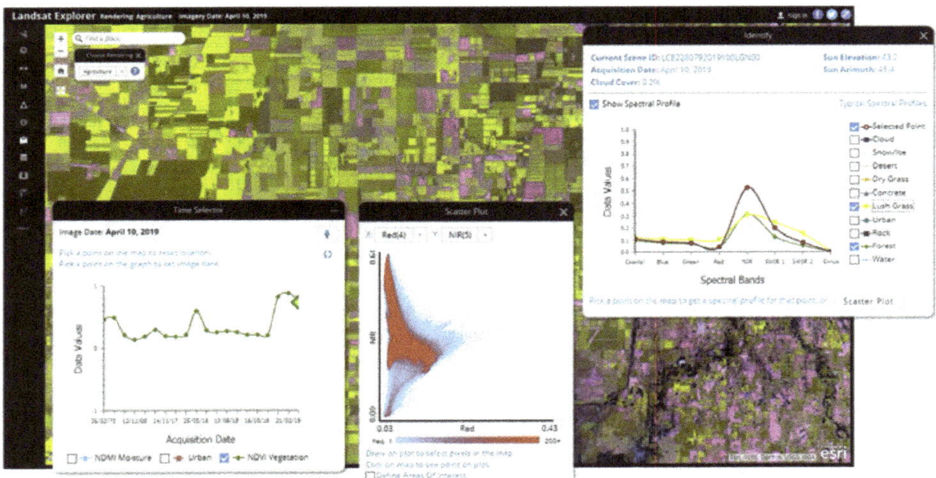

Figure 6. Landsat Explorer [28] is a free web application for interactively exploring the spectral and temporal characteristics of the world. It was created with the Web AppBuilder imagery widgets and utilizes image services published with the methodology described in this paper.

Whereas the traditional approach to an image archive would start with first searching and selecting individual scenes to view, these services and web application provide an interface that enable users to pan and zoom to any location on earth and see the best available Landsat scenes in a wide range of different band combinations and enhancements. The web client also allows searching based on geographic extent, date, cloud cover, and other criteria. A timeline slider control is used to refine the date selection.

The application supports a variety of interactive analyses from common band indices for evaluating vegetation, water, and burn scares, as well as image thresholding and change detection. All image processing functions are applied on the fly by the server using the dynamic processing capabilities described in Section 2. The application also includes graphing widgets for exploring spectral signatures of areas, or of individual locations through time, Figure 6. With a few clicks, anyone with internet access can examine vegetation health through time, and map burn scars or water bodies, anywhere in the world.

A more traditional approach to Landsat data would be to search for individual images, download images to a local desktop computer, then use a desktop image analysis software application to process the data, all of which requires more skill, and significantly more time.

Since the initial release, the Landsat collection in Amazon has grown to over a million scenes. Each day new scenes are added to the collection and an automated script updates the mosaic dataset referencing the scenes, making new imagery available as image services accessible to client applications. Python scripts are triggered as new scenes become available and use a raster type to read the metadata from USGS. MTL files to set up the appropriate functions in the mosaic datasets. The resulting mosaic dataset defines a virtual data cube that can then be accessed for analysis or served as dynamic image services.

The application and services are deployed on scalable commercial cloud infrastructure, allowing them to grow with demand over the last 4 years. They now regularly handle tens of thousands of requests per hour and maintain response times suitable for interactive web application use.

In 2018, Esri developed similar web services to access and share the Sentinel 2 available from the AWS Open Data registry [29]. These services use similar scripts that automate the ingestion of the metadata into the mosaic datasets from the Sentinel 2 DIMAP files. It can be accessed via a corresponding Sentinel Explorer web application [30].

These Landsat and Sentinel 2 services are also available in a single Earth Observation Explorer application [31]. It provides access to both collections in a single application and allows easy side-by-side comparisons of proximal image dates from different sensors, providing useful insight when used with the interactive image swipe tool.

3.2. Digital Earth Australia

The Australian GeoScience Data Cube was developed by Geoscience Australia and hosted on the National Computing Infrastructure at Australia National University. It was a 25-meter resolution, pixel aligned collection of over 300,000 Landsat scenes which were geometrically and spectrally calibrated to surface reflectance [8,16]. This was the first continental Landsat data cube with a large number of overlapping scenes in time, enabling new types of analysis to be envisioned and developed. A new version of this data cube has been created and is now known as Digital Earth Australia (DEA).

Geoscience Australia replicated the DEA cube in Amazon S3 and provided access to Esri Australia to prototype a capability serving DEA data from S3 using ArcGIS Image Server to enable hosted visualization and analysis from external client applications. The DEA data cube utilizes a YAML file to store metadata about each image tile. A Python raster type was created to allow the system to parse the YAML file of the data cube to generate an Esri mosaic dataset which holds the metadata and references to images stored in the cloud. The mosaic dataset makes all the imagery and metadata easily accessible through multiple APIs, directly in desktop applications or through image services published from ArcGIS Image Server, and therefore web clients.

This prototype also developed server-side analytics for common DEA data products such as Water Observations from Space (WOfS) [32] Figure 7. WOfS is a relevant test case because it benefits from the data preprocessing such as surface reflectance calibration and pixel alignment that is part of the data cube construction preprocessing. The analysis runs in Amazon near the DEA imagery. These calculations can be computed on-the-fly for the current display extent and resolution, or at full resolution to a persistent output dataset using a distributed computation service. Data visualization and analytics were shown to be accessible and responsive through Javascript web clients and ArcGIS Pro. Additional analytics and web applications are being considered for future development.

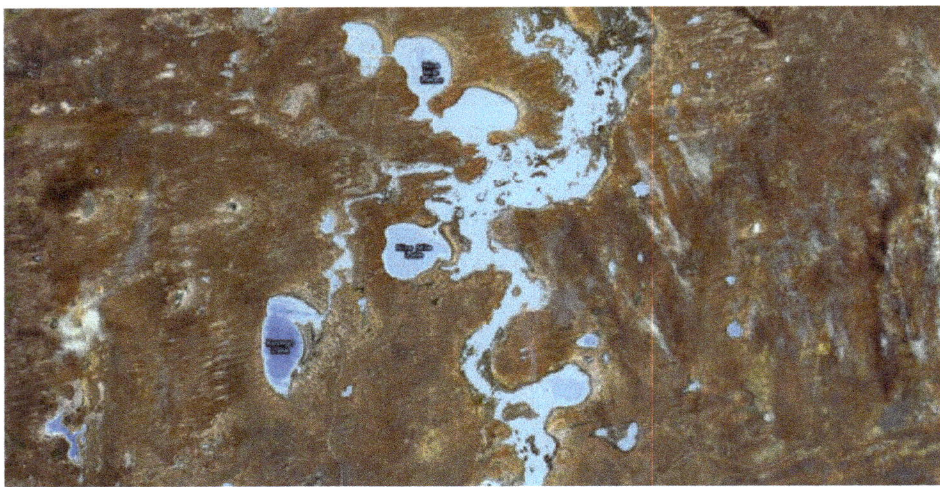

Figure 7. Water Observations from Space (WOfS) [32] calculation in New South Wales, Australia. Dark blue is persistent water, light blue is intermittent water, as infrequent as once every 10 years.

3.3. Digital Earth Africa

The Open Data Cube [9] is a NASA-CEOS initiative in collaboration with Geoscience Australia to create open source software for creating, managing, and analyzing Earth observation data cubes. The largest project is a data cube for the continent of Africa known as Digital Earth Africa. The initial prototype of this project was the Africa Regional Data Cube (ARDC) created in 2018 which covers Kenya, Senegal, Sierra Leone, Ghana, and Tanzania [10]. NASA-CEOS provided access to the image data cube of Ghana stored in Amazon S3, which contains Landsat and Sentinel 1 data. The Sentinel 1 data is a collection of 44 monthly mosaics created by Norwegian Research Centre (NORCE).

Building upon experience gained from the Digital Earth Australia collaboration described above, the Python raster type for reading the YAML files was modified for the characteristics of the Africa data collections and a mosaic dataset created to reference the images.

A collection of image services and web applications were published for exploring spectral indices through time, as well as performing interactive change detection of spectral and radar data, Figure 8. Additional analytic workflows are under development. This ongoing collaborative research and prototyping exercise is developing Python workflows and tools to publish similar open capabilities from the Digital Earth Africa data cube when it becomes available.

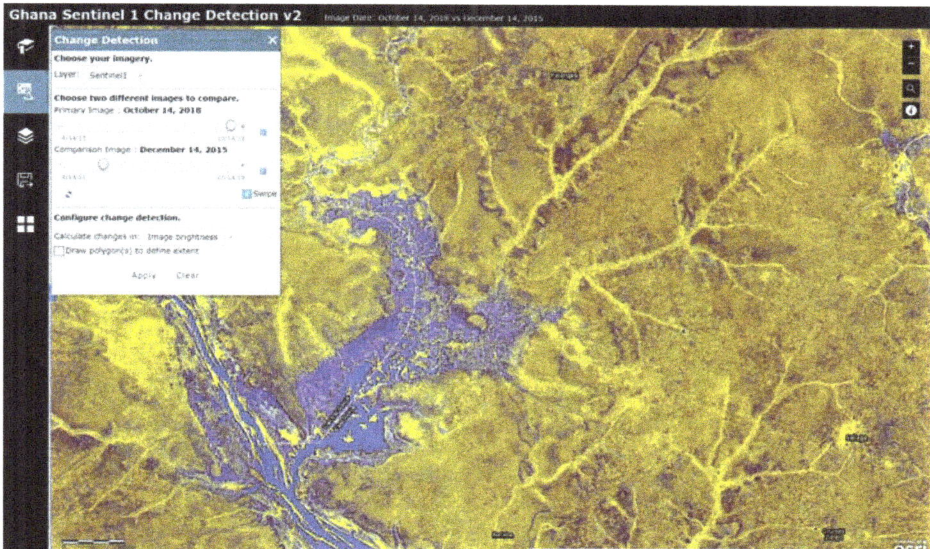

Figure 8. Change detection web application showing Sentinel 1 radar data of the Volta River area in central Ghana. The RGB is created as VV, VH, VV/VH. This application was created with the Change Detection web application template.

The image services, analytic services, and web applications will be made openly accessible through the Africa GeoPortal [33], a no-cost open portal for geospatial data and applications in Africa. The Africa GeoPortal is a fully cloud hosted SaaS, enabling users to create and share their own data, web maps, analysis, and custom hosted web applications. These open services and applications for Africa will be included into the search of other geospatial portals such as ArcGIS Online, which is part of the federated search of the GEOSS Portal and others supporting the appropriate OGC web service standards.

The three use cases described above use a common approach and software platform for accessing, analyzing, and sharing imagery from four different image data cubes. These data collections were built by four different groups of people in different locations, using different formats, compression, coordinate systems, and tiling schemes. The data is hosted on three different continents, yet is accessible from any internet connection through easy to use web applications and standard APIs, Figure 9. All applications presented maintain response times suitable for interactive exploration and analysis through web applications, and required no data download or image replication. Interested readers are encouraged to visit the Earth Observation Explorer application [31] for their area of interest. This approach is applicable to nearly any spatially and temporally organized collection of earth observation imagery.

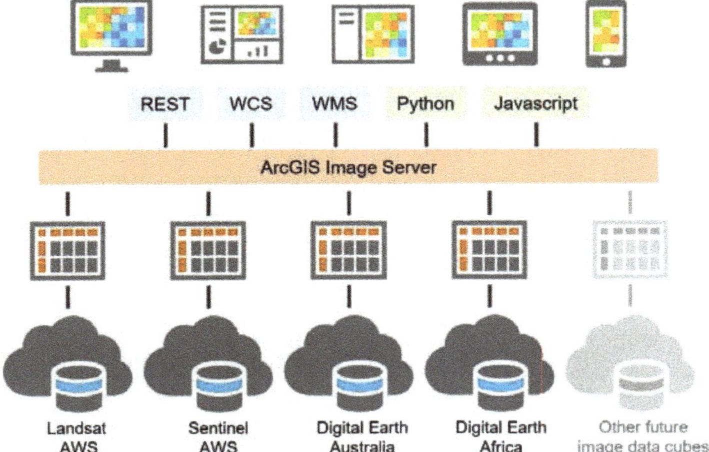

Figure 9. Conceptual overview of three use case examples illustrating cloud hosted image cubes from multiple providers (at the bottom), each accessed through a mosaic dataset referencing image metadata and pixel locations, served with ArcGIS Image Server in a variety of common APIs, consumed by a range of desktop, web, and mobile applications.

4. Conclusions

We described here a platform for accessing, analyzing, and sharing earth observation imagery from a variety of data cubes. Free and low-cost open data sharing and cloud hosting of large collections of imagery as data cubes is improving access and increasing usage of earth observation data. ARD preprocessing is an important and beneficial step in building data cubes but includes compromises that should be understood by data cube creators. Cloud hosted image analysis transforms image data into actionable information, increasing its value and potential for impact. Analytic web applications tailored to specific stakeholders and problems are growing in use and make multispectral and multitemporal imagery more approachable to a larger audience.

Within the next 5 years, we expect that nearly all earth observation data will exist in some form of ARD image data cube hosted in the cloud and provide hosted analytic web services. Therefore, it is important that software be capable of efficiently accessing and using data without replication from the variety of data cubes which will be created.

All python scripts, custom raster types, and sample applications developed for and described in this paper are freely available as examples to follow for anyone interested in connecting to or building other similar data cube integrations. New online training materials for how to build and integrate data cubes are being developed, as well as new training materials [34] illustrating how to use image data cubes and other data for addressing United Nations Sustainable Development Goals.

Author Contributions: Conceptualization, P.P. and A.D.; Data curation, A.D., K.Z., and H.X.; Methodology, P.P. and A.D.; Project administration, S.K.; Software, A.D., S.K., K.Z., and H.X.; Visualization, S.K.; Writing—original draft, S.K. and P.P.; Writing—review & editing, S.K., P.P., and D.J.W.

Funding: This research received no external funding.

Acknowledgments: The authors would like to acknowledge the contributions of the NASA-CEOS data cube team, Geoscience Australia, and Esri Australia for sharing their expertise, time and data for this collaboration. The comments and suggestions provided by anonymous reviewers are appreciated in creating a more clear and comprehensive narrative.

Conflicts of Interest: The authors declare no conflict of interest.

References

1. USGS. *Landsat 5 Landsat Science, 1 March 1984*; USGS: Reston, VA, USA. Available online: https://landsat.gsfc.nasa.gov/landsat-5/ (accessed on 31 May 2019).
2. Popkin, G. US Government Considers Charging for Popular Earth-Observing Data. *Nature* **2018**, *556*, 417–419. Available online: https://www.nature.com/articles/d41586-018-04874-y (accessed on 24 April 2018).
3. Wulder, M.A.; Masek, J.G.; Cohen, W.B.; Loveland, T.R.; Woodcock, C.E. Opening the Archive: How Free Data Has Enabled the Science and Monitoring Promise of Landsat. *Remote Sens. Environ.* **2012**, *122*, 2–10. [CrossRef]
4. Faundeen, J.L.; Kanengieter, R.L.; Buswell, M.D. U.S. Geological Survey Spatial Data Access. *J. Geospat. Eng.* **2002**, *4*, 145–152.
5. Meißl, S. Top 5 Trends in EO Data Usage—EOX. Available online: https://eox.at/2015/09/top5/ (accessed on 12 September 2017).
6. Larraondo, P.R.; Pringle, S.; Guo, J.; Antony, J.; Evans, B. GSio: A Programmatic Interface for Delivering Big Earth Data-as-a-Service. *Big Earth Data* **2017**, *1*, 173–190. [CrossRef]
7. Appel, M.; Lahn, F.; Buytaert, W.; Pebesma, E. Open and Scalable Analytics of Large Earth Observation Datasets: From Scenes to Multidimensional Arrays Using SciDB and GDAL. *ISPRS J. Photogramm. Remote Sens.* **2018**, *138*, 47–56. [CrossRef]
8. Lewis, A.; Lymburner, L.; Purss, M.B.; Brooke, B.; Evans, B.; Ip, A.; Dekker, A.G.; Irons, J.R.; Minchin, S.; Mueller, M.; et al. Rapid, High-Resolution Detection of Environmental Change over Continental Scales from Satellite Data—the Earth Observation Data Cube. *Int. J. Digit. Earth* **2016**, *9*, 106–111. [CrossRef]
9. Killough, B. Overview of the Open Data Cube Initiative. In Proceedings of the IGARSS 2018—2018 IEEE International Geoscience and Remote Sensing Symposium, Valencia, Spain, 23–27 July 2018; pp. 8629–8632. [CrossRef]
10. Killough, B.D. Satellite Analysis Ready Data for the Sustainable Development Goals. In *Earth Observation Applications and Global Policy Frameworks*; AGU Geophysical Monograph Series: Washington, DC, USA, 2019; in press.
11. Frye, C.; Paige, D.; Mead, R. Democratizing GIS: Are We There Yet? In Proceedings of the 1997 Esri Users Conference, Palm Springs, CA, USA, 14 July 1997. Available online: http://proceedings.esri.com/library/userconf/proc97/proc97/to700/pap664/p664.htm (accessed on 4 July 2019).
12. Smith, H. Data Democratization: Finally Living up to the Name. *InfoWorld.* 10 January 2018. Available online: https://www.infoworld.com/article/3246632/data-democratization-finally-living-up-to-the-name.html (accessed on 4 July 2019).
13. Wulder, M.A.; Loveland, T.R.; Roy, D.P.; Crawford, C.J.; Masek, J.G.; Woodcock, C.E.; Allen, R.G.; Anderson, M.C.; Belward, A.S.; Cohen, W.B.; et al. Current Status of Landsat Program, Science, and Applications. *Remote Sens. Environ.* **2019**, *225*, 127–147. [CrossRef]
14. Lewis, A.; Lacey, J.; Mecklenburg, S.; Ross, J.; Siqueira, A.; Killough, B.; Szantoi, Z.; Tadono, T.; Rosenavist, A.; Goryl, P.; et al. CEOS Analysis Ready Data for Land (CARD4L) Overview. In Proceedings of the IGARSS 2018—2018 IEEE International Geoscience and Remote Sensing Symposium, Valencia, Spain, 23–27 July 2018; pp. 7407–7410. [CrossRef]
15. Roy, D.P.; Li, J.; Zhang, H.K.; Yan, L. Best Practices for the Reprojection and Resampling of Sentinel-2 Multi Spectral Instrument Level 1C Data. *Remote Sens. Lett.* **2016**, *7*, 1023–1032. [CrossRef]
16. Lewis, A.; Oliver, S.; Lymburner, L.; Evans, B.; Wyborn, L.; Mueller, N.; Raevksi, G.; Hooke, J.; Woodcock, R.; Sixsmith, J.; et al. The Australian Geoscience Data Cube—Foundations and Lessons Learned. *Remote Sens. Environ.* **2017**, *202*, 276–292. [CrossRef]
17. Holmes, C. Cloud Native Geospatial Part 2: The Cloud Optimized GeoTIFF. Available online: https://medium.com/planet-stories/cloud-native-geospatial-part-2-the-cloud-optimized-geotiff-6b3f15c696ed (accessed on 10 October 2017).
18. Becker, P.; Plesea, L.; Maurer, T. Cloud Optimized Image Format and Compression. *ISPRS Int. Arch. Photogramm. Remote Sens. Spat. Inf. Sci.* **2015**, *7*, 613–615. [CrossRef]
19. Deutsch, P. RFC 1951 DEFLATE Compressed Data Format Specification Version 1.3, RFC Editor May 1996. Available online: https://www.rfc-editor.org/rfc/rfc1951.pdf (accessed on 4 July 2019).

20. Plesea, L. *Meta Raster Format (MRF) User Guide*; Reprint, Global Imagery Browse Services: Washington, DC, USA, 2014. Available online: https://github.com/nasa-gibs/mrf (accessed on 20 May 2019).

21. Warmerdam, F. The Geospatial Data Abstraction Library. In *Open Source Approaches in Spatial Data Handling*; Springer: Berlin, Germany, 2008; pp. 87–104. Available online: https://link.springer.com/chapter/10.1007/978-3-540-74831-1_5 (accessed on 4 July 2019).

22. Xu, H.; Becker, P. ArcGIS Data Models for Managing and Processing Imagery. *ISPRS Int. Arch. Photogramm. Remote Sens. Spat. Inf. Sci.* **2012**, *39*, 97–101. [CrossRef]

23. Xu, H.; Abdul-Kadar, F.; Gao, P. *An Information Model for Managing Multi-Dimensional Gridded Data in a GIS. IOP Conference Series: Earth and Environmental Science*; IOP Publishing: Bristol, UK, 2016. [CrossRef]

24. Strobl, P.; Baumann, P.; Lewis, A.; Szantoi, Z.; Killough, B.; Purss, M.; Craglia, M.; Nativi, S.; Held, A.; Dhu, T. The Six Faces of the Data Cube. In Proceedings of the 2017 Conference on Big Data from Space, Toulouse, France, 28–30 November 2017. [CrossRef]

25. Wagemann, J.; Clements, O.; Figuera, R.M.; Rossi, A.P.; Mantovani, S. Geospatial Web Services Pave New Ways for Server-Based on-Demand Access and Processing of Big Earth Data. *Int. J. Digit. Earth* **2018**, *11*, 7–25. [CrossRef]

26. *Gartner Forecasts Worldwide Public Cloud Revenue to Grow 17.3 Percent in 2019*; Gartner: Stanford, CT, USA, 1979. Available online: https://www.gartner.com/en/newsroom/press-releases/2018-09-12-gartner-forecasts-worldwide-public-cloud-revenue-to-grow-17-percent-in-2019 (accessed on 12 September 2018).

27. Sundwall, J. Start Using Landsat on AWS. AWS News Blog. 19 May 2015. Available online: https://aws.amazon.com/blogs/aws/start-using-landsat-on-aws/ (accessed on 4 July 2019).

28. Esri. *Landsat Unlock Earth's Secrets*; Esri: Redlands, CA, USA, 1969. Available online: https://www.esri.com/en-us/arcgis/landsat (accessed on 31 May 2019).

29. Sentinel-2-Registry of Open Data on AWS. Registry of Open Data on AWS. Available online: https://registry.opendata.aws/sentinel-2/ (accessed on 31 May 2019).

30. Esri. *Sentinel Explorer*; Esri: Redlands, CA, USA, 1969. Available online: https://sentinel2explorer.esri.com/ (accessed on 31 May 2019).

31. Esri. *Earth Observation Explorer*; Esri: Redlands, CA, USA, 1969. Available online: https://livingatlas2.arcgis.com/eoexplorer/ (accessed on 31 June 2019).

32. Mueller, N.; Lewis, A.; Roberts, D.; Ring, S.; Melrose, R.; Sixsmith, J.; Lymburner, L.; McIntyre, A.; Tan, P.; Curnow, S.; et al. Water Observations from Space: Mapping Surface Water from 25 Years of Landsat Imagery across Australia. *Remote Sens. Environ.* **2016**, *174*, 341–352. [CrossRef]

33. Cozzens, T. Esri's Africa GeoPortal to Help with Urgent Development Challenges. Geospatial Solutions: Cleveland, OH, USA, 2018. Available online: http://geospatial-solutions.com/tag/africa-geoportal/ (accessed on 11 May 2018).

34. Esri. *Solve Problems for Sustainable Development Goals*; Esri: Redlands, CA, USA, 1969. Available online: https://learn.arcgis.com/en/paths/solve-problems-for-sustainable-development-goals/ (accessed on 31 June 2019).

Article

Snow Cover Evolution in the Gran Paradiso National Park, Italian Alps, Using the Earth Observation Data Cube

Charlotte Poussin [1,2,3,*], **Yaniss Guigoz** [1,3], **Elisa Palazzi** [4], **Silvia Terzago** [4], **Bruno Chatenoux** [1,3] **and Gregory Giuliani** [1,3]

[1] Institute for Environmental Sciences, University of Geneva, Bd Carl-Vogt 66, CH-1205 Geneva, Switzerland; yaniss.guigoz@unige.ch (Y.G.); bruno.chatenoux@unige.ch (B.C.); gregory.giuliani@unepgrid.ch (G.G.)

[2] Department of F.-A. Forel for Environment and Water Sciences, Faculty of Sciences, University of Geneva, Bd Carl-Vogt 66, CH-1205 Geneva, Switzerland

[3] UNEP/GRID-Geneva, 11 ch. des Anémones, CH-1219 Châtelaine, Switzerland

[4] Institute of Atmospheric Sciences and Climate, National Research Council (ISAC-CNR), corso Fiume 4, 10133 Torino, Italy; e.palazzi@isac.cnr.it (E.P.); s.terzago@isac.cnr.it (S.T.)

[*] Correspondence: charlotte.poussin@unige.ch; Tel.: +41-22-510-56-85

Received: 13 June 2019; Accepted: 3 October 2019; Published: 9 October 2019

Abstract: Mountainous regions are particularly vulnerable to climate change, and the impacts are already extensive and observable, the implications of which go far beyond mountain boundaries and the environmental sectors. Monitoring and understanding climate and environmental changes in mountain regions is, therefore, needed. One of the key variables to study is snow cover, since it represents an essential driver of many ecological, hydrological and socioeconomic processes in mountains. As remotely sensed data can contribute to filling the gap of sparse in-situ stations in high-altitude environments, a methodology for snow cover detection through time series analyses using Landsat satellite observations stored in an Open Data Cube is described in this paper, and applied to a case study on the Gran Paradiso National Park, in the western Italian Alps. In particular, this study presents a proof of concept of the preliminary version of the snow observation from space algorithm applied to Landsat data stored in the Swiss Data Cube. Implemented in an Earth Observation Data Cube environment, the algorithm can process a large amount of remote sensing data ready for analysis and can compile all Landsat series since 1984 into one single multi-sensor dataset. Temporal filtering methodology and multi-sensors analysis allows one to considerably reduce the uncertainty in the estimation of snow cover area using high-resolution sensors. The study highlights that, despite this methodology, the lack of available cloud-free images still represents a big issue for snow cover mapping from satellite data. Though accurate mapping of snow extent below cloud cover with optical sensors still represents a challenge, spatial and temporal filtering techniques and radar imagery for future time series analyses will likely allow one to reduce the current cloud cover issue.

Keywords: data cube; optical remote sensing; snow cover; Gran Paradiso National Park; climate change

---

1. Introduction

The latest scientific observations [1,2] highlight a clear warming of the global climate system in recent decades, directly affecting the atmosphere, land, oceans and the cryosphere. Mountainous environments are among the regions most sensitive to and most affected by climate change [3,4]. Several studies using both measured and modelled data show evidence that warming rates are amplified at higher elevations (e.g., [3] and references therein). Among the major effects of such warming are the shrinking of glaciers; reductions of snow cover extension, quantity and duration; and permafrost thawing [1,2,5], bringing on consequences from water availability; e.g., [6], biodiversity and ecosystem functions and services.

To assess and understand environmental changes in mountains and to support decision-making and adaptation, regular and continuous monitoring is required. One of the Essential Climate Variables [7] deserving particular attention is snow cover. The reduction of snow cover causes an amplification of warming rates through snow-albedo feedback. Moreover, snow represents a crucial seasonal reserve of water for downstream areas and has a key role in many ecological processes.

Owing to the difficulty of monitoring high-altitude environments with in-situ station networks, satellite Earth observation (EO) data can be considered as an appropriate source to complement scattered in-situ measurements [8]. EO are well suited to map snow cover because of the good contrast of snow with most other natural surfaces, except some clouds [9,10]. Moreover, the global coverage and regular repeatability of measurements offered by satellite images allow experts to monitor the vast temporal and spatial variability of snow cover where ground measurements may be insufficient, expensive or even dangerous [10–12]. For more than 40 years, snow has been successfully mapped from space using a variety of sensors [9,10,13–17]. Since the 1970s, the long-term data records make Landsat datasets frequently used to study snow cover around the world at medium resolution (15 to 100 m) [18–21]. With the advent of the European Space Agency's (ESA's) Sentinel constellation, it is now possible to have images every 5 days at 10-m resolution. This further enhances monitoring capabilities to provide nearly real-time information on several geophysical parameters [22].

To efficiently exploit the increasing availability of satellite EO data, Earth observations data cubes (EODC) have recently emerged [23,24]. They represent a solution to store, organize, manage and analyze large amounts of multi-sensor EO data. The ambition is to allow scientists, researchers and other possible users to harness big EO data, facilitating the access and use of analysis ready data (ARD) [25–27], which are consistently processed using the highest scientific standards for immediate analysis in applications and for time-series exploitation. The interest in that objective has been proven by various implementations that already exist, such as the Open Data Cube (ODC) [25], the EarthServer [28], the e-sensing platform [29], the JRC Earth Observation Data and Processing Platform (JEODPP) [30], the Copernicus Data and Information Access Services (DIAS) [31,32] and the Google Earth Engine (GEE) [33].

In this paper we show, for the first time, an application of the EO Swiss Data Cube (SDC), an EODC specifically developed for Switzerland, based on Landsat imagery. We apply a preliminary version of an algorithm for snow detection, hereafter referred to as snow observations from space (SOfS), to the SDC to extract information on snow cover area (SCA) in the period 1984–2018, focusing on the Gran Paradiso National Park (GPNP) in the western Italian Alps. Snow cover information for the Gran Paradiso National Park based on in-situ stations is very limited and not easily available. In this case remote sensing data can provide valuable additional information for park management.

The first objective of this study is to assess the characteristics of this new snow dataset, such as the frequency of available images and of the cloud-free observations per month. Owing to significant cloud obstruction in many satellite images, we then test an approach, part of the SOfS algorithm, to combine images into monthly aggregations, providing maximum snow cover area products for a given month. A similar approach to mitigate the impact of clouds has already been applied to MODIS data by combining the information provided by MODIS Terra and Aqua sensors at the weekly time scale (i.e., [34–36]). The SOfS algorithm also allows one to evaluate, on a seasonal (winter) scale, the number of cloud-free observations and the number of times snow was observed in the corresponding cloud-free scenes. Combining the information of the number of snow observations relative to the number of cloud-free observations, we produce "snow cover summaries" for each winter season from 1984 to 2018 and for the aggregation of all 34 winters seasons. Third, a tentative analysis of the temporal evolution of the percent area of the GPNP with high/low probability to observe snow in winter is also presented, though this has to be regarded with caution owing to the non-homogeneous frequency of the underlying observations. We finally identify and discuss the main benefits and limitations of the snow data cube and of the SOfS algorithm employed for snow discrimination. Possible strategies to further enhance the capabilities of the SOfS algorithm exploiting the EODC infrastructure are presented and discussed.

The paper is structured as follows: Section 2 describes the area of study and its climatic and environmental characteristics; Section 3 presents the Swiss Data Cube; Section 4 describes the methodology used to generate the snow cover dataset, including the procedure to identify clouds and the preliminary version of the SOfS algorithm, and describes how the snow cover monthly and seasonal summaries are generated; Section 5 shows the application of that methodology, discussing the characteristics of the snow cover data cube obtained for the GPNP area and presents the main results of the paper; Section 6 discusses the limitations and perspectives of this study and Section 7 concludes the paper.

2. Study Area and Its Climatic Characteristics

This study focuses on the Gran Paradiso National Park (GPNP) protected area, located in the western Italian Alps, encompassing the Aosta Valley and Piedmont regions, as shown in Figure 1. Established in 1922, the Italian national park covers a surface of about 720 km^2 [37], with elevations ranging from about 700 m a.s.l. to about 4000 m a.s.l. More than 75% of the park area lies above 2000 m a.s.l., and it is covered by snow for several months per year. A total of 60% of the park is covered by areas with scarce or no vegetation (rocks, screes and glaciers), while 20.2% is characterized by alpine vegetation (woods and shrubs) [38]. The pastures and meadows represent 17% of the Italian park and only 0.8% of the park is dedicated to urban areas and cultivated lands. The GPNP is home to almost 15 altitude lakes and to great richness and diversity of fauna and flora [38].

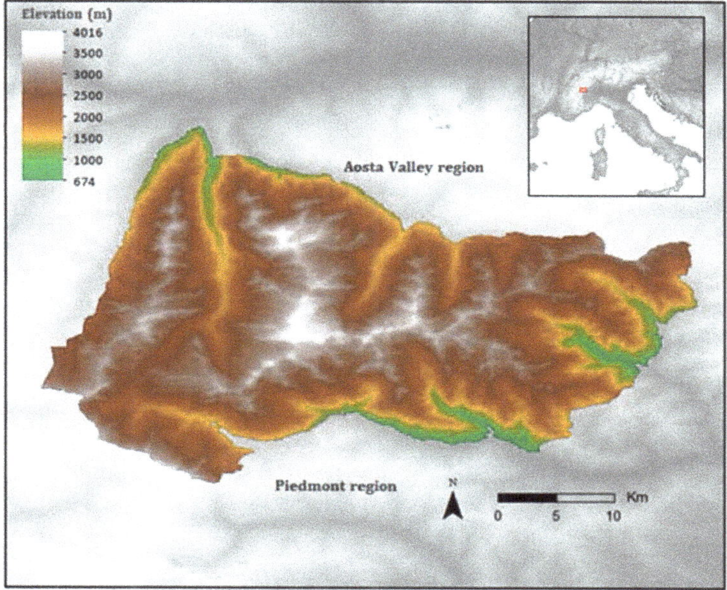

Figure 1. Location and topographic map of the Gran Paradiso National Park, northwestern Italian Alps.

Though differences in elevation, slope and aspect between valleys can give rise to diverse microclimate conditions inside the park, the average climate of the area is characterized by a relatively low mean temperature, scarce precipitation and well-defined seasons. Figure 2 shows the climatological annual cycle of temperature and precipitation in the GPNP area obtained from an observation-based gridded dataset specifically developed for the Greater Alpine Region, HISTALP [39,40]. This dataset provides monthly temperature and (total and solid) precipitation at a spatial resolution of 0.08° latitude-longitude, corresponding to about 10 km, from 1780 to 2014 (for temperature) and from 1801 to 2014 (for precipitation). Based on this dataset and considering averages over the period 1950–2014 (Figure 2), the coldest months in the GPNP are December to March, with an average temperature of about −6 °C. The warmest months are July and August, with an average temperature of about 9 °C. Precipitation has maxima in spring/early summer (April, May and June, 110 mm/month) and autumn (particularly in November with about 180 mm/month), while the driest conditions along the year are found in July–August (71 mm/month) and in December to March (76 mm/month).

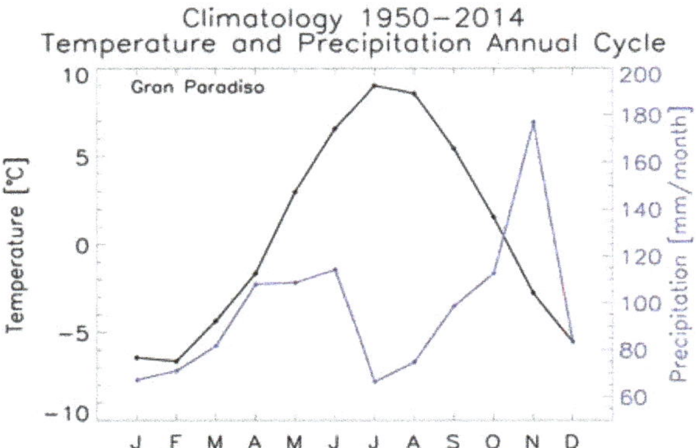

Figure 2. Climatological annual cycle of temperature (left y-axis, black) and total precipitation (right y-axis, blue) spatially averaged over the Gran Paradiso National Park area, obtained from the HISTALP observation-based dataset. The climatology was obtained after temporally averaging the data over the period 1950–2014.

The temporal extent of the HISTALP dataset allows one to study changes in temperature and precipitation that occurred over the last few decades; for example, considering the averages in the period 1951–1980 (Figure 3, left column plots), the averages in the period 1985–2014 (middle column) and the difference between the latter and the former (right column) of temperature (panels c, d, e), total precipitation (panels f, g, h) and solid precipitation (panels i, l, m). We considered only the winter season, including the months from December to February (DJF), which are particularly relevant for the subsequent snow analysis. The Gran Paradiso National Park area experienced positive winter temperature changes in 1985–2014 compared to 1951–1980, ranging from about 1 °C to about 1.7 °C (Figure 3e). The spatial distribution of the observed warming is positively correlated with the elevation distribution (Figure 3b) with a Pearson's correlation coefficient greater than 0.9. Precipitation changes, for both total (Figure 3h) and solid (Figure 3m) precipitation are negative everywhere in the considered domain, indicating a decrease of precipitation in 1985–2014 compared to 1951–1980. The spatial pattern of change is similar for the total and solid precipitation, showing the largest negative values in the western and north-western part of the box encompassing the GPNP; overall, snowfall decreased slightly more than total precipitation.

In addition, Table 1 summarizes the temporal trends of winter (DJF) temperature and precipitation from 1950–2014 and for the more recent sub period (1985 to 2014), for which satellite data are available. For completeness, results for the months of February and April are also reported in the table, since these are key months for understanding snow dynamics in the Alps [41]. In particular, February is representative of a cold winter snowpack while April roughly corresponds to a time of maximum snowpack in the Alps [42].

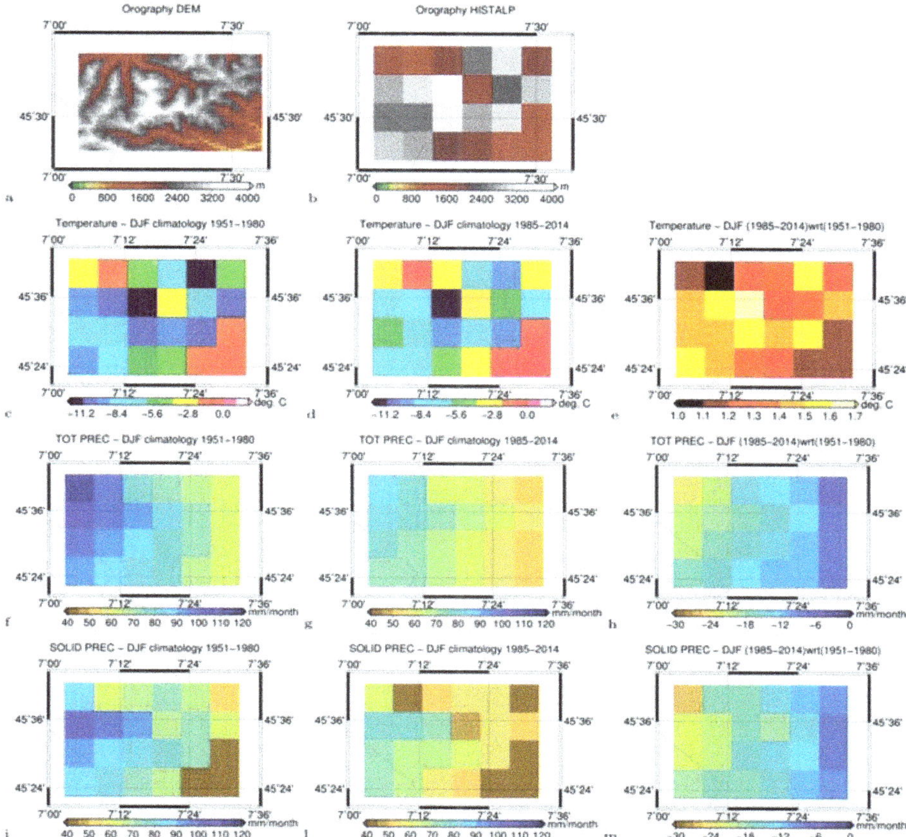

Figure 3. (**a**,**b**): Topographic map of the study area from a Digital Elevation Model at 0.008° (~1 km) resolution (**a**) and from the HISTALP dataset at 0.08° (~10 km) resolution (**b**). (**c–e**): Winter (December–February, DJF) spatial maps of surface air temperature averaged (**c**) over the periods 1951–1980 and (**d**) 1985–2014. (**e**) Spatial map of the difference between the (1985–2014) and (1951–1980) climatologies. (**f–h**): The same as (**c–e**) for total precipitation. (**i–m**): the same as (**c–e**) for solid precipitation.

Table 1. Trends of temperature, total and solid precipitation, from 1985 to 2014 and from 1950 to 2014, for DJF, February and April. The (*) indicates statistically significant values (p-value < 0.05).

		Temperature (°C/decade)	Total Precipitation (mm/month/decade)	Solid Precipitation (mm/month/decade)
DJF	1985–2014	−0.07	7.88	5.93
	1950–2014	0.31 (*)	−1.22	−1.93
FEB	1985–2014	−0.24	9.11	7.75
	1950–2014	0.29	−2.79	−2.93
APR	1985–2014	0.30	−18.34	−16.45
	1950–2014	0.20 (*)	−2.91	−2.78

3. The Swiss Data Cube

The Swiss Data Cube (SDC) is the second operational national Earth Observation Data Cube (EODC) worldwide, after Australia's [25,43,44]. It is an initiative supported by the Swiss Federal Office for the Environment and jointly developed, implemented and operated by the United Nations Environment Programme (UNEP)/GRID-Geneva and the University of Geneva. The SDC aims at supporting the Swiss government and institutions for their environmental monitoring and reporting mandates, and supports national scientific institutions to benefit from EO data for research and education. The SDC is built upon the ODC, a storage and analytical open source framework supported by Geoscience Australia, the Commonwealth Scientific and Industrial Research Organization, the United States Geological Survey (USGS), the National Aeronautics and Space Administration (NASA) and the Committee on Earth Observations Satellites (CEOS) [26,45,46]. The SDC currently holds 35 years of Landsat 5,7 and 8 data from the NASA Landsat program [47], covering the period from 1984 to 2019, updated daily, along with the entire Sentinel-2 archives [48,49] and a part of the Sentinel-1 archives [50,51].

The data available in the SDC are in an analysis ready form (after LEDAPS/LaSRC processing by USGS), meaning that all radiometric, geometric, solar and atmospheric corrections have already been applied and the data are spatially segmented into 30×30 m resolution grids that cover the area considered. All Landsat data used have undergone the same transformations to become analysis ready data (ARD) in order to get a consistent and harmonized multi-sensor dataset. ARD ensures that EO measurements are radiometrically comparable and geometrically aligned. Indeed, as mentioned by the CEOS [52], ARD is expected to limit as far as possible, barriers to interoperability both through time and with data from different sensors. Therefore, this multi-sensor dataset can be considered to be derived from a single sensor [53]. The unique ARD archive of Switzerland accounts for a total data volume of approximately 6 TB and more than 150 billion observations over the entire country that can be analyzed both in the spatial and temporal dimensions.

The Swiss Data Cube was originally developed employing eight Landsat tiles following the tiling notation system for Landsat data (the Worldwide Reference System-2 paths and rows) to cover the whole of Switzerland, also including parts of northern Italy, such as the Gran Paradiso National Park. Sentinel data, incorporated subsequently into the data cube, also cover the same domain. Despite the availability of Sentinel data in the archive, this proof of concept study focuses on Landsat sensors only. Indeed, for reasons of consistency and homogeneity, it is preferable for an initial test study to use data from the same satellite series with the same spatial resolution and having undergone the same ARD process. This study used the Collection 1 Level 1 data from Landsat 5 Thematic Mapper (TM) available from March 1984 to November 2011, Landsat 7 Enhanced Thematic Mapper Plus (ETM+) available since June 1999 and Landsat 8 Operational Land Imager (OLI)/Thermal Infrared Sensor (TIRS), available since March 2013. These datasets are characterized by a spatial resolution of 30 m and a revisit time of 16 days. Spectral bands of Landsat series satellites useful for snow detection and used in this study are the green and the shortwave Infrared (SWIR1) bands processed to orthorectified surface reflectance. Moreover, we employed the Collection 1 Level-1 Quality Assessment (QA) 16-bit Band, which provides information on the surface characteristics, cloud cover and sensor conditions (Table 2) that can be used to generate masks according to the application and user needs. In our case, we used information from the QA product to create a cloud mask. The main characteristics of the dataset used in this study are summarized in Table 3.

For the generation of snow cover products over the GPNP, we processed more than 480 Landsat images for the months of December to April in the period 1984–2018.

Table 2. Values assumed by the Surface Reflectance Pixel Quality Assessment (QA) band values [54].

Landsat-5 and Landsat-7		Landsat-8
Attribute	**Pixel Values**	
Fill (no data values in the pixel)	1	1
Clear	66, 130	322, 286, 834, 898, 1346
Water	68, 132	324, 388, 836, 900, 1348
Cloud shadow	72, 136	328, 392, 840, 904, 1350
Snow/Ice	80, 112, 144, 176	336, 368, 400, 432, 848, 880, 912, 944, 1352
Cloud	96, 112, 160, 176, 224	352, 368, 416, 432, 480, 864, 880, 928, 944, 992
Low confidence * cloud	66, 68, 72, 80, 96, 112	322, 324, 328, 336, 352, 368, 834, 836, 840, 848, 864, 880
Medium confidence * cloud	130, 132, 136, 144, 160, 176	386, 388, 392, 400, 416, 432, 900, 904, 928, 944
High confidence * cloud	-	480, 992
Low confidence * cirrus	-	322, 324, 328, 336, 352, 368, 386, 388, 392, 400, 416, 432, 480
High confidence * cirrus	-	834, 836, 840, 848, 864, 880, 898, 900, 904, 912, 928, 944, 992
Terrain occlusion	-	1346, 1348, 1350, 1352

* Low confidence = 0–33% confidence; Medium confidence = 34–66% confidence; High confidence = 67–100% confidence.

Table 3. Characteristics of Landsat data used in this study.

Optical Satellite Platform		Landsat-5	Landsat-7	Landsat-8
Sensor		TM	ETM+	OLI/TIRS
Period (start–end)		March 1984–November 2011	June 1999–	March 2013–
Revisit time (day)		16	16	16
Spatial resolution (m)		30	30	30
Bands used		Green, SWIR1, QA band	Green, SWIR1, QA band	Green, SWIR1, QA band
Wavelenght (μm)	**Green**	0.52–0.60	0.52–0.60	0.53–0.59
	SWIR1	1.55–1.75	1.55–1.75	1.57–1.65

4. Snow Observation from Space (SOfS) Algorithm

This study presents a preliminary version of the snow observation from space (SOfS) algorithm, which exploits consolidated techniques to detect snow cover from satellite images in the SDC and provides further tools to combine snow cover information over different temporal scales, for example, monthly and seasonal. A detailed description of the SOfS procedure is provided in the following sub-sections.

4.1. Cloud and Water Masks

Satellite observations are affected by many factors which can lead to poor observational quality, such as instrument failure or cloud interference [55]. During the winter season and in mountain regions especially, the presence of clouds reduces the capability of snow cover mapping through optical remote sensing [56,57]. Additionally, since clouds can have spectral signatures similar to snow, it is sometimes difficult to discriminate them from spaceborne multispectral sensors [16,21,58,59]. In order to exclude non-clear sky pixels, a cloud mask can be applied before evaluating the presence of snow. In this paper, to discriminate cloud covered pixels, we considered the cloud cover information included in the Landsat Collection 1 Level-1 Quality Assessment Band, derived using the C Function of Mask (CFMask) algorithm [60]. This algorithm provides information on the presence of clouds, along with a measure of the level of confidence that a pixel is actually cloud covered. The CFMask is very conservative and may have issues over bright targets, such as snow and ice surfaces, leading to possible overestimation of cloud cover, thus a loss of useful information for snow cover mapping. Therefore, we reprocessed the pixels and we considered as "cloud-covered," those previously identified as cloudy with a medium (34%–66%) or high (67%–100%) level of confidence, and pixels with cirrus clouds (at any level of confidence). More precisely, we considered as cloud covered, all pixels having the following attributes (see Table 2): "fill" (no data value), "low confidence cirrus," "high confidence cirrus," "medium confidence cloud" and "high confidence cloud".

Next, water pixels were filtered using a water mask from the Global Surface Water Explorer [61].

4.2. Snow Cover Detection and the Generation of Monthly Products

After filtering the dataset with a cloud and water mask, in order to detect the presence of snow in each pixel, we employed the well-known normalized difference snow index (NDSI) test [62–64] which has proven to be effective for the binary (i.e., for distinguishing between "snow" or "not snow") monitoring of snow cover [9,15,16,58,65–67]. NDSI is defined as the reflectance difference between visible (green) and shortwave infrared (SWIR) wavelengths:

$$\text{NDSI} = (r_{\text{green}} - r_{\text{SWIR}})/(r_{\text{green}} + r_{\text{SWIR}}) \tag{1}$$

where r_{green} and r_{SWIR} are the surface reflectances in the green and in the SWIR bands, respectively (see Table 3). The index ranges from −1 to +1. A pure snow pixel is characterized by a high NDSI, while a pixel with mixed elements (e.g., snow, water, vegetation, bare ground, etc.) is characterized by lower NDSI values [9]. A threshold test is then required to classify a pixel as snow covered or snow free. Based on a previous study performed using Landsat TM data in Sierra Nevada [64], pixels with at least 50% snow cover are found to have a NDSI value greater than 0.4. Though this value is commonly used [12], it may vary according to landscape conditions [15,20,68–70]. For the Gran Paradiso National Park, we decided to apply the same threshold as the one adopted for a study over Switzerland (in preparation), i.e., 0.45, given the similarity of land cover and altitude characteristics between the two areas.

The low temporal resolution of Landsat satellites combined with the presence of clouds in mountainous environments makes it difficult to get cloud-free observations. Therefore, numerous techniques, such as multi-date composite and a combination of different sensors, have been developed to aggregate multiple observations over time [15,21,71,72]. In this paper, we combine the single NDSI maps into monthly aggregations (for each month from December to April) using a mosaicking method. As shown in Figure 4, this method takes the maximum NDSI value per pixel from any available image independently from the sensor during the considered month. The NDSI threshold test is finally applied to the maximum NDSI value. For each month (December to April) we then obtained one snow cover area (SCA) map, showing the maximum snow cover extent for that month. Figure 5 shows an example of the SCA map for February 2000, for which cloud-free and water-free pixels were classified as snow-covered (NDSI > 0.45) or snow-free (NDSI ≤ 0.45).

This technique minimizes, though without completely removing, the number of cloud-covered pixels in the monthly composite. Indeed, the percentage of scenes with less than 30% cloud cover increases from 51% to 63% after considering the maximum NDSI value by month. This multi-temporal and multi-sensor approach allows one to reduce limitations due to Landsat temporal resolution and to increase the number of pixels tested for the presence of snow [67,73].

A flow diagram of the SOfS algorithm which we set up is shown in Figure 6.

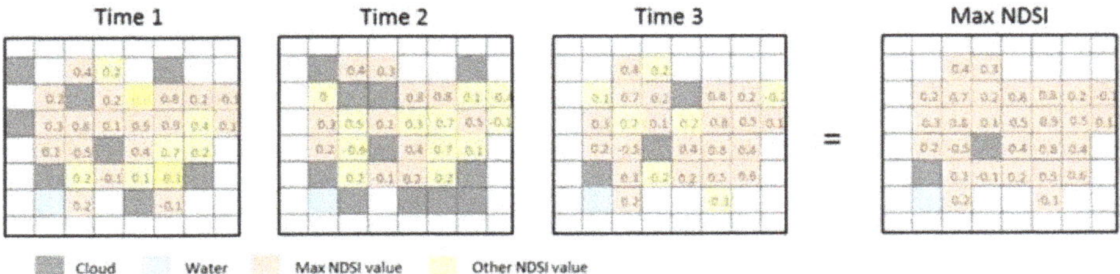

Figure 4. An example of the mosaicking technique employed to derive the maximum normalized difference snow index (NDSI) for a given month. From the available images (three in this example), NDSI was derived and the highest NDSI value was retained for the NDSI threshold test.

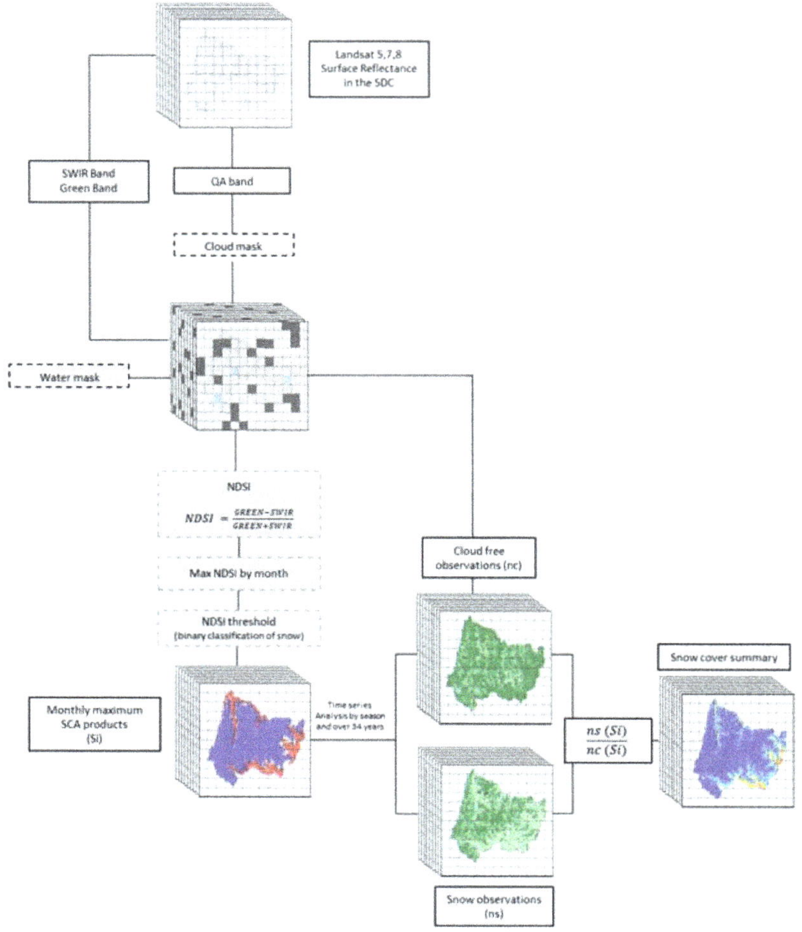

Figure 5. Monthly maximum snow covered area in February 2000, displaying snow-covered (blue), snow-free (red) and cloud-covered (grey) areas. Water surfaces are shown in white.

Figure 6. Flow chart of the preliminary version of the snow observation from space algorithm, applied to the Landsat images referring to the Gran Paradiso National Park.

4.3. Seasonal Snow Cover Summaries

Besides monthly aggregations, we also performed a seasonal-scale analysis, focusing on the winter season (DJF). For each winter season at the pixel level, we calculated the seasonal snow cover summary, Si, representing the relative frequency of snow cover observations in the corresponding monthly products, with respect to the total number of available cloud-free observations:

$$Si = ns/nc \times 100, \qquad (2)$$

where nc (from 0 to 3) is the number of cloud-free observations in the three-monthly snow cover products and ns (0 to 3) is the number of snow occurrences in the three snow cover products.

Figure 7 shows two examples of seasonal snow cover summaries for winters 1988/89 (upper panels) and 2001/02 (lower panels), the former characterized by substantial cloud cover in each of the monthly snow cover products, and the latter characterized by mostly cloud-free conditions. The number of cloud-free observations (nc) are displayed in Figure 7a,d: a white color corresponds to no observation available (0 out of 3 months), and a dark green color to the maximum number of observations available; i.e., three out of three over the winter season. The percentage of occurrences in each of the four considered classes is summarized in the pie chart insets in Figure 7a,d: for example, for the 2001/02 winter season (Figure 7d), 78% of observations are cloud-free in all the three images, while 14% (8%) are cloud-free in two (one) out of three images. The number of snow observations (ns) is displayed in Figure 7b,e: that corresponds to the number of times one pixel was classified as snow during the three winter months, with possible values of 0 (indicating absence of snow or no data either because of clouds or sensor technical issues), 1 (snow in only one out of three months), 2 (snow in two out of three months) or 3 (snow in all three months). Figure 7c,f shows the snow summaries (Si), expressing the percentage of snow cover observations with respect to the total number of cloud-free observations. In that case, each pixel can take one of the following values: 0%, (no snow cover in none of the three-monthly products), 33%, 50%, 66% (both snow cover and snow-free conditions in the three-monthly products) or 100% (snow cover in each of the cloud-free monthly products). The white pixels in Figure 7c,f are those with no data, or water or cloud covered in all three-monthly products.

The winter snow cover summary maps were employed to evaluate each season individually and explore the temporal evolution of the winter snow cover in the period 1984–2018. In addition to those time series, a snow cover summary considering all winter months in the period of study (1984–2018) was calculated. In the following, snow cover summary maps were interpreted in terms of the probability of observing snow cover in a given period of time in the GPNP.

To summarize, the preliminary version of the snow observation from space algorithm based on the NDSI approach, widely used to map snow cover in mountain regions, was applied to derive (i) monthly (December, January, February and April) SCA maps, defined as the maximum snow cover extent in the study area for that month; (ii) the winter (DJF) SCA analysis representing the percentage of snow observations with respect to the total number of cloud-free observations in all December, January and February in the period 1984–2018; (iii) winter (DJF) SCA time series, defined as the percentages of snow observations with respect to the total cloud-free observations during each winter season in the period 1984–2018. Implemented in an Earth Observation Data Cube environment, the algorithm could process a large amount of remote sensing data ready for analysis and benefited from all Landsat series since 1984 being put into one single multi-sensor dataset. Inspired by a methodology often used for MODIS sensors data, a multi-date composite of one month was applied to the snow dataset to reduce the impact of cloud cover. This proof-of-concept study tested, for the first time, the potential of the data cube infrastructure to monitor snow cover evolution in a small region of the Alps. In this context, Landsat images, with their long historical records and finer spatial resolution, become very attractive for multi-decade time series analyses, despite a 16-days repeat cycle. Applying the data cube methodology to EO datasets attempts to address new computation capabilities and to shift paradigm from scene-based analysis to pixel-based time series analysis. Time series analysis, with the aim to assess and monitor the variability and trends in snow cover extent in the GPNP.

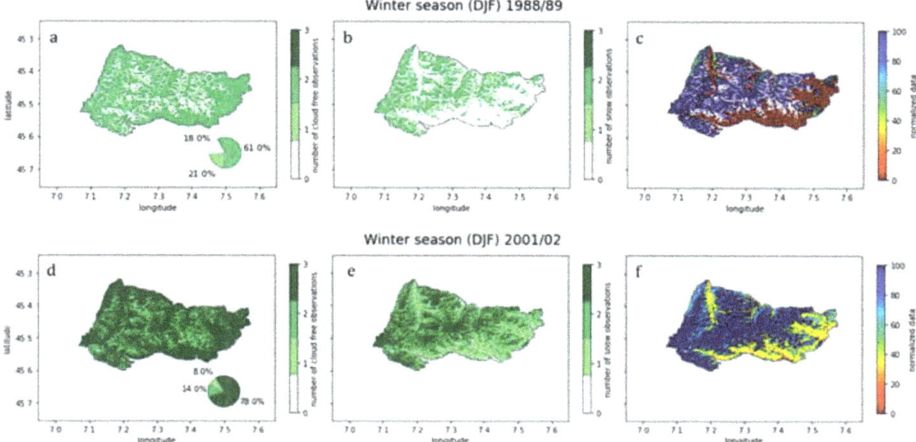

Figure 7. Example of aggregation of monthly maximum snow cover information for December, January and February (DJF) one a seasonal time scale, for the 1988/89 (upper panels) and the 2001/02 (bottom panels) winter seasons in Gran Paradiso National Park: (**a,d**) number of cloud-free observations ranging from 0 to 3 in the three-month period, (**b,e**) number of snow observations ranging from 0 to 3 in the same three-month period, and (**c,f**) snow cover summaries. The pie charts in (**a,d**) summarize the statistics for each of the four considered classes.

4.4. Snow Cover and Climatic Data Correlation

As temperature and precipitation are the two principal drivers of snow [74], we found it interesting to evaluate their trends over the past three decades in the Gran Paradiso National Park and analyze them jointly with the snow cover information resulting from the application of the SOfS algorithm. We used the HISTALP climate data already described in Section 2. We calculated the regression and correlation coefficients between climatic and snow cover data by least-squares linear fitting over the time period 1984–2014, for the winter season and for the months of February and April separately. Correlations were calculated excluding months with a cloud cover exceeding 30% (15%) of the entire park's surface in the monthly (seasonal) analyses.

5. Results

5.1. Availability of Cloud-Free Observations

Figure 8 shows in green colors, for every December, January, February, March and April in the period 1984–2018, the availability of Landsat scenes with less than 30% of cloud cover, while a cross indicates months with cloud cover exceeding 30% of the total GPNP area. The month of December is particularly under-represented with only 41% of months containing one or more scenes with less than 30% of cloud cover. Spring months (March and April) were also frequently affected by clouds. More than half of the cloudy scenes refer to the years before 2000 when only Landsat-5 (dark green in Figure 8) data were available. Since June 1999, the availability of two sensors (Landsat-5 and Landsat-7 for the period 2000–2012, and Landsat-7 and Landsat-8 for the period 2013–2018) considerably increased the number of scenes with less than 30% of cloud cover. The multi-temporal and multi-sensor approach employed in this study, i.e., the mosaicking method, allowed us to increase the number of valid monthly products with cloud cover of <30%. The last two columns of Figure 8 shows the percentage of monthly

SCA products with less than 30% of cloud cover before and after the application of the mosaicking method. The percentage values referring to the monthly aggregations are always greater than the corresponding ones referring to the single satellite scenes. As an example, for April, the percentage of valid monthly SCA products (cloud cover <30% of the total area) increased from 47% to 56%, gaining three additional months with less than 30% of cloud cover out of the 34 years studied.

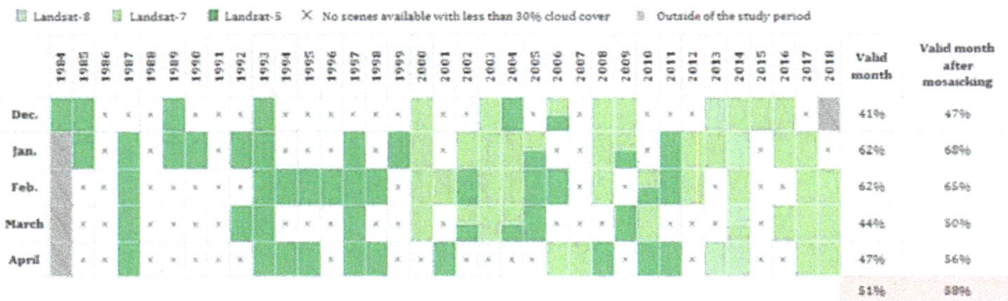

Figure 8. Green cells show the availability of Landsat (TM, ETM+ and OLI) scenes for the Gran Paradiso National Park with less than 30% cloud cover (after application of the CFMask) for December, January, February, March and April since 1984. Cells filled with more than one color indicate that valid scenes are available for more than one sensor. The last two columns represent the percentages of months over the full period (1984–2018) with at least one valid scene per month before and after the application of the mosaicking procedure respectively.

For completeness, Figure 9 shows the percentage of cloud cover which still remains in the monthly SCA products after application of the mosaicking procedure. Despite the improvements brought by the applied methodology, on average, 58% of all considered months were valid using a cloud cover threshold of 30% (see Figure 9). The percentage of cloud cover varied widely from year to year and from month to month. However, since 2000, and particularly for the months of January, February and April, we observed a decrease in the percentage of cloud cover, owing to the availability of two different Landsat sensors. For example, for the month of April (February) the percentage of cloudy months (cloud cover >30% of the total area) decreased from 62% (50%) for the period 1985–2000 to 28% (22%) for the period 2001–2018.

5.2. Winter Snow Cover Probability Map

Individual monthly maximum SCA products for December, January and February were employed to evaluate the "snow cover summary" over 34 years in the period 1984–2018 in the Gran Paradiso National Park area. Figure 10a shows the percentage of cloud-free observations per pixel in all December, January and February products in the period 1984–2018 (102 images considered in total; i.e., three winter months times 34 years). The percentage of cloud-free observations with respect to the total number of images available is at best equal to 88%, and rarely above 80%, as shown in Figure 10a in dark green colors. Most of the area (72%) had between 60% and 80% cloud-free observations, 27% of the area had between 40% and 60% cloud-free observations, while a small part of the park (0.63%) had less than 40% of cloud-free observations. This last class includes high-altitude areas (>3000 m a.s.l.) in the center of the park, that are particularly prone to cloud cover and showed few cloud-free observations available (20% to 40%) during the considered period. For those areas, highlighted in light green colors in Figure 10a, the uncertainty of snow observations was high and did not allow for a robust analysis. The percentage of occurrence in each of the five considered classes (0%–20%, 20%–40%, 40%–60%, 60%–80%, 80%–100%) is summarized in the pie chart inset into Figure 10a.

The percentage of snow observations in all December, January and February products in the period 1984–2018, with respect to the total number of cloud-free observations, is shown in Figure 10b.

More than 80% of the park surface (about 512 km^2 SCA) is usually covered by snow during winter (the probability to observe snow is greater than 80%), while only 1.8% of the park surface (about 11 km^2) is mainly snow-free in winter (probability to observe snow < 20%). Active areas, where ephemeral snow covers the ground (probability to observe snow between 20% and 80%), can be considered transitional zones between snow-covered and snow-free areas, covering about 17% of the Gran Paradiso park (corresponding to 109 km^2). The probability to observe snow in winter months was evaluated as well, for different 500 m-thick elevation bins between 500 m and 4500 m, as shown in Figure 10c. For each elevation bin, the frequency of the five different probability classes (0%–20%, 20%–40%, 40%–60%, 60%–80% and 80%–100%) were expressed in terms of percent area in that class with respect to the total area of that elevation bin. Areas above 2500 meters, corresponding to about 58% of the total area of the GPNP, are characterized by a high probability (80%–100%) of observing snow cover during winter. Between 1500 m and 2000 m about 50% of the area still has a high (80%–100%) probability to be snow-covered, while below 1500 m the area with high snow probability rapidly decreases. At elevations higher than 1500 m, the area where snow cover is rarely observed is negligible.

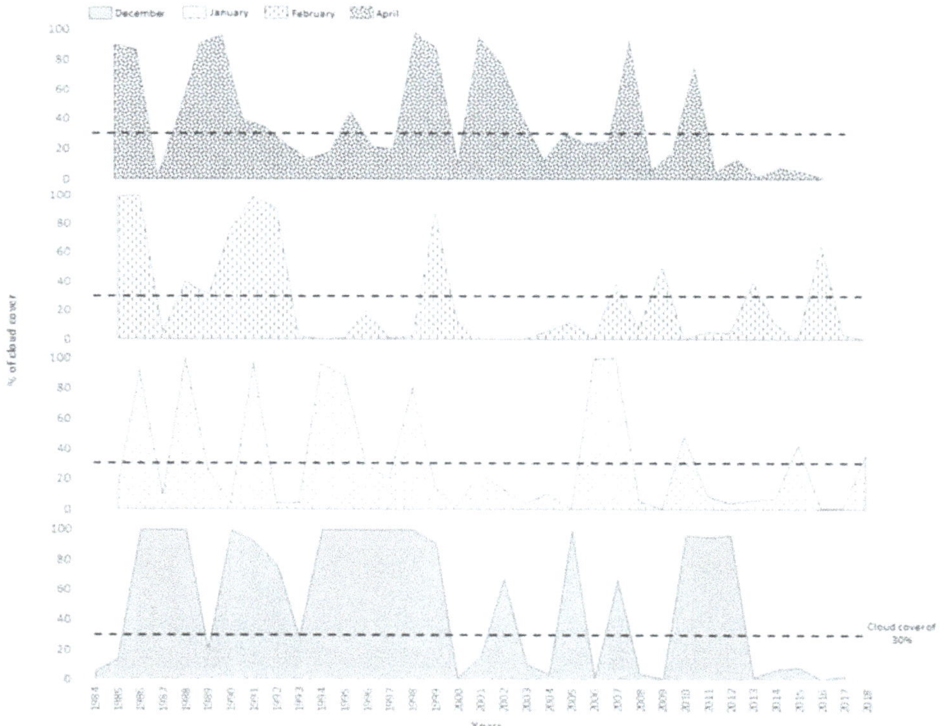

Figure 9. Percentage of cloud cover for December, January, February and April monthly snow cover area (SCA) products since 1984. The dashed line in each panel shows the 30% cloud cover threshold employed in this study, as described in the text.

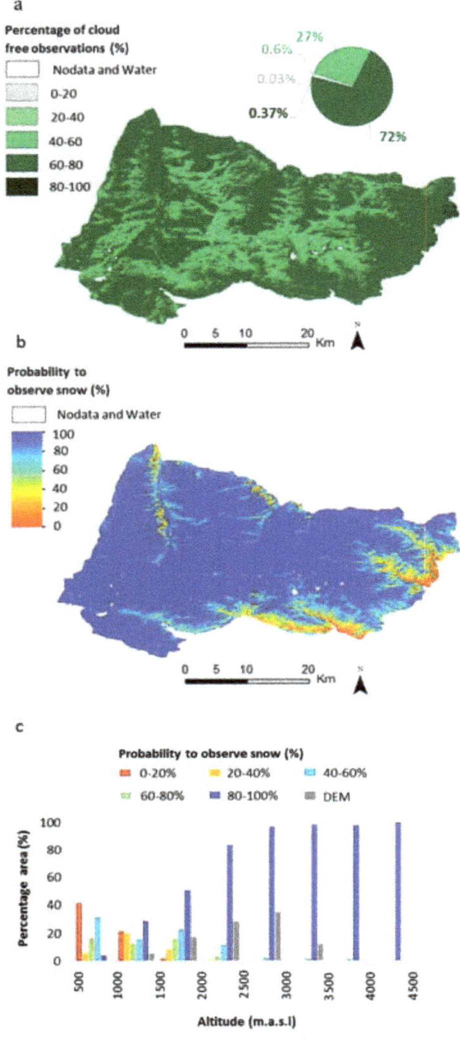

Figure 10. (**a**) Percentage of cloud-free observations for December, January and February maximum SCA products in the period 1984 to 2018 for the Gran Paradiso National Park. Dark green areas indicate that 60% to 100% of cloud-free observations were available for the pixel, while light green areas indicate that 20% to 60% of cloud-free observations were available for the pixel. The pie chart in panels (**a**) summarize the statistics for each of the five considered classes (0%–20%, 20%–40%, 40%–60%, 60%–80% and 80%–100%). (**b**) Winter snow cover probability map, for December, January and February maximum SCA products in the period 1984 to 2018 for the Gran Paradiso National Park (GPNP). Blue areas indicate presence of snow in 80%–100% of cloud-free monthly aggregations, while red areas indicate presence of snow cover in 0%–20% of the cloud-free monthly aggregations. (**c**) Frequency, for each 500 m elevation bins, of the different probability classes (0%–20%, 20%–40%, 40%–60%, 60%–80% and 80%–100%) are expressed in terms of percentage of the total area of that bin. The grey bars show the percentage area in each elevation band, according to the ASTER Global Digital Elevation Model at 30 m spatial resolution [75].

5.3. Winter Snow Cover Evolution

The snow cover summary referring to individual winter seasons in the period 1984–2018 (two examples in Figure 7) provides information on which pixels were classified as snow covered (i) in all the cloud-free images in that winter (high probability to observe snow cover); (ii) in none of the cloud-free images in that winter (low probability to observe snow cover). For each season we calculated the number of observations in those two classes (for which the probability to observe snow cover was high and low, respectively) and the corresponding area, expressing it as a percentage of the total GPNP area. The temporal evolution of the percent snow cover area, along with the cloud cover area, is shown in Figure 11c. The representativeness of those measures is displayed in Figure 11d, showing, for each winter season, the percent of pixels with three, two, one or no cloud-free observations available in the corresponding snow cover summary. For the first 15 winter seasons, cloud-free observations were generally not available for the three winter months considered (dark green bars), but rather on two (medium green bars) or one (light green bars) winter months only. Since 2000/01, the availability of cloud-free observations for the three winter months increased considerably, with almost half of the winter seasons having more than 50% of their observations being cloud-free for the three months. The uneven consistency of the data hampers a proper long-term trend analysis, for which more advanced techniques to reconstruct missing data are required. The information currently available seems to suggest, although with a high level of uncertainty, relatively large interannual variations in the percentage of the area with a high probability of observing snow (blue bars), varying from 58% to 98%, with the largest value being in winter 2005/06 and the smallest in 2015/16. There was a small negative, though not statistically significant, trend (decrease of about 0.5 km^2/season, y = −0.0695x + 86.157 and R = 0.0044). Please note that for the trend analysis, only the seasons with less than 15% cloud cover were retained (seasons 1987/88, 1988/89, 1999/91 and 1995/96 exceeding this threshold have been excluded). The percentage of the area with a low probability of observing snow (red bars) was generally very small, varying between 0% and 18%, the latter being observed in 1991/92 and 2011/12 winters.

For completeness, winter mean temperature and (total and solid) precipitation time series from the HISTALP dataset are displayed in the upper panels of Figure 11a,b, showing no statistically-significant trend in the period 1985–2014 (see Table 1). Winter mean temperature varied from about 8 °C (during the 2009/10 season) to about −3 °C (during the 2006/07 season). Figure 11b highlights some relatively dry winters in 1991/92, 1992/93, 1994/95 and 1999/00 with less than 22 mm/month. Inversely, the 2003/04 winter season was characterized by large quantities of precipitation, exceeding 150 mm/month, most being snow. We jointly analyzed the snow cover information presented in the previous section with the climatic data from the HISTALP dataset, by quantifying the time correlation between the meteorological variables and the percentages of SCA (Figure 12). The percentage of SCA with a high probability of observing snow was positively and significantly correlated with solid precipitation (R = 0.46), while the percentages of SCA with low probability of observing snow and solid precipitation were significantly anticorrelated (R = −0.54). The role of winter temperature on the extent snow is more complex. The percentage of SCA with a high probability to observe snow cover shows a weak, not statistically-significant, negative correlation with temperature; and the percentage of SCA with low probability to observe snow cover shows a weak, not statistically-significant, positive correlation with temperature. The weak correlation could be attributed to the fact that surface air temperature is often below the freezing point at such altitudes during winter, with limited effects on the precipitation phase.

Table 4 provides a general summary of the statistics (minimum and maximum values, means and standard deviations) of the various variables, for DJF and for the months of February and April. For the minimum and maximum values also, the year of occurrence is reported in parentheses.

The same analysis performed for the winter season and shown in Figures 11 and 12 was repeated for the months of February and April, since they are key months to understand the dynamics and the changes of snow cover in mountainous areas. The results are displayed and commented on in Annex 5 (Figures S4 and S5) and are briefly summarized in the following text. February's percentage of SCA exhibited a large interannual variability, with some years (1994, 2002 and 2006) in which almost the entire extent of the Gran Paradiso National Park appeared snow-covered, and others (1993 and 2000) in which the park appeared poorly snow-covered, according to the available satellite observations in that month. April's percentages of SCA and precipitation varied widely over the period 1984–2014 (Annex 5, Figure S5). April 1987 and 2006 were particularly snow-covered with more than 90% of the Gran Paradiso territory covered by snow. For both February and April, the analysis of the temporal percentage of SCA's evolution and trend was seriously hampered by the high number of years excluded from the analysis, owing to the presence of clouds.

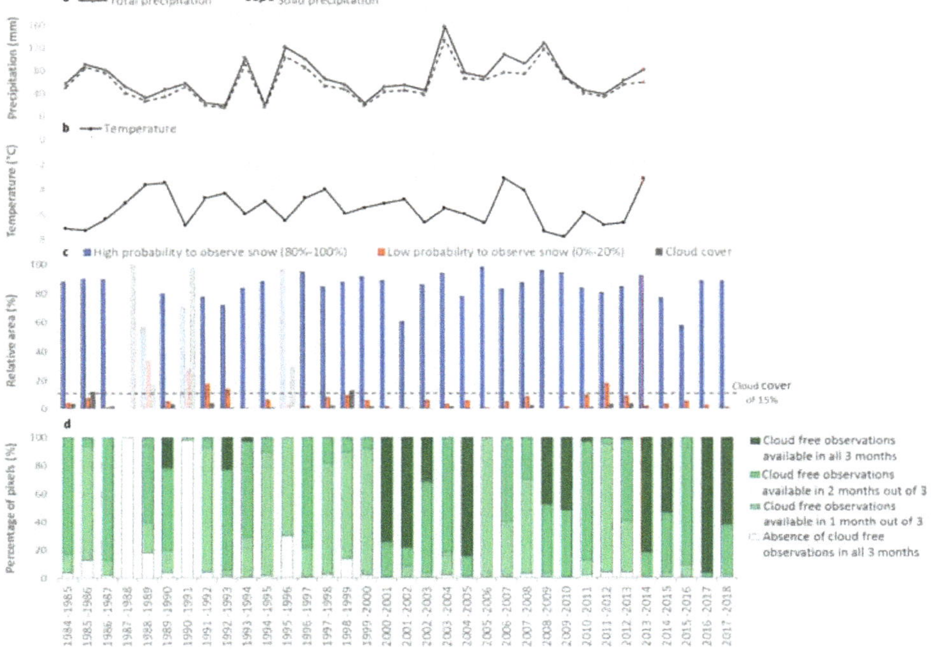

Figure 11. Mean winter (December–January–February) (**a**) total and solid precipitation, and (**b**) temperatures, obtained from the HISTALP observation-based dataset from 1984 to 2014 in the Gran Paradiso National Park. (**c**) Relative area with high (>80%) and low (<20%) probabilities of observing snow cover (in blue and red respectively), and the relative areas with cloud cover (dark gray) derived from the winter time series analyses from 1984 to 2018 in the Gran Paradiso National Park. Light grey bars illustrate winter seasons where cloud cover exceeded 15% of the entire park surface. (**d**) Percentage of pixels with cloud-free observations in none of the three months (white), in one month (light green) out of three months, in two months (medium green) out of three months and in all three months (dark green).

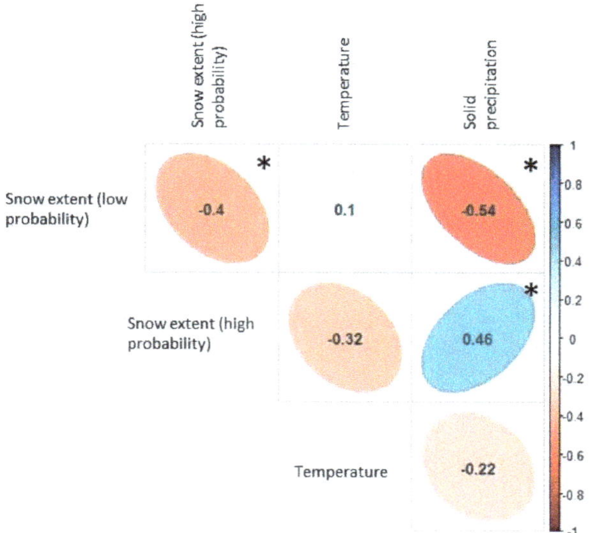

Figure 12. Correlation matrix between seasonal (DJF) snow extent and climate variables (temperature and solid precipitation). The shape, color and the orientation of the ellipse visually describe the correlation coefficients, whose value are also reported. The (*) indicate statistically significant correlations at a 95% confidence level (p-value < 0.05).

Table 4. Mean, minimum (Min), maximum (Max) and standard deviation (SD) for winter (DJF), February and April's percentage of SCA, air temperature, total precipitation and solid precipitation. Statistics were calculated over the periods 1984–2018 for snow cover and 1984–2014 for temperature, and total and solid precipitation.

	Variables	Mean	Min (year)	Max (year)	SD
WINTER (DJF)	%SCA high probability	86.15	58.13 (2015/016)	98.3 (2005/2006)	9.26
	%SCA low probability	6.35	0.17 (2008/09)	18.53 (1991/92)	4.84
	Mean air temperatures (°C)	−5.53	−7.82 (2009/10)	−3.08 (2006/07)	1.34
	Mean total precipitation (mm/month)	66.58	18 (1993/94; 1994/95)	156 (2003/04)	33.41
	Mean solid precipitation (mm/month)	56.92	14 (1992/93)	133 (2003/04)	29.20
FEB	%SCA	83.82	70.87 (2000)	99.15 (1994)	16.99
	%No-SCA	16.17	0.85 (1994)	29.12 (2000)	16.99
	Mean air temperature (°C)	−5.98	−11.2 (2012)	−1.58 (1998)	2.51
	Mean total precipitation (mm/month)	59.20	4 (2000)	140 (2002)	38.54
	Mean solid precipitation (mm/month)	53.53	4 (2000)	128 (1994)	35.41
APR	%SCA	79.10	65.61 (1997)	92.16 (1987)	8.34
	%No-SCA	20.90	7.83 (1987)	34.39 (1997)	8.34
	Mean air temperature (°C)	−1.16	−2.9 (1991)	3.54 (2007)	1.32
	Mean total precipitation (mm/month)	129	4 (2006)	333 (1986)	87.49
	Mean solid precipitation (mm/month)	86.93	1 (2006)	288 (1986)	71.97

6. Discussion

In this paper, we presented a methodology to exploit Landsat series satellite data stored in the Swiss Data Cube (SDC) for snow cover detection, applied to the Gran Paradiso National Park (GPNP). The frequency of available images and the cloud cover analysis over the analyzed 34 years showed that the availability of cloud-free observations remains a major limitation for mapping snow with optical remote sensing data. Indeed, almost half of the Landsat scenes available in the SDC for the period 1984–2018 in the GPNP are cloud-covered by more than 30% of the total area.

Built in an EODC environment, the preliminary version of the SOfS algorithm implemented in the SDC can take advantage from the entire Landsat satellite series from 1984 to 2018 to combine different satellite scenes and generate maximum snow cover area products on a monthly time scale. Temporal filtering is the most common approach employed, particularly with Moderate Resolution Imaging Spectroradiometer (MODIS) satellites, to mitigate cloud contamination (i.e., [34,36,59]). In our study, the mosaicking method used to generate the monthly SCA products allowed to reduce cloud obstruction by approximately 7% and to gain about two additional monthly observations with cloud cover less than 30% of the total area. Besides the multi-day combination, we created a multi-sensor snow dataset combining observations from three different Landsat missions into a single dataset. This multi-sensor data fusion allowed us to go back further in time with the Landsat-5 mission, but the 2000s benefited from two Landsat missions at the same time (doubling the number of observations per month). By comparing the period 1984–2000, for which only observations from one satellite are available (two observations per month), with the period 2001–2018, for which the data are available from two satellites (four observations per month, Landsat-7 and Landsat-8 for the period 2013 to 2018), we observed that monthly cloud contamination was reduced by half after the 2000s. Using the potential of an EODC infrastructure to overcome the spatiotemporal constraints associated with the conventional single-sensor satellite mission, the combination of different sensors data into a unique dataset seemed to be an effective approach to increase the number of available/usable data. This new way of handling satellites data storage and analysis allows users to access the same ARD for different purposes and facilitates data analyses using a multitude of observations.

Monthly maximum snow cover products for December to February were aggregated to create "snow cover summaries" over the full 34 year-period and over each winter season. Snow cover summaries represented the percentages of snow cover observations with respect to the number of the total cloud-free observations in the considered period. Given that, we have to keep in mind that winter snow summaries are generally not representative of the three winter months for 34 years, particularly before the year 2000 when only the Landsat-5 sensor was active. However, the map showing the number of cloud-free observations available per pixel in the park over the full 34 year-period illustrates that cloud-free observations are relatively homogeneous, except in higher altitude areas where less cloud-free observations are found. As expected, the winter snow cover probability map shows that regions lying above 2000 m a.s.l. have high probability to be snow-covered in winter months. The probability to observe snow cover in winter is lowest in the south-eastern part of the park, where snowfall is also less abundant (from about 40 to 60 mm/month, see Figure 31) compared to the rest of the park. Those are the areas lying at relatively low elevations. At higher elevations, lower air temperature and more abundant solid precipitation create favorable conditions for the formation and maintenance of snow cover (as shown in Supplementary Material, Annex 4). Using historical snow cover observations since 1984, our analysis highlights where snow cover is usually present and where it is seldom observed. Information on snow cover distribution and its evolution is essential for studying Alpine ecosystems, as well as highlighting the opportunities and risks for winter tourism's development. For example, according to a study carried out in Switzerland, it has been estimated that under warmer conditions, only ski areas at high elevations (above 2000 m a.s.l.) and with installations for producing artificial snow, will be able to open before Christmas time [76]. With more than one million people visiting the GPNP every year [38], information on the snow cover distribution during winter is essential to manage the local economy in a sustainable way.

The seasonal "snow cover summaries" produced in this study are based mostly on one or two monthly products per season, thus providing only partial information on the presence of snow during that period. As mentioned before, the month of December is more frequently covered by clouds than January and February, and thus it was less represented in our winter snow cover summary analysis. Despite this, winter SCA products showed some consistency with the climatic variables (particularly with solid precipitation) derived from the HISTALP dataset specifically developed for the alpine region. These preliminary results encourage further investigations using the proposed methodology.

6.1. Limitations

One of the major problems of snow detection using optical sensors, such as Landsat, is their inability to deliver surface information under cloudy conditions [77]. Our analysis has shown that, despite a monthly mosaicking methodology and a multi-sensors analysis, the cloud issue persists and limits the number of usable observations for snow cover mapping. On the monthly scale, abundant cloud cover strongly limits snow observations and the possibility to evaluate long-term trends and significant correlations between snow cover and climatic data.

The multi-day composite methodology tested in this study is subject to several other limitations. Using data from Landsat, which has a revisit time of 16-days, forced us to create a one-month composite with only two dates available, at least for the first 15 years (1984–2000). Previous studies have confirmed that the overall accuracy of multi-day combined snow cover products improves as the number of composition-days increase. In other words, the higher the number of images in the composite, the higher the overall accuracy. Similarly, the number of composition days was not homogeneous over the 34 years. Indeed, for the period 1984–2000 where only Landsat-5 was active, we used two images to calculate the maximum NDSI per month, while for the period 2001–2018, we use four images in a month to create the maximum NDSI products (with the combination of Landsat-5 and 7 and Landsat-7 and 8). Moreover, as mentioned by [78], the overall accuracy of a mosaicking method decreases when increasing the compositing time period window (one month here). Based on the assumption that snow will persist on the surface for a certain period of time [31], a one-month compositing period based on two or four images might be insufficient to assess snow cover change. It would thus be beneficial to exploit all satellites currently available in the SDC (Sentinel-1 and Sentinel-2, besides the Landsat series) and MODIS data to increase the number of remote sensing observations per month, and hence reduce the time window used in the mosaicking procedure. A previous study using MODIS observations over Austria has demonstrated that seven days of temporal filtering can remove more than 95% of clouds while maintaining an accuracy exceeding 92% [79].

6.2. Perspectives

In order to mitigate the influence of cloud cover, several methods have been proposed, such as spatial and temporal filtering or interpolation techniques [66,78,80,81]. Spatial filtering is designed to replace cloud-contaminated pixels by using information on surrounding adjacent pixels that are not obscured by clouds on the same date. According to the spatial position of the pixel, the estimation of the regional snow line (defined as the elevation above which all pixels are covered by snow) can help when reclassifying cloudy pixels as snow-covered or snow-free [66]. The study undertaken by Parajka et al. [79] using the daily MODIS snow cover products shows that the snow line method provides robust snow cover mapping, even if cloud cover is as large as 90%. In our case, with only Landsat-5 data available in the Swiss Data Cube from 1984 to 2000, that spatial filtering could be used to effectively reduce the impact of clouds. Similarly, as suggested by Dietz et al. [82], digital elevation model (DEM) information can be used to infer information on cloudy pixels. For example, Qobilov et al. [83] suggested that if a pixel is cloud covered, the nearest cloud-free pixel with the same elevation, azimuth and slope angle can be used instead. Another possibility to get information on cloud covered areas is to combine different optical datasets (e.g Landsat, Sentinel-2 and MODIS) and radar-based (Sentinel-1) sensors [84]. In fact, the synthetic aperture radar (SAR) from the first Sentinel satellite in the European Copernicus program, Sentinel-1, demonstrates an effective ability to map snow in all-weather and day-and-night conditions [10,15]. This will allow us to combine potentialities and limit their individual drawbacks. EODC seems to be an appropriate environment to conduct such multi-sensor data fusion and analysis and to improve the SOfS algorithm.

7. Conclusions

In this paper, we presented a first study on the snow cover evolution in the Gran Paradiso National Park, north-western Italian Alps, from 1984 to 2018, using the preliminary version of the snow observation from space (SOfS) algorithm benefiting from the Swiss Data Cube technology. The SOfS algorithm was applied to all Landsat series images (Landsat-5, Landsat-7 and Landsat-8) available in the period 1984–2018 in an analysis-ready form from the SDC, which corresponds to more than 480 Landsat scenes, to study the snow cover evolution in the GPNP.

The Earth observation data cube environment allowed to generate a monthly snow cover area time series in a totally automated-way for the Gran Paradiso National Park. The SOfS algorithm was employed for the first time in a multi-date (monthly composite) and a multi-sensor (combination of Landsat-5, 7 and 8 in the same dataset) mosaicking approach for increasing the number of cloud-free observations. Indeed, by using three different Landsat sensors incorporated into the same dataset, cloud cover contamination in the monthly SCA products after the 2000s was reduced by more than half.

However, despite the improvements deriving from our approach, the availability of cloud-free scenes remains a major limitation for mapping snow cover with optical remote sensing. There is a need to further investigate the detection of snow under cloud conditions using alternative approaches, such as spatial and temporal filtering techniques, and a combination of additional satellite data, such as Sentinel-1 and Sentinel-2. The integration and combination of consistent and comparable data provides new avenues and opportunities, increasing information density to minimize cloud contamination and to look further back in the past [52].

The preliminary version of the snow detection tool contained in the SDC and described in this paper needs to be further enhanced, in order to increase the quality, availability and accessibility of snow cover information across the Alps. With those improvements, the data cube technology might support a full analysis of the snow cover evolution, compared to a more traditional, scene-based sampling approach that might limit our ability to detect changes [26], allowing us better understand them and their consequences. As mentioned by Guo et al. [85] "multi-temporal Earth observation data may reveal large-scale processes and features that are not observable via traditional methods." Climate change must be taken as an opportunity to gain a more comprehensive understanding of the past and present evolution of our environment, for rethinking our mountainous regions from both environmental and economic perspectives through the establishment of adaptation measures to ensure the sustainable use of alpine resources [86].

Supplementary Materials: Supplementary materials can be accessed at: http://www.mdpi.com/2306-5729/4/4/138/s1.

Author Contributions: C.P.: conceptualization, methodology and writing—original draft preparation; Y.G.: conceptualization, methodology and writing—original draft preparation; E.P.: conceptualization and writing—original draft preparation; S.T.: conceptualization and writing—original draft preparation; B.C.: conceptualization, methodology and writing—review and editing; G.G.: writing—review and editing.

Funding: This research was funded by European Commission "Horizon 2020 Program" ECOPOTENTIAL project, grant number 641762; and with the financial support of the Swiss Federal Office for the Environment (FOEN) who supports the Swiss Data Cube.

Acknowledgments: The authors would like to thank the Swiss Federal Office for the Environment (FOEN) for their financial support to the Swiss Data Cube. Results of this publication partly rely on the Swiss Data Cube (http://www.swissdatacube.org), operated and maintained by UN Environment/GRID-Geneva, the University of Geneva, the University of Zurich and the Swiss Federal Institute for Forest, Snow and Landscape Research WSL. This work was supported by the European Union's Horizon 2020 research and innovation program under grant agreement number 641762 (ECOPOTENTIAL). The views expressed in the paper are those of the authors and do not necessarily reflect the views of the institutions they belong to.

Conflicts of Interest: The authors declare no conflict of interest.

References

1. IPCC. *Climate Change 2014: Synthesis Report. Contribution of Working Groups I, II and III to the Fifth Assessment Report of the Intergovernmental Panel on Climate Change*; Core Writing Team, Pachauri, R.K., Meyer, L.A., Eds.; 9291691437; IPCC: Geneva, Switzerland, 2014; p. 151.
2. IPCC. Global Warming of 1.5 °C: Summary for Policy Makers. In *Global Warming of 1.5 °C. An IPCC Special Report on the Impacts of Global Warming of 1.5 °C above Pre-Industrial Levels And Related Global Greenhouse Gas Emission Pathways, in the Context of Strengthening the Global Response to the Threat of Climate Change, Sustainable Development, and Efforts to Eradicate Poverty*; Masson-Delmotte, V., Zhai, H.-O.P., Pörtner, D., Roberts, J., Skea, P.R., Shukla, A., Pirani, W., Moufouma-Okia, C., Péan, R., Pidcock, S., et al., Eds.; IPCC: Geneva, Switzerland, 2018.
3. Mountain Research Initiative EDW Working Group; Pepin, N.; Bradley, R.S.; Diaz, H.F.; Baraer, M.; Caceres, E.B.; Forsythe, N.; Fowler, H.; Greenwood, G.; Hashmi, M.Z.; et al. Elevation-dependent warming in mountain regions of the world. *Nat. Clim. Chang.* **2015**, *5*, 424. [CrossRef]
4. Rangwala, I.; Sinsky, E.; Miller, J.R. Variability in projected elevation dependent warming in boreal midlatitude winter in CMIP5 climate models and its potential drivers. *Clim. Dyn.* **2016**, *46*, 2115–2122. [CrossRef]
5. Global Climate Report for January 2019. Available online: https://www.ncdc.noaa.gov/sotc/global/201901 (accessed on 12 March 2019).
6. Donnelly, C.; Greuell, W.; Andersson, J.; Gerten, D.; Pisacane, G.; Roudier, P.; Ludwig, F. Impacts of climate change on European hydrology at 1.5, 2 and 3 degrees mean global warming above preindustrial level. *Clim. Chang.* **2017**, *143*, 13–26. [CrossRef]
7. Bojinski, S.; Verstraete, M.; Peterson, T.C.; Richter, C.; Simmons, A.; Zemp, M. The Concept of Essential Climate Variables in Support of Climate Research, Applications, and Policy. *Bull. Am. Meteorol. Soc.* **2014**, *95*, 1431–1443. [CrossRef]
8. Salzano, R.; Salvatori, R.; Valt, M.; Giuliani, G.; Chatenoux, B.; Ioppi, L. Automated Classification of Terrestrial Images: The Contribution to the Remote Sensing of Snow Cover. *Geosciences* **2019**, *9*, 97. [CrossRef]
9. Hall, D.K.; Riggs, G.A.; Salomonson, V.V.; Barton, J.; Casey, K.; Chien, J.; DiGirolamo, N.; Klein, A.; Powell, H.; Tait, A.J.N.G. *Algorithm Theoretical Basis Document (ATBD) for the MODIS Snow and Sea Ice-Mapping Algorithms*; NASA: Washington, DC, USA, 2001; pp. 1–45.
10. Nolin, A.W. Recent advances in remote sensing of seasonal snow. *J. Glaciol.* **2010**, *56*, 1141–1150. [CrossRef]
11. Chokmani, K.; Bernier, M.; Royer, A. A merging algorithm for regional snow mapping over eastern Canada from AVHRR and SSM/I data. *Remote Sens.* **2013**, *5*, 5463–5487. [CrossRef]
12. Zhang, H.; Zhang, F.; Zhang, G.; Che, T.; Yan, W.; Ye, M.; Ma, N. Ground-based evaluation of MODIS snow cover product V6 across China: Implications for the selection of NDSI threshold. *Sci. Total Environ.* **2019**, *651*, 2712–2726. [CrossRef]
13. Barnes, J.C.; Smallwood, M.D. *Synopsis of Current Satellite Snow Mapping Techniques, with Emphasis on the Application of Near-Infrared Data*; NASA: Washington, DC, USA, 1975.
14. Dozier, J.; Painter, T.H. Multispectral and hyperspectral remote sensing of alpine snow properties. *Annu. Rev. Earth Planet. Sci.* **2004**, *32*, 465–494. [CrossRef]
15. Hall, D.K. *Satellite Snow-Cover Mapping: A Brief Review*; NASA: Washington, DC, USA, 1995.
16. Kaur, R.; Saikumar, D.; Kulkarni, A.V.; Chaudhary, B.J.C.S. Variations in snow cover and snowline altitude in Baspa Basin. *Curr. Sci.* **2009**, *96*, 1255–1258.
17. Lemke, P.; Ren, J.; Alley, R.B.; Allison, I.; Carrasco, J.; Flato, G.; Fujii, Y.; Kaser, G.; Mote, P.; Thomas, R.H. *Observations: Changes in Snow, Ice and Frozen Ground*; IPCC: Geneva, Switzerland, 2007.
18. Paul, F.; Kääb, A.; Maisch, M.; Kellenberger, T.; Haeberli, W. The new remote-sensing-derived Swiss glacier inventory: I. Methods. *Ann. Glaciol.* **2002**, *34*, 355–361. [CrossRef]
19. Selkowitz, D.J.; Forster, R.; Sensing, R. Automated mapping of persistent ice and snow cover across the western US with Landsat. *ISPRS J. Photogramm. Remote Sens.* **2016**, *117*, 126–140. [CrossRef]
20. Yin, D.; Cao, X.; Chen, X.; Shao, Y.; Chen, J. Comparison of automatic thresholding methods for snow-cover mapping using Landsat TM imagery. *Int. J. Remote Sens.* **2013**, *34*, 6529–6538. [CrossRef]
21. Zhu, Z.; Woodcock, C.E. Continuous change detection and classification of land cover using all available Landsat data. *Remote Sens. Environ.* **2014**, *144*, 152–171. [CrossRef]

22. Claverie, M.; Ju, J.; Masek, J.G.; Dungan, J.L.; Vermote, E.F.; Roger, J.-C.; Skakun, S.V.; Justice, C. The Harmonized Landsat and Sentinel-2 surface reflectance data set. *Remote Sens. Environ.* **2018**, *219*, 145–161. [CrossRef]

23. Baumann, P.; Misev, D.; Merticariu, V.; Huu, B.P.; Bell, B. Datacubes: A Technology Survey. In Proceedings of the IGARSS 2018 IEEE International Geoscience and Remote Sensing Symposium, Valencia, Spain, 22–27 July 2018; pp. 430–433.

24. Baumann, P.; Rossi, A.P.; Bell, B.; Clements, O.; Evans, B.; Hoenig, H.; Hogan, P.; Kakaletris, G.; Koltsida, P.; Mantovani, S.; et al. Fostering Cross-Disciplinary Earth Science Through Datacube Analytics. In *Earth Observation Open Science and Innovation*; Mathieu, P.-P., Aubrecht, C., Eds.; Springer International Publishing: Cham, Switzerland, 2018; pp. 91–119. [CrossRef]

25. Killough, B. Overview of the Open Data Cube Initiative. In Proceedings of the IGARSS 2018 IEEE International Geoscience and Remote Sensing Symposium, Valencia, Spain, 22–27 July 2018; pp. 8629–8632.

26. Lewis, A.; Oliver, S.; Lymburner, L.; Evans, B.; Wyborn, L.; Mueller, N.; Raevksi, G.; Hooke, J.; Woodcock, R.; Sixsmith, J.; et al. The Australian Geoscience Data Cube—Foundations and lessons learned. *Remote Sens. Environ.* **2017**, *202*, 276–292. [CrossRef]

27. Strobl, P.; Baumann, P.; Lewis, A.; Szantoi, Z.; Killough, B.; Purss, M.; Craglia, M.; Nativi, S.; Held, A.; Dhu, T. The Six Faces of The Datacube. In Proceedings of the Conference on Big Data from Space (BIDS'2017), Toulouse, France, 28–30 November 2017; pp. 28–30.

28. Baumann, P.; Mazzetti, P.; Ungar, J.; Barbera, R.; Barboni, D.; Beccati, A.; Bigagli, L.; Boldrini, E.; Bruno, R.; Calanducci, A.; et al. Big Data Analytics for Earth Sciences: The EarthServer approach. *Int. J. Dig. Earth* **2016**, *9*, 3–29. [CrossRef]

29. Camara, G.; Assis, L.F.; Ribeiro, G.; Ferreira, K.R.; Llapa, E.; Vinhas, L. Big earth observation data analytics: Matching requirements to system architectures. In Proceedings of the 5th ACM SIGSPATIAL International Workshop on Analytics for Big Geospatial Data, Burlingame, CA, USA, 31 October–3 November 2016; pp. 1–6.

30. Soille, P.; Burger, A.; De Marchi, D.; Kempeneers, P.; Rodriguez, D.; Syrris, V.; Vasilev, V. A versatile data-intensive computing platform for information retrieval from big geospatial data. *Future Gener. Comput. Syst.* **2018**, *81*, 30–40. [CrossRef]

31. Lehmann, A.; Nativi, S.; Mazzetti, P.; Maso, J.; Serral, I.; Spengler, D.; Niamir, A.; McCallum, I.; Lacroix, P.; Patias, P.; et al. GEOEssential—Mainstreaming workflows from data sources to environment policy indicators with essential variables. *Int. J. Dig. Earth* **2019**, 1–17. [CrossRef]

32. Sudmanns, M.; Tiede, D.; Lang, S.; Bergstedt, H.; Trost, G.; Augustin, H.; Baraldi, A.; Blaschke, T. Big Earth data: Disruptive changes in Earth observation data management and analysis? *Int. J. Dig. Earth* **2019**, 1–19. [CrossRef]

33. Gorelick, N.; Hancher, M.; Dixon, M.; Ilyushchenko, S.; Thau, D.; Moore, R. Google Earth Engine: Planetary-scale geospatial analysis for everyone. *Remote Sens. Environ.* **2017**, *202*, 18–27. [CrossRef]

34. Hall, D.; Riggs, G.A.; Salomonson, V.V. Development of Methods for Mapping Global Snow Cover Using Moderate Resolution Imaging Spectroradiometer Data. *Remote Sens. Environ.* **1995**, *54*, 127–140. [CrossRef]

35. Hou, J.; Huang, C.; Zhang, Y.; Guo, J.; Gu, J. Gap-Filling of Modis Fractional Snow Cover Products Via Non-Local Spatio-Temporal Filtering Based on Machine Learning Techniques. *Remote Sens.* **2019**, *11*, 90. [CrossRef]

36. López-Burgos, V.; Gupta, H.V.; Clark, M. Reducing cloud obscuration of MODIS snow cover area products by combining spatio-temporal techniques with a probability of snow approach. *Hydrol. Earth Syst. Sci.* **2013**, *17*, 1809–1823. [CrossRef]

37. Gran Paradiso National Park—Italy. Available online: https://deims.org/e33c983a-19ad-4f40-a6fd-1210ee0b3a4b (accessed on 31 May 2019).

38. Ecopotential Project. Available online: https://www.ecopotential-project.eu/ (accessed on 15 May 2019).

39. Auer, I.; Böhm, R.; Jurkovic, A.; Lipa, W.; Orlik, A.; Potzmann, R.; Schöner, W.; Ungersböck, M.; Matulla, C.; Briffa, K. HISTALP—historical instrumental climatological surface time series of the Greater Alpine Region. *Int. J. Climatol.* **2007**, *27*, 17–46. [CrossRef]

40. Chimani, B.; Böhm, R.; Matulla, C.; Ganekind, M. Development of a long-term dataset of solid/liquid precipitation. *Adv. Sci. Res.* **2011**, *6*, 39–43. [CrossRef]

41. Crawford, C.J.; Manson, S.M.; Bauer, M.E.; Hall, D. Multitemporal snow cover mapping in mountainous terrain for Landsat climate data record development. *Remote Sens. Environ.* **2013**, *135*, 224–233. [CrossRef]

42. Marty, C.; Tilg, A.-M.; Jonas, T. Recent Evidence of Large-Scale Receding Snow Water Equivalents in the European Alps. *J. Hydrometeorol.* **2017**, *18*, 1021–1031. [CrossRef]

43. Dhu, T.; Dunn, B.; Lewis, B.; Lymburner, L.; Mueller, N.; Telfer, E.; Lewis, A.; McIntyre, A.; Minchin, S.; Phillips, C. Digital earth Australia—Unlocking new value from earth observation data. *Big Earth Data* **2017**, *1*, 64–74. [CrossRef]

44. Nativi, S.; Mazzetti, P.; Craglia, M. A view-based model of data-cube to support big earth data systems interoperability. *Big Earth Data* **2017**, *1*, 75–99. [CrossRef]

45. Rizvi, S.R.; Killough, B.; Cherry, A.; Gowda, S. The Ceos Data Cube Portal: A User-Friendly, Open Source Software Solution for the Distribution, Exploration, Analysis, and Visualization of Analysis Ready Data. In Proceedings of the IGARSS 2018 IEEE International Geoscience and Remote Sensing Symposium, Valencia, Spain, 22–27 July 2018; pp. 8639–8642.

46. Woodcock, R.; Paget, M.; Wang, P.; Held, A. Accelerating Industry Innovation Using the Open Data Cube in Australia. In Proceedings of the IGARSS 2018 IEEE International Geoscience and Remote Sensing Symposium, Valencia, Spain, 22–27 July 2018; pp. 8636–8638.

47. Landsat Missions. Available online: https://www.usgs.gov/land-resources/nli/landsat (accessed on 31 May 2019).

48. Giuliani, G.; Chatenoux, B.; De Bono, A.; Rodila, D.; Richard, J.-P.; Allenbach, K.; Dao, H.; Peduzzi, P. Building an Earth Observations Data Cube: Lessons learned from the Swiss Data Cube (SDC) on generating Analysis Ready Data (ARD). *Big Earth Data* **2017**, *1*, 100–117. [CrossRef]

49. Giuliani, G.; Chatenoux, B.; Honeck, E.; Richard, J. Towards Sentinel-2 Analysis Ready Data: A Swiss Data Cube Perspective. In Proceedings of the IGARSS 2018 IEEE International Geoscience and Remote Sensing Symposium, Valencia, Spain, 22–27 July 2018; pp. 8659–8662.

50. Sentinel-1 Satellites Observe Snow Melting Processes. Available online: https://earth.esa.int/web/sentinel/missions/sentinel-1/news/-/article/sentinel-1-satellites-observe-snow-melting-processes (accessed on 15 May 2019).

51. Nagler, T.; Rott, H.; Ossowska, J.; Schwaizer, G.; Small, D.; Malnes, E.; Luojus, K.; Metsämäki, S.; Pinnock, S. Snow Cover Monitoring by Synergistic Use of Sentinel-3 Slstr and Sentinel-L Sar Data. In Proceedings of the IGARSS 2018 IEEE International Geoscience and Remote Sensing Symposium, Valencia, Spain, 22–27 July 2018; pp. 8727–8730.

52. Committee on Earth Observations Satellites (CEOS). Available online: http://ceos.org/ard/users.html (accessed on 15 July 2019).

53. Masek, J.; Ju, J.; Roger, J.-C.; Skakun, S.; Claverie, M.; Dungan, J. Harmonized Landsat/Sentinel-2 Products for Land Monitoring. In Proceedings of the IEEE International Geoscience and Remote Sensing Symposium, Valencia, Spain, 22–27 July 2018.

54. Landsat Collection 1 Level-1 Quality Assessment Band. Available online: https://www.usgs.gov/land-resources/nli/landsat/landsat-collection-1-level-1-quality-assessment-band?qt-science_support_page_related_con=0-qt-science_support_page_related_con (accessed on 31 May 2019).

55. Mueller, N.; Lewis, A.; Roberts, D.; Ring, S.; Melrose, R.; Sixsmith, J.; Lymburner, L.; McIntyre, A.; Tan, P.; Curnow, S.; et al. Water observations from space: Mapping surface water from 25 years of Landsat imagery across Australia. *Remote Sens. Environ.* **2016**, *174*, 341–352. [CrossRef]

56. Dedieu, J.P.; Lessard-Fontaine, A.; Ravazzani, G.; Cremonese, E.; Shalpykova, G.; Beniston, M. Shifting mountain snow patterns in a changing climate from remote sensing retrieval. *Sci. Total Environ.* **2014**, *493*, 1267–1279. [CrossRef] [PubMed]

57. Tong, J.; Déry, S.J.; Jackson, P.L. Topographic control of snow distribution in an alpine watershed of western Canada inferred from spatially-filtered MODIS snow products. *Hydrol. Earth Syst. Sci.* **2009**, *13*, 319–326. [CrossRef]

58. König, M.; Winther, J.-G.; Isaksson, E. Measuring snow and glacier ice properties from satellite. *Rev. Geophys.* **2001**, *39*, 1–27. [CrossRef]

59. Stillinger, T.; Roberts, D.A.; Collar, N.M.; Dozier, J. Cloud Masking for Landsat 8 and MODIS Terra Over Snow-Covered Terrain: Error Analysis and Spectral Similarity Between Snow and Cloud. *Water Resour. Res.* **2019**, *55*, 1–16. [CrossRef]

60. CFMask Algorithm. Available online: https://www.usgs.gov/land-resources/nli/landsat/cfmask-algorithm (accessed on 31 May 2019).

61. Global Surface Water Explorer. Available online: https://global-surface-water.appspot.com/ (accessed on 31 May 2019).

62. Kyle, H.; Curran, R.; Barnes, W.; Escoe, D. A cloud physics radiometer. In Proceedings of the 3rd Conference on Atmospheric Radiation, Davis, CA, USA, 28–30 June 1978; pp. 107–109.

63. Dozier, J.; Sensing, R. Snow reflectance from Landsat-4 thematic mapper. *IEEE Trans. Geosci. Remote Sens.* **1984**, *GE-22*, 323–328. [CrossRef]

64. Dozier, J. Spectral signature of alpine snow cover from the Landsat Thematic Mapper. *Remote Sens. Environ.* **1989**, *28*, 9–22. [CrossRef]

65. Klein, A.G.; Barnett, A.C. Validation of daily MODIS snow cover maps of the Upper Rio Grande River Basin for the 2000-2001 snow year. *Remote Sens. Environ.* **2003**, *86*, 162–176. [CrossRef]

66. Gascoin, S.; Grizonnet, M.; Bouchet, M.; Salgues, G.; Hagolle, O. Theia Snow collection: High-resolution operational snow cover maps from Sentinel-2 and Landsat-8 data. *Earth Syst. Sci. Data* **2019**, *11*, 493–514. [CrossRef]

67. Kulkarni, A.V.; Singh, S.K.; Mathur, P.; Mishra, V.D. Algorithm to monitor snow cover using AWiFS data of RESOURCESAT-1 for the Himalayan region. *Int. J. Remote Sens.* **2006**, *27*, 2449–2457. [CrossRef]

68. Burns, P.; Nolin, A. Using atmospherically-corrected Landsat imagery to measure glacier area change in the Cordillera Blanca, Peru from 1987 to 2010. *Remote Sens. Environ.* **2014**, *140*, 165–178. [CrossRef]

69. Grumman, N.J.R.B. *VIIRS Snow Cover Algorithm Theoretical Basis Document (ATBD)*; Northrup Grumman Aerospace Systems: Redondo Beach, CA, USA, 2010.

70. Härer, S.; Bernhardt, M.; Siebers, M.; Schulz, K. On the need for a time- and location-dependent estimation of the NDSI threshold value for reducing existing uncertainties in snow cover maps at different scales. *Cryosphere* **2018**, *12*, 1629–1642. [CrossRef]

71. Riggs, G.A.; Hall, D.K.; Salomonson, V.V. A snow index for the Landsat thematic mapper and moderate resolution imaging spectroradiometer. In Proceedings of the IGARSS'94-1994 IEEE International Geoscience and Remote Sensing Symposium, Pasadena, CA, USA, 8–12 August 1994; pp. 1942–1944.

72. Wang, X.; Xie, H.; Liang, T. Evaluation of MODIS snow cover and cloud mask and its application in Northern Xinjiang, China. *Remote Sens. Environ.* **2008**, *112*, 1497–1513. [CrossRef]

73. Dietz, A.J.; Kuenzer, C.; Dech, S. Analysis of Snow Cover Time Series–Opportunities and Techniques. In *Remote Sensing Time Series*; Springer: Berlin/Heidelberg, Germany, 2015; pp. 75–98.

74. Beniston, M. Climate change in mountain regions a review of possible impacts. *Clim. Chang.* **2003**, *59*, 5–31. [CrossRef]

75. ASTER Global Digital Elevation Map. ASTER GDEM is a Product of METI and NASA. Available online: https://asterweb.jpl.nasa.gov/gdem.asp (accessed on 31 May 2019).

76. Schmucki, E.; Marty, C.; Fierz, C.; Weingartner, R.; Lehning, M. Impact of climate change in Switzerland on socioeconomic snow indices. *Theor. Appl. Climatol.* **2017**, *127*, 875–889. [CrossRef]

77. Hüsler, F.; Jonas, T.; Riffler, M.; Musial, J.P.; Wunderle, S. A satellite-based snow cover climatology (1985–2011) for the European Alps derived from AVHRR data. *Cryosphere* **2014**, *8*, 73–90. [CrossRef]

78. Gao, Y.; Xie, H.; Yao, T.; Xue, C. Integrated assessment on multi-temporal and multi-sensor combinations for reducing cloud obscuration of MODIS snow cover products of the Pacific Northwest USA. *Remote Sens. Environ.* **2010**, *114*, 1662–1675. [CrossRef]

79. Parajka, J.; Blöschl, G. Spatio-temporal combination of MODIS images-potential for snow cover mapping. *Water Resour. Res.* **2008**, *44*. [CrossRef]

80. Gafurov, A.; Bárdossy, A. Cloud removal methodology from MODIS snow cover product. *Hydrol. Earth Syst. Sci.* **2009**, *13*, 1361–1373. [CrossRef]

81. Parajka, J.; Pepe, M.; Rampini, A.; Rossi, S.; Blöschl, G. A regional snow-line method for estimating snow cover from MODIS during cloud cover. *J. Hydrol.* **2010**, *381*, 203–212. [CrossRef]

82. Dietz, A.J.; Kuenzer, C.; Gessner, U.; Dech, S. Remote sensing of snow—A review of available methods. *Int. J. Remote Sens.* **2012**, *33*, 4094–4134. [CrossRef]

83. Qobilov, T.; Pertziger, F.; Vasilina, L.; Baumgartner, M. *Operational Technology for Snow-Cover Mapping in the Central Asian Mountains Using NOAA-AVHRR Data*; NOAA: Silver Sprin, ML, USA, 2001; pp. 76–80.

84. Dietz, A.J.; Hu, Z.; Tsai, Y.-L. Remote Sensing of Snow Cover in The Alps-an Overview of Opportunities and Constraints. In Proceedings of the EO4Alps on the Alps from Space Workshop, Innsbruck, Austria, 27–29 June 2018.
85. Guo, H.-D.; Zhang, L.; Zhu, L.-W. Earth observation big data for climate change research. *Adv. Clim. Chang. Res.* **2015**, *6*, 108–117. [CrossRef]
86. Bavay, M.; Grünewald, T.; Lehning, M.J.A. Response of snow cover and runoff to climate change in high Alpine catchments of Eastern Switzerland. *Adv. Water Resour.* **2013**, *55*, 4–16. [CrossRef]

 *data*

Article

Land Cover Mapping Using Digital Earth Australia

Richard Lucas [1,*], Norman Mueller [2], Anders Siggins [2], Christopher Owers [1], Daniel Clewley [3], Peter Bunting [1], Cate Kooymans [2], Belle Tissott [2], Ben Lewis [2], Leo Lymburner [2] and Graciela Metternicht [4]

1   Department of Geography and Earth Sciences, Aberystwyth University, Aberystwyth, Ceredigion, Wales SY233DB, UK; cho18@aber.ac.uk (C.O.); pfb@aber.ac.uk (P.B.)
2   Geoscience Australia/Digital Earth Australia (DEA), GPO Box 378, Canberra, ACT 2601, Australia; norman.mueller@ga.gov.au (N.M.); anders.siggins@gmail.com (A.S.); cate.kooymans@ga.gov.au (C.K.); belle.tissott@ga.gov.au (B.T.); ben.lewis@ga.gov.au (B.L.); leo.lymburner@ga.gov.au (L.L.)
3   Plymouth Marine Laboratory, Plymouth PL1 3DH, UK; dac@pml.ac.uk
4   School of Biological, Earth and Environmental Sciences, the University of New South Wales, High Street, Kensington, NSW 2052, Australia; g.metternicht@unsw.edu.au
*   Correspondence: rml2@aber.ac.uk

Received: 31 May 2019; Accepted: 29 August 2019; Published: 1 November 2019

Abstract: This study establishes the use of the Earth Observation Data for Ecosystem Monitoring (EODESM) to generate land cover and change classifications based on the United Nations Food and Agriculture Organisation (FAO) Land Cover Classification System (LCCS) and environmental variables (EVs) available within, or accessible from, Geoscience Australia's (GA) Digital Earth Australia (DEA). Classifications representing the LCCS Level 3 taxonomy (8 categories representing semi-(natural) and/or cultivated/managed vegetation or natural or artificial bare or water bodies) were generated for two time periods and across four test sites located in the Australian states of Queensland and New South Wales. This was achieved by progressively and hierarchically combining existing time-static layers relating to (a) the extent of artificial surfaces (urban, water) and agriculture and (b) annual summaries of EVs relating to the extent of vegetation (fractional cover) and water (hydroperiod, intertidal area, mangroves) generated through DEA. More detailed classifications that integrated information on, for example, forest structure (based on vegetation cover (%) and height (m); time-static for 2009) and hydroperiod (months), were subsequently produced for each time-step. The overall accuracies of the land cover classifications were dependent upon those reported for the individual input layers, with these ranging from 80% (for cultivated, urban and artificial water) to over 95% (for hydroperiod and fractional cover). The changes identified include mangrove dieback in the southeastern Gulf of Carpentaria and reduced dam water levels and an associated expansion of vegetation in Lake Ross, Burdekin. The extent of detected changes corresponded with those observed using time-series of RapidEye data (2014 to 2016; for the Gulf of Carpentaria) and Google Earth imagery (2009–2016 for Lake Ross). This use case demonstrates the capacity and a conceptual framework to implement EODESM within DEA and provides countries using the Open Data Cube (ODC) environment with the opportunity to routinely generate land cover maps from Landsat or Sentinel-1/2 data, at least annually, using a consistent and internationally recognised taxonomy.

Keywords: land cover classification; change; Digital Earth Australia; open data cube; Landsat; Australia

---

1. Introduction

To date, there have been few land cover maps that provide consistent coverage for all of Australia. Those that exist have mostly focused on only part of the land cover classification spectrum. For example, the National Vegetation Information System (NVIS; Australian Department of Environment and Energy,

2017) describes 32 broad categories linked to grasslands, shrublands, woodlands and forests, and natural and artificial bare and water classes. However, as with most Australian land cover mapping systems, NVIS is a combination of inputs from separate mapping efforts by governments at the state and national levels, resulting in an output that has detailed content in some areas but not in others. To create a consistent land cover map for Australia, Geoscience Australia (GA) and the Australian Bureau of Agricultural and Resource Economics and Sciences (ABARES) developed the National Dynamic Land Cover Dataset (DLCD) [1] using data from the Moderate Resolution Imaging Spectroradiometer (MODIS). The DLCD was generated from 250 m resolution 16-day Enhanced Vegetation Index (EVI) composites, from which 12 coefficients (based on statistical, phenological and seasonal characteristics for each pixel) were clustered using a support vector clustering algorithm. Each class was then labelled using a combination of catchment-scale land-use mapping and the NVIS to provide a consistent land cover map for Australia [2]. However, the 250 m scale spatial resolution and methodology employed to create DLCD limited its ability to provide high accuracy across its classes and discriminate features at a scale useful for policy and land management.

A number of international projects have classified land cover across Australia according to global taxonomies, including Landcover 2000 and the European Space Agency's (ESA) Climate Change Initiative (CCI) 300 m resolution land cover product, which employed a subset of the Food and Agriculture Organisation's (FAO) Land Cover Classification System (LCCS), a supervised approach, and temporal information for some biophysical variables (e.g., canopy cover). The value of the FAO LCCS was recognised for Australia by Atyeo and Thackway [3], who stated that this taxonomy provided a comprehensive and flexible system for remapping existing Australian State and Territory vegetation and land cover types and highlighted its potential for providing comprehensive descriptions and maps of land cover for national and international reporting. With the development of Digital Earth Australia (DEA) [4], containing the Australian archive of public good Earth observation (EO) data, such as from the Landsat and Copernicus programs, the LCCS is seen as a candidate to provide ongoing national land cover data for Australia on the DEA platform.

An advantage of the FAO LCCS is that the classes generated closely align with habitat taxonomies that are widely used by ecologists. For example, the taxonomy was adopted within the EU FP7 Biodiversity Multi-Souce Monitoring System (BIO_SOS) [5] project, which applied ecologically-based rules (based on an approach developed for Wales [6] to progressively classify Very High Resolution (VHR) satellite data acquired during pre-flush (e.g., temperate spring or tropical/subtropical dry season) and/or peak flush (e.g., temperate summer or wet season) periods. The BIO_SOS approach followed the LCCS dichotomous hierarchy [6]. During the subsequent FP7 Horizon 2020 Ecopotential Project [7], ecological rules derived from single or dual images were replaced by inputs from a defined set of environmental variables (EVs; e.g., canopy cover, water turbidity) that aligned with those used by the FAO LCCS Version 2 [8]. A diverse and expandable set of additional EVs external to the LCCS (e.g., plant species, woody/herbaceous biomass, sea surface temperature, snow depth) were also integrated. Change events and processes were then identified by accumulating comparisons of LCCS component codes (e.g., for canopy cover) and EVs internal and external to the taxonomy. This led to the establishment of a flexible evidence-based historical and near real-time change alert system. This integrated system of land cover and evidence-based change detection is termed the Earth Observation Data for Ecosystem Monitoring (EODESM) [9].

The EODESM system is currently implemented in Python and makes use of the functionality of the RSGISLib software [10] and KEA file format [11]—specifically, raster attribute tables (RAT). A major advantage of EODESM is that it is applicable to any site worldwide, and it can provide land cover and change classifications at any scale and temporal separation. The system primarily generates land cover and change from available EVs, many of which are obtained from EO data. The approach also focuses on and encourages the standardisation of scientifically robust protocols for recording land cover information and EVs.

2. Aims

The aim of this research was to test the use of EODESM for generating moderate (~25 m) spatial resolution land cover and evidence-based change maps within the Open Data Cube (ODC) environment and using the framework of Geoscience Australia's (GA) Digital Earth Australia (DEA). Focus was therefore on sites located on the Australian mainland. Specific objectives were to (a) review the nature and current availability of EVs held within or accessible through DEA, their relevance at a national level, the requirements for their translation as inputs for EODESM and any gaps that might exist; and (b) outline an approach for integrating EODESM within the broader ODC environment.

3. Background to EODESM

3.1. Land Cover Classification

The FAO LCCS Version 2 taxonomy is hierarchical, consisting of a decision tree structure (Figure 1). Each level in the hierarchy consists of one or more binary decisions. The first three levels of the LCCS tree classify, in sequence, vegetated, aquatic, cultivated/managed, urban and artificial water. More detail can then be provided through what is known as the Level 4 classification. These resulting descriptors (such as canopy cover or leaf type) are then combined to produce a cumulated land cover class.

The EODESM system mirrors this hierarchical and modular classification approach by combining products from EO data that (a) assign values of 1 or 0 to Level 1–3 raster inputs and (b) thematic values (e.g., 1, 2 and 3 for water, ice and snow categories, respectively) to layers relevant to the Level 4 hierarchy, with several derived from continuous layers (e.g., water hydroperiod). Additional descriptors external to the LCCS classification (e.g., above ground biomass (AGB)) can also be included. To ensure comparability, scalability and consistency over time, all continuous EVs are accepted only if they can be quantified using pre-defined and recognisable units. Examples include vegetation height (m) and cover (%), water depth (m) and AGB (Mg ha^{-1}). Thematic layers (e.g., plant species composition) also need to be associated with pre-defined and standardised lists or categories. Indices (e.g., the Normalised Difference Vegetation Index (NDVI)) are avoided within EODESM but are instead used to generate descriptors such as Net Primary Productivity (NPP) or the start of leaf flush (in days) that retain the same meaning (and units) over time. This ensures the longevity of the system, interpretability and objective classification.

3.2. Change Detection

To capture the varying nature and rates of change across landscapes, EODESM compares both temporal thematic classifications (e.g., leaf type and water extent) and quantitative (continuous) information on EVs (including those external to the LCCS) retrieved from EO data to build up evidence of changes identified within a defined change taxonomy (e.g., sea level rise, harvesting of crops, wildfires). The same EVs used to generate a land cover classification for a single date or period (e.g., annual; t_1) are generated for a second time step (t_2), and the resulting component codes are compared to determine changes in land cover. The comparison is augmented by comparisons of EVs external to the LCCS. The time interval between t_1 and t_2 can vary between days, weeks and months to decades, and land cover and EV comparisons can be undertaken before and after events (e.g., fires, floods) or processes (e.g., regrowth), which can be indicated using more traditional change detection methods such as the Breaks for Additive Season and Trend (BFAST) algorithm [12] or cross correlation analysis [13]. As well as the LCCS classes, the EODESM system also compares an expandable list of EVs that are external to the classification but provide additional descriptors.

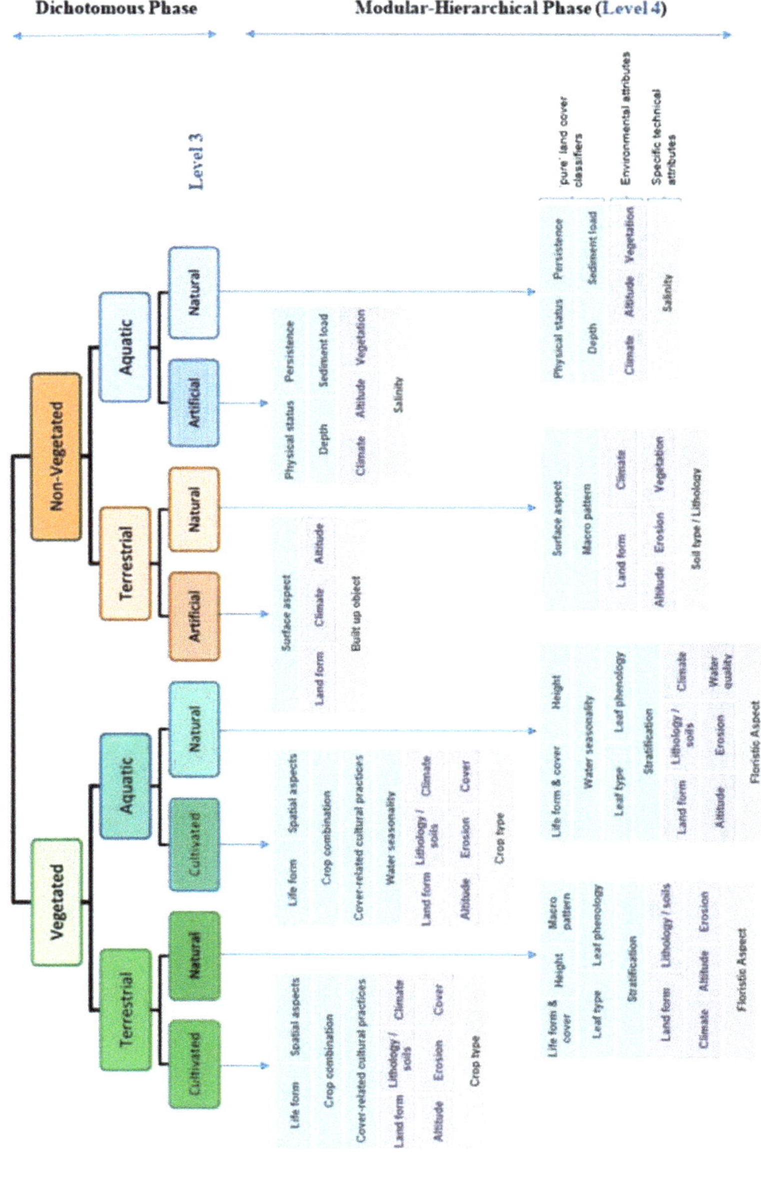

Figure 1. The Food and Agriculture Organisation (FAO) Land Cover Classification System (LCCS) Taxonomy; consisting of the dichotomous and modular–hierarchical phase.

4. Study Sites

The study focused initially on four areas in Queensland and New South Wales (Figure 2). Those in Queensland were the lower Burdekin catchment (coastal semi-natural/natural and agriculture/urban; including the townships of Ayr and Townsville), the Diamantina River (inland riverine natural) and the south eastern Gulf of Carpentaria, including part of the Leichhardt River catchment (coastal dry natural). The Gwydir catchment was selected in New South Wales as it supported inland wetlands and agriculture. These areas collectively were known to experience changes in mangrove extent, hydroperiod and/or agricultural use. Within these, the full range of FAO LCCS Level 3 classes (cultivated or semi-natural, terrestrial or aquatic vegetation, natural and artificial bare surfaces and water) and a diverse range of potential Level 4 classes were present.

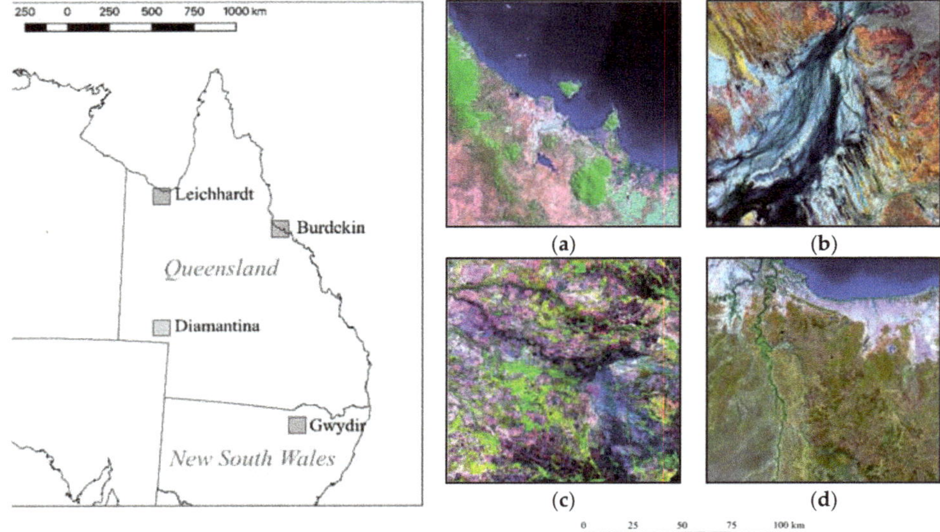

Figure 2. Landsat annual false colour composites for (**a**) Ayr (Queensland), (**b**) Diamantina (Queensland), (**c**) Gwydir (New South Wales) and (**d**) the Leichhardt River (northern Queensland).

Initially, focus was on 100 × 100 km tiles, with these associated with the spatial storage units used within DEA. However, the classification was extended to several adjoining tiles (200 × 200 km for Leichhardt, with this encompassing the original 100 × 100 km area) to demonstrate consistency in classification across a larger region. Higher spatial resolution (< 2–6 m) and temporal change data were available to support the interpretation of the Level 3 change products for the lower Norman River catchment (Leichhardt) and Lake Ross (Townsville, Ayr), and hence these two case study areas were selected to demonstrate the value and potential of the approach. The following subsections provide an overview of the four sites.

4.1. Lower Burdekin (Townsville and Ayr), Queensland

Two 100 × 100 km tiles centred around the towns of Ayr and Townsville in northern Queensland support a diverse range of cultivated/managed lands with extensive irrigation (primarily for sugar cane). The natural vegetation includes grass, shrub and woodlands. Extensive wetlands (including mangroves and intertidal mudflats) are present along the coast and in some areas inland. Major changes are associated with water inundation, with the hydroperiod varying annually and impacting the dynamics of the wetlands but also controlling the amount of water available for irrigation and general use by the population. Agricultural and urban changes also occur over a range of time steps.

4.2. Diamantina, Queensland

The region is remote and dominated by the Diamantina River, which (because of the flat topography) supports numerous river channels that form a dense interconnected network across the region. Many of the river channels present combinations of herbaceous and woody vegetation, and flooding is highly variable, both intra- and inter-annually. The surrounding landscape is semi-arid with extensive areas of bare ground or sparse vegetation. No cultivated or actively managed areas occur.

4.3. Gwydir, New South Wales

The Gwydir River is a perennial river within the Murray Darling River Basin and the catchment is a major agricultural production area. The availability of water has led to the establishment of irrigated agriculture, with particular focus on cotton, although non-irrigation farming also occurs. Within this region, the Gwydir Wetlands support combinations of both herbaceous and woody aquatic vegetation, although much of the non-cultivated area supports dry grass, shrub and woodlands.

4.4. Southwest Gulf of Carpentaria, Queensland and the Leichhardt River

The Gulf of Carpentaria is a remote region that supports extensive areas of mud and sandflats, largely because most of the coastal plan is only a few meters above mean sea level and is hence subject to tidal inundation as well as extensive flooding from rivers (particularly the Flinders River). Agricultural areas occupy a very small proportion of the landscape, and the main urban centers are at Karumba and Burketown. The coastline of the Gulf of Carpentaria is macrotidal and supports extensive tracts of mangrove and saltmarsh. However, A substantive dieback of mangroves was experienced in late 2015.

5. Methods

5.1. Available Data

For the classification of land cover according to the LCCS, a number of EV data layers were available from DEA products (Table 1), whilst others were obtained from other sources but were accessible within the DEA environment (Table 2). All layers listed were available at a 25 m spatial resolution and provided continental coverage.

5.2. Input Layers for EODESM

To support classification within the dichotomous phase (to LCCS Level 3), reference was made to the annual fractional cover layers held within DEA [14]. A threshold of 10% (i.e., where annual observation summaries were greater than 10% for the green and non-photosynthetic vegetation fractions) was determined to define vegetated extent (Level 1). The Level 2 aquatic class was generated from a combination of layers with user determined thresholds. The Inter-Tidal Extent model (ITEM) was used to differentiate between non-vegetated intertidal areas such as mud and sand flats. This dataset represents the tidal range of the Australian coastline, generated from the frequency of inundation as quantified from all cloud free Landsat observations of the coastal margin over a 28-year period. Observations were validated with reference to corresponding tide height data at the time of image capture [15]. This study applied ITEM to extract aquatic areas; that is, areas with greater than 10% and less than 80% of inundated observations for the Landsat archive. The national (Australian) mangrove maps in [16] were then used to identify areas of tidally inundated areas not captured by ITEM because of the obscuration of water by the mangrove canopies. The Water Observations from Space (WOfS) [17] identified inland and coastal water bodies. An area was considered aquatic when the presence of water detected was greater than 10% over the annual observations. An ocean mask (defined as water on the seaward margin of the minimum extent of the ITEM intertidal zone) was also added. All four data layers were then used in combination to identify areas associated with the Level 2 aquatic class. To generate the Level 3 classification (8 classes), the Level 1 and Level 2 classes were cross-tabulated against data on the extent of agricultural cultivation and artificial land covers, namely urban and water (artificial), with these extracted from the Catchment Scale Land Use of Australia [18] layer. Areas not classified as cultivated or managed were considered to be semi-naturally or naturally vegetated or naturally bare. A summary of the input layers and a description of how these are relevant to EODESM in generating land cover classifications, but also more advanced descriptions (through the inclusion of thematic and continuous EV layers, used internally or external to the LCCS taxonomy), are given in Table 3.

Table 1. Thematic (TM) and continuous (CO) layers from Digital Earth Australia (DEA) products.

Data Layer	Type	Derivation	Reference/Source
Fractional cover (photosynthetic and non-photosynthetic vegetation and bare surface)	CO	Spectral unmixing	Gill et al. (2018) [14]
Water Observations from Space (WOfS)	CO	Classification	Mueller et al. (2016) [17]
Inter-Tidal Extent model (ITEM)	TM	Classification	Sagar et al. (2017) [15]
National Mangroves	TM	Classification	Lymburner et al. (2018) [16]

Table 2. Thematic (TM) and continuous (CO) layers currently external to DEA.

Data Layer	Type	Derivation	Reference/Source
TERN Continental Vegetation Height (CVH)	CO	Generated through integration of LiDAR, L-band SAR and Landsat	Scarth et al. (2019) [19]
National Vegetation Information System (NVIS)	C)	Collation of State and Territory vegetation maps	Australian Department of the Environment and Energy DOEE [20]
Catchment Scale Land Use of Australia	TM	Cultivated areas from cadastral information	Australian Bureau of Agricultural and Resource Economics and Sciences ABARES (2016) [18]
	TM	Buildings and infrastructure from cadastral information	
Australian Hydrological Geospatial Fabric (Geofabric)	TM	Artificial water (dams and reservoirs)	Bureau of Meteorology

Table 3. Use of different inputs to the Earth Observation Data for Ecosystem Monitoring (EODESM).

Input Layers	Level 3	Level 4	EVs
Fractional Cover	Vegetated	Canopy cover	Canopy cover (%)
WOfS	Aquatic	Hydroperiod	Hydroperiod (days)
ITEM	Aquatic/bare [1]	Tidal extent	Relative tidal inundation frequency (%)
GMW Mangroves	Aquatic	Tidal extent	
TERN CVH		Lifeform, vegetation (canopy) height	Canopy height (m)
NVIS			Dominant genus
ABARE	Cultivated	Field size	Field size (ha) Crop type
	Urban	Density, geometry	Area (%)
	Artificial water	Water depth	Water depth (m)

[1] Classification at Level 3 depends upon the date or period of observation.

In the modular–hierarchical Phase (LCCS Level 4), more detailed descriptions of the LCCS Level 3 vegetated category were obtained by again referencing the Landsat-derived fractional cover (indicating canopy cover percentage) and canopy height in meters [19], with the latter generated using a combination of Advanced Land Observing Satellite (ALOS) Phased Arrayed L-band Synthetic Aperture Radar (PALSAR) and Landsat sensor data acquired in 2009. Woody vegetation was then associated with objects for which the average canopy height and cover exceeded 2 m and 20%, respectively, with trees and shrubs distinguished as being ≥5 m and <5 m, respectively. All forests were assumed to be evergreen (as is typical in Australia). NVIS data can also be used to broadly discriminate those dominated by broadleaved (primarily Eucalyptus and Acacia) and needle-leaved types (e.g., primarily *Callitris* and *Casuarina* species). The areas defined as aquatic were then further described using WOfS according to the hydroperiod classes defined within LCCS Level 4, namely 1–3, 4–6, 7–9 and >9 months. ABARES data can also be used to indicate field size and crop type and metrics describing urban areas.

For all four areas, a comparison of the LCCS land cover classifications was undertaken between 2009 (associated with the structural classification of Scarth et al. [19]) and 2016 (the latest year, at the time of study, with coverages for all DEA layers) to assess the potential of EODESM for detecting change in EVs (namely hydroperiod and canopy cover). However, for the Gulf of Carpentaria, including the Leichardt River, comparisons were made between 2014 and 2016 as a substantive dieback of mangroves was observed from late 2015 [22].

5.3. Implementation of the Land Cover Classification

The overall mechanism for applying the classification was based on the system outlined in Clewley et al. [23] using Python to combine functionality from a number of different packages, primarily RSGISLib [10], and using the KEA file format [11] for data storage.

5.3.1. Segmentation

For each area, objects were generated by applying the algorithm of Shepherd et al. [24] (available within RSGISLib) to annual (geomedian; [25]) composites (for 2016) of multi-spectral Landsat sensor data and for each tile through DEA. Once proven at the object level, the analysis was applied at a per-pixel level to take advantage of the full resolution of the data.

5.3.2. Classification

The classification procedure applied rules to each environmental dataset and for each time period (i.e., t_1 and t_2). For both periods, and to generate the Level 3 classes, each object (which can be a segment or pixel) was populated with a value of 1 or 0, respectively, to indicate the presence or absence of the feature in question (i.e., vegetation, aquatic environments, cultivation and artificial surfaces or water). The values for each were then cross-tabulated to assign each object to the appropriate LCCS Level 3 category (e.g., vegetated aquatic cultivated). Once the classification system progressed to the more detailed modular levels of LCCS in Level 4, values were assigned separately on the basis of the numerical values associated with each LCCS code specific to each modular layer. For example, leaf types are assigned to broadleaved (D1), needle-leaved (D2) and aphyllous (D3) in the FAO LCCS taxonomy, and the raster layer is accordingly given values of 1, 2 and 3. In the case of hydroperiod, canopy cover and canopy height, the continuous layers were converted to codes according to the LCCS-2 classification scheme, with each category assigned a value associated with specific LCCS codes (e.g., perennial water (7–9 months) was assigned a value of 7 representing the code B7).

As well as the component codes, the original numeric values derived from the EVs (e.g., canopy height) for each object were integrated within the Raster Attribute Table (RAT), as was additional information external to the classification (e.g., NVIS dominant species type, as a numeric code value). Once all layers were integrated and RAT columns populated, the component codes were combined to produce the final combined string prior to translation to a meaningful Level 4 class name. Standardised colour schemes developed by Lucas and Mitchell [8] for the Level 3 and Level 4 classifications were then applied.

5.3.3. Evidence-Based Change Detection

To demonstrate the capacity of EODESM to generate descriptions of land cover change alerts, two of the four test sites were chosen (the lower Norman River catchment near Karumba in Queensland

and within the Leichhardt River scene) and Lake Ross (within the Townsville–Ayr scene), given the availability of suitable field and/or higher resolution EO data for validation (Figure 3).

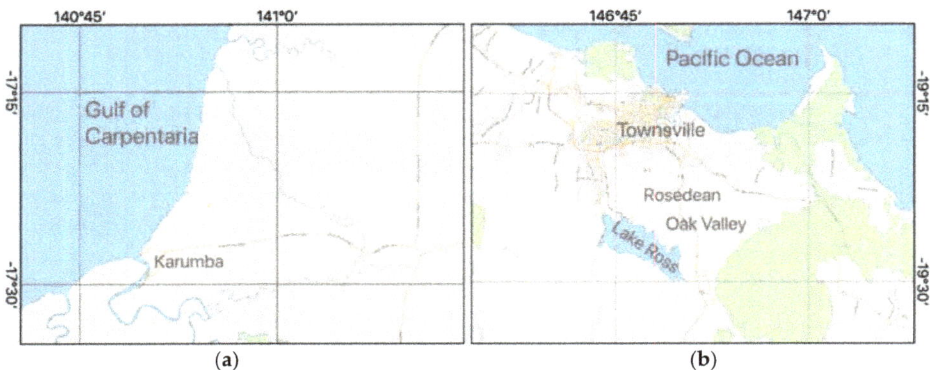

Figure 3. Location maps for detailed assessment of changes in (**a**) mangroves on the coastal margin near Karumba, Gulf of Carpentaria between 2014 and 2016 and (**b**) water extent in Lake Ross, Burdekin Catchment, near Townsville between 2009 and 2016.

In late 2015, a substantive dieback of mangroves occurred along the coastline of the Gulf of Carpentaria, with this attributed to a combination of a substantive drop (20–30 cm) in sea level, high temperatures and low rainfall [22]. The dieback was most rapid and noticeable within mangrove communities dominated by lower stature species (e.g., dominated by *Avicennia marina*) on the landward margin, but there was some evidence of dieback several years earlier (2012) within mangroves dominated by *Rhizophora stylosa* on the landward margin. Airborne and field observations in dieback areas [26,27] indicated that the dieback in mangroves did not immediately result in a loss of mangrove extent nor a change in canopy height but rather a change in canopy cover. Hence, attention focused on establishing changes in the Level 3 classes but also Level 4 canopy cover between 2014 and 2016, with these representing the pre-dieback and post-dieback periods. Comparisons were made against time-series of RapidEye data and the derived Normalised Difference Vegetation Index (NDVI) data provided by Planet [28] for 2014 and 2016. Changes in the WOfS hydroperiod product were also noted in the lower catchment of the Leichhardt River. For Lake Ross, a progressive loss of water area was noted between 2014 and 2016 within the WOfS product. Reservoir volume data (percentage of maximum capacity), Google Earth Imagery (GEI) and RapidEye data were available for the validation of these changes.

6. Results

6.1. Land Cover Classifications

The EODESM land cover classification procedure was applied to the four study areas, for time periods t_1 and t_2, with these being 2009 and 2016, respectively, for Townsville–Ayr, Diamantina and Gwydir and 2014 to 2016 for the Leichhardt. The latter time interval was selected to capture the mangrove dieback event. Level 3 and 4 outputs for 2016 are shown in Figures 4 and 5, respectively, with the latter showing an extended area of 200 × 200 km (for Leichhardt) for the Level 4 classification.

Each map was generated from objects (segments as a first test and then pixels as a final product) that were populated with LCCS components and combined codes and then translated to a full taxonomic description plus thematic and continuous EVs (e.g., % canopy cover or hydroperiod in days). The set of descriptive attributes to complement the classification were also retained within the RAT row for each object. The legend for the classifications was built from the available layers and provided a comprehensive description of land cover classes.

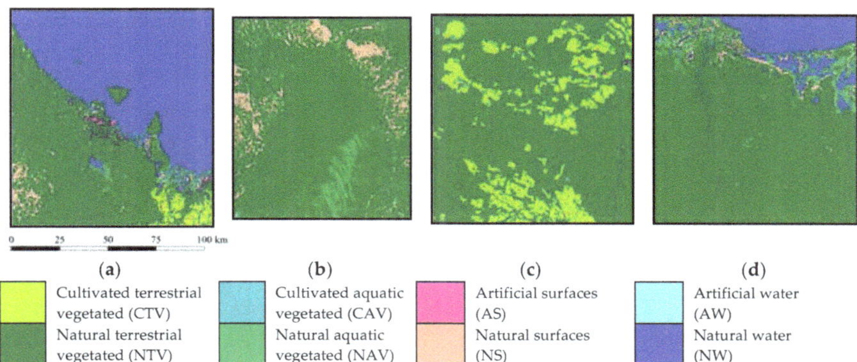

Figure 4. Land cover classifications for (**a**) Ayr (Queensland), (**b**) Diamantina (Queensland), (**c**) Gwydir (New South Wales) and (**d**) the Leichhardt River (northern Queensland) according to the LCCS-3 taxonomy. Each area represents 100×100 km. Acronyms are also provided.

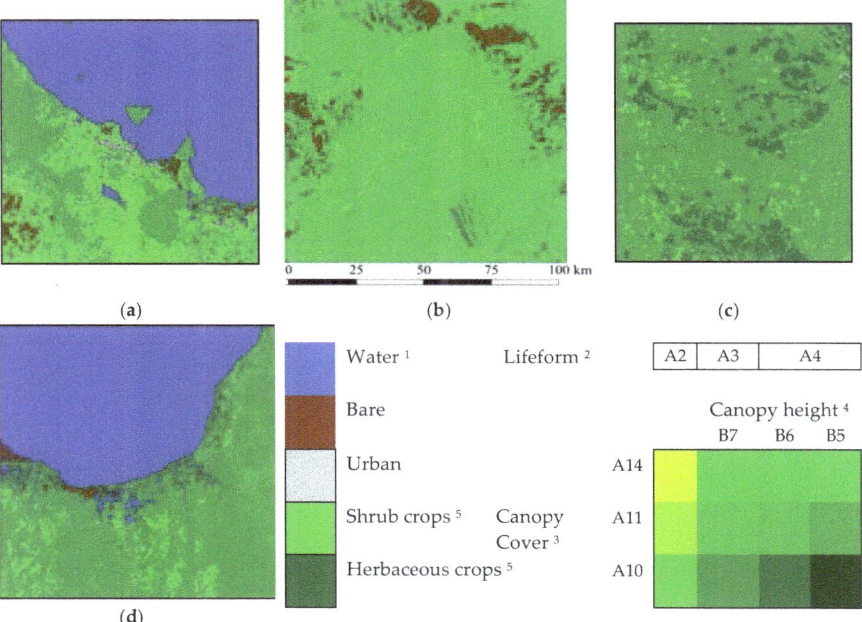

[1] Water areas include hydroperiod classes (> 9 months (B1), 7–9 months (B7), 4–6 months (B8) and 1–3 months (B9)).

[2] Lifeform classes of herbaceous (graminoids and/or forbs (A2), trees (A3) and shrubs (A4)).

[3] Canopy cover classes of sparse (1 to 10–20%; A14), open (10–20 to 60–70%; A11) and closed (> 60–70%; A10) for both aquatic and terrestrial vegetation.

[4] Canopy height classes of 3–7 m (B7), 7–14 m (B6) and > 14 m (B5) for both aquatic and terrestrial vegetation.

[5] Gwydir only.

Figure 5. Land cover classifications for (**a**) Ayr (Queensland), (**b**) the Diamantina River (Queensland), (**c**) Gwydir catchment (New South Wales) and (**d**) the Leichhardt River (northern Queensland; 200×200 km) according to the LCCS-4 taxonomy.

The accuracy of the land cover classifications is dependent upon that of the input layers used for their generation and, where available, these are listed in Table 4. Those reported for the individual input layers ranged from between 80% (for cultivated, urban and artificial) to over 95% (for hydroperiod and fractional cover). A current limitation is that these accuracies are reported using different mechanisms, and hence a future challenge will be to develop an effective approach for standardizing and combining these within a common framework.

Table 4. Reported accuracies of the EV layers used as input to EODESM.

EV	Derived Layer	Validation
Fractional Cover	Vegetated	The fractional cover product has an overall RMSE of 11.8%. The error margins for photosynthetic vegetation, non-photosynthetic vegetation and bare soil fractions are 11.0%, 17.4% and 12.5%, respectively [14].
WOfS	Aquatic	Based on 3.4 million validation points; overall accuracy of 97%; with water identified 93% of time.
ITEM	Aquatic/Bare	Mean absolute height difference between (non-inundated) estimated and actual surface elevation of 0.57 m at the continental level. Based on Real Time Kinematic (RTK) Global Positioning Systems (GPS) ground data, with this being indicative of tidal water depth.
GMW Mangroves	Aquatic/Vegetated [1]	Users' and producers' accuracies from 92–93 and 97–99%, respectively [16].
TERN CVH		Close correspondence with airborne LIDAR profiles from TERN sites [19].
NVIS	Dominant species [2]	Final accuracy of 85% in the delineation of vegetation map units based on aerial photography at 1:20,000.
ABARE [3]	Cultivated	Composite product generated from State and Territory land cover maps with stated overall accuracies above 80% at the catchment scale.
	Urban	As above
	Artificial water	As above

[1] Generated through GMW. [2] Undertaken by experienced interpreters (final accuracy of 85%). [3] Australian Bureau of Agricultural and Resource Economics and Sciences (2011), guidelines for land-use mapping in Australia: principles, procedures and definitions, fourth edition, Australian Bureau of Agricultural and Resource Economics and Sciences, Canberra.

6.2. Land Cover Change Maps

By comparing the FAO LCCS component codes between t_1 and t_2, maps of changes in Level 3 categories were obtained for all four sites (Figure 6). Between-class changes in Level 3 categories from t_1 and t_2 can indicate multiple transitions (e.g., bare ground to terrestrial vegetation, natural water to artificial cultivated vegetation or vice versa) or relative stability in the landscape (e.g., water stays as water). Where major transitions occur at Level 3, the Level 4 classifications can provide more detailed information just before and immediately following the transitions. Where the Level 3 class between t_1 and t_2 does not change, a comparison of the Level 4 classifications provides insights into within-class changes, such as the annual hydroperiod (e.g., from 1–3 months to 4–6 months) or fractional vegetation cover (percentage change). Within EODESM, changes can be represented at both levels, with this indicated in Figure 7, whereby both changes between LCCS Level 3 categories and within Level 4 (hydroperiod) are overlain on the t_2 Level 4 classification. Multiple layers of change (annual, monthly or daily) can also be viewed simultaneously and for different change groups (e.g., canopy cover as well as hydroperiod).

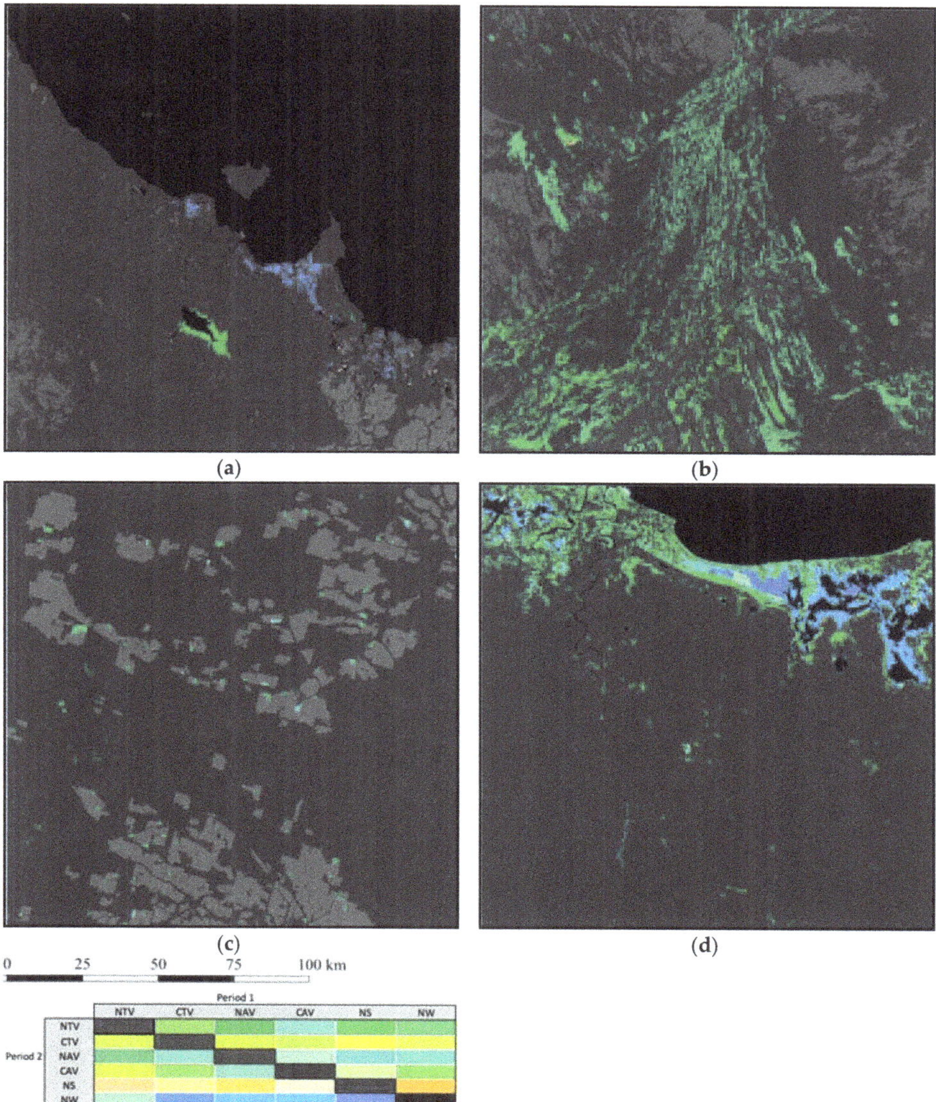

Figure 6. Land cover (LCCS Level 3) change maps for (**a**) Ayr (Queensland), (**b**) Diamantina (Queensland), (**c**) Gwydir (New South Wales) and (**d**) the Leichhardt River (northern Queensland) according to the LCCS-3 taxonomy. The dominant transition is from natural bare surfaces (NS) and natural aquatic vegetation (NAV) to natural water (NW).

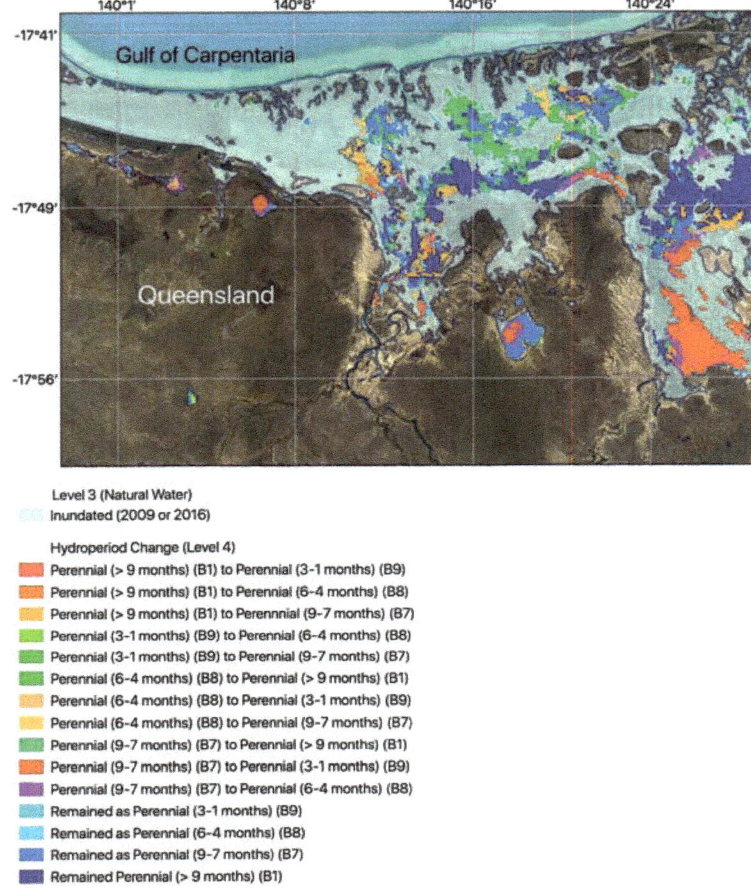

Figure 7. Classification of LCCS Level 4 transitions (in hydroperiod) between t_1 and t_2 for the Leichhardt area (100 × 100 km) and only for areas where the LCCS Level 3 classes remained stable. Hydroperiod classes are >9 months (B1), 7–9 months (B7), 4–6 months (B8) and 1–3 months (B9). The total area that was inundated in either 2009 or 2016 is indicated.

6.3. Evidence-Based Change Descriptions

A fuller description of changes between t_1 and t_2 was demonstrated by combining transitions in the two taxonomic levels for the lower Norman River catchment (Leichhardt) and Lake Ross (Townsville, Ayr). In both cases, the LCCS component codes (for Levels 3 and 4) and the original thematic and continuous information for each LCCS input variable were maintained in the RAT and, when used in combination, allowed changes to be detected and described on the basis of evidence.

In the lower Norman River, the dieback of mangroves between 2014 and 2016 was identified by a rapid decline in the Landsat-derived Normalised Difference Vegetation Index (NDVI) from values typically exceeding 0.6 to those that were < 0.1 (Figure 8a). This decline was captured by a Level 3 change from natural aquatic vegetation (i.e., mangroves) to natural water but also a loss of canopy cover (typically from 70–100%), as determined by comparisons of the Landsat-derived fractional (green) cover and derived Level 4 component codes for canopy cover between these periods (Figure 8b). Whilst time-series data on canopy height were not available, as the height layer of Scarth et al. [19]

was generated for 2009, airborne LIDAR data were acquired in 2016, with these indicating degraded forests with loss of structural integrity. The change classification indicated a close correspondence with the area of dieback observed within mangroves (which had been mapped previously by applying a random forest classifier to 2014 Rapideye spectral data—Figure 8c—and quantified by comparing time-series of RapidEye NDVI data from 2014 and 2016—Figure 8d)).

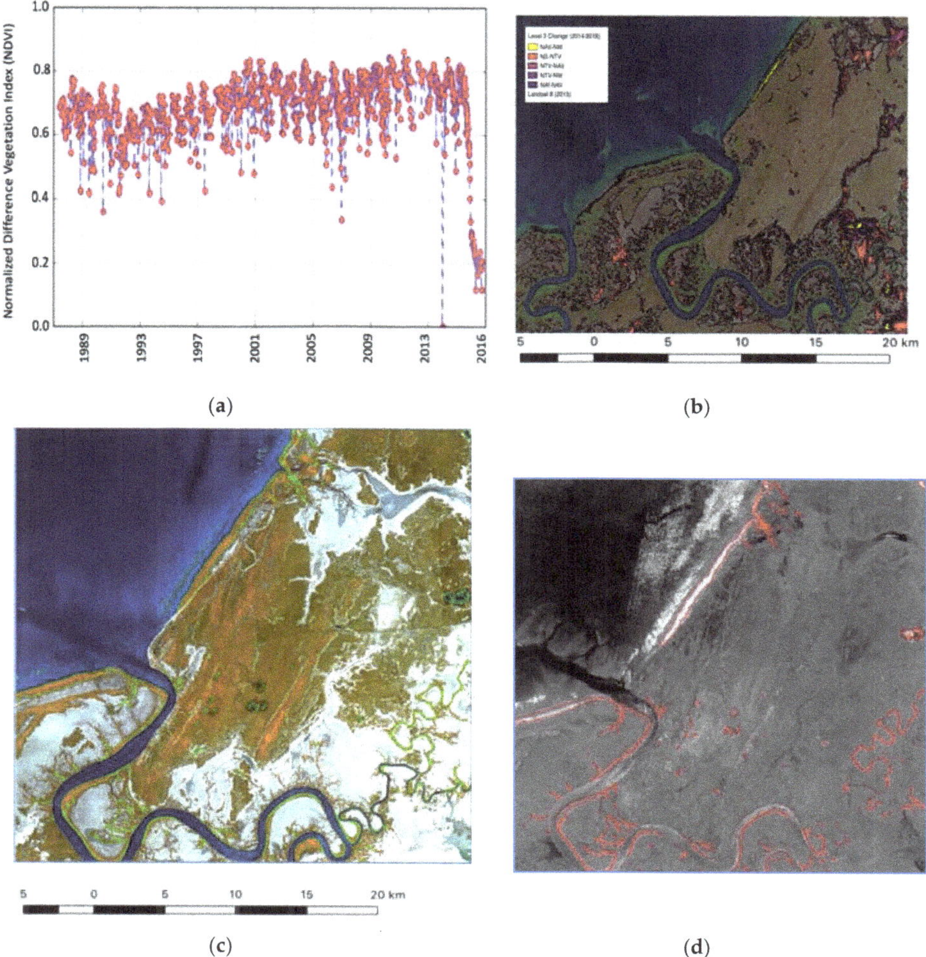

(**a**)

(**b**)

(**c**)

(**d**)

Figure 8. (**a**) Time-series of Landsat-derived Normalised Difference Vegetation Index (NDVI) data extracted from DEA, indicating declines between 2014 and 2016 associated with mangrove dieback along the Gulf of Carpentaria. (**b**) Changes in land cover detected based on a transition between the Level 3 classes, including mangrove dieback (from natural aquatic vegetation to natural water; NAV to NW). Other changes, including flooding of natural terrestrial vegetation (NTV) and vegetation encroachment onto previously naturally bare surfaces (NS), are also indicated. (**c**) RapidEye image from 2014 showing the extent of mangroves (green line) as mapped using a random forest classifier and (**d**) differences in the RapidEye-derived NDVI between 2014 and 2016 showing the extent of mangrove dieback (decreases indicated in white; mangrove area classified in 2014 overlain in red).

The change detection example over the Townsville–Ayr study area indicated that, between 2014 and 2016, the area experienced a change in hydroperiod and a subsequent decrease of water within Lake Ross. This decrease was accompanied by an increase in aquatic vegetation, with this restricting the area of open water. The evidence for this change was based on a change in the LCCS Level 3 category from artificial water to vegetated aquatic (semi) natural indicated by a change in fractional cover from 0% (as a water class) to over 80% vegetated (Figure 9a). The change indicated herbaceous vegetation instead of open water, providing evidence of falling water levels within the dam. This was attributed to low rainfall during this period and water extraction for use by population and irrigated agriculture in proximal areas. This change was also observed within RapidEye data (Figure 9b) and agreed with records of water extent produced by Townsville County Council (Figure 9c). Over this period, dam levels have decreased progressively to a minimum (in 2016).

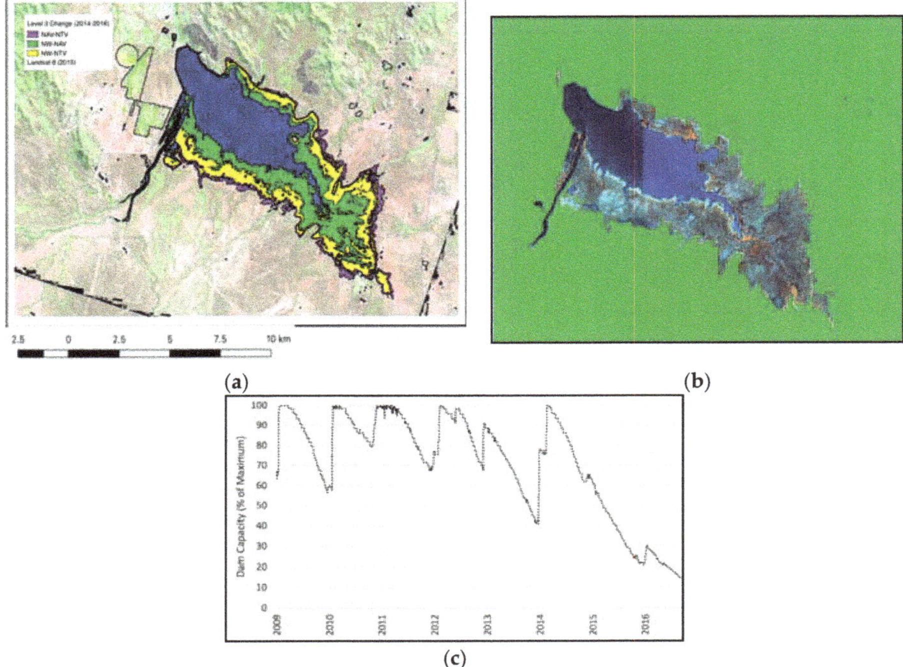

Figure 9. (**a**) Transitions between the Level 3 classes of natural aquatic vegetation (NAV), natural terrestrial vegetation (NTV) and natural water (NW) for Lake Ross between 2014 and 2016. Such changes were associated with a progressive decrease in hydroperiod between 2009 and 2016 and an associated increase in the extent of both aquatic and terrestrial vegetation. (**b**) RapidEye image (near infrared (NIR), red edge and red in RGB) from 2016 highlighting the reduced extent of water and replacement by aquatic and terrestrial vegetation. (**c**) Changes in dam capacity (%) between 2009 and 2016.

For Lake Ross (Burdekin catchment), the Landsat sensor imagery within the Google Earth Engine (GEE; Figure 10) further confirmed the progressive loss of open water, the retreat of aquatic (wet) vegetation and a transition to drier vegetation on the outer margins of the lake's basin. A confusion matrix was difficult to generate because of the lack of a field survey at the time of the image acquisitions, with this highlighting the requirement for the near real-time measurement of equivalent land covers.

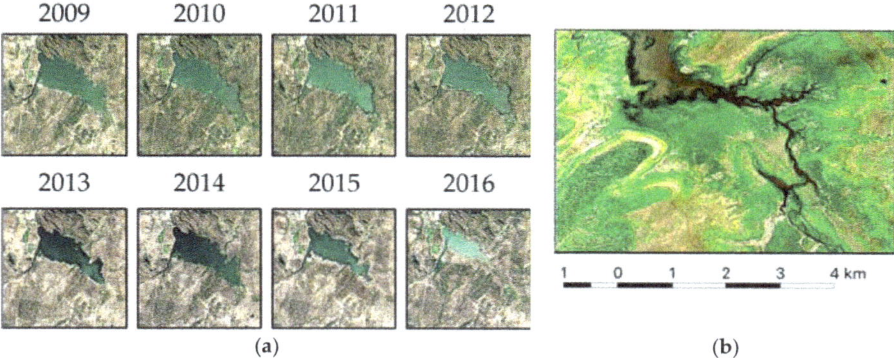

Figure 10. (a) Decrease in the extent of water in Lake Ross near Townsville, Queensland, between 2009 and 2016 and the associated increase in aquatic vegetation, as observed within Google Earth Engine. (b) The distribution of aquatic (wet; primarily green) and drier (brown) vegetation in the high-resolution Google Image of 2016 (southeast section).

7. Discussion

7.1. Overview of Approach

The classifications of land cover according to the FAO LCCS were generated from multiple thematic and continuous EV layers, published within or accessible by DEA. Whilst the focus initially was on several 100 × 100 km areas relating to the four sites, all data layers used were also available at a national level, with this indicating the capacity for generating similar classifications across Australia for different points in time through DEA. All of the input layers were also associated with unit measures (e.g., m, %, number of days) or pre-defined thematic categories (e.g., numeric codes representing urban, cultivated areas, dominant genus). All data were provided at a 25 m spatial resolution, which provided a consistent resolution and spatial reference for comparison. The initial testing was undertaken using segments but then focused at the pixel level to allow the capture of greater detail within the landscape.

In this study, attention was focused on the use of annual summaries of EVs (including fractional cover and persistent green fraction), with these providing a consistent time reference for comparison and a basis for linking concurrent or dependent changes between EVs. Several, but all, datasets were also associated with assessments of retrieval accuracy. Some datasets were only available for one point in time (e.g., a specific year, as in the case of the ABARES layers and also the TERN CVH. Whilst these can be construed as limiting, the current implementation of EODESM highlighted the need for the more regular production of input datasets. For example, the production of annual vegetation CVHs would provide significant support for the mapping of vegetation change, particularly when integrated with canopy cover. This could be achieved by using, for example, the Global Ecosystem Dynamics Investigation (GEDI) LiDAR or interferometric SAR (Tandem-X) data. Algorithms for routinely retrieving these parameters across Australia could also be developed or improved. A limitation of the current CVH layer was that heights were assigned to segments [19], and so the area and outline of the segments was evident within the classification of vegetated categories at LCCS Level 4.

The EODESM system compares input layers from t_1 and t_2, and this comparison aligns well with annual change assessments. The time between t_1 and t_2 can, however, be varied, allowing the comparison of images acquired from consecutive image overpasses (e.g., daily, weekly, monthly) as well as inter-annual and even decadal periods (based on derived summaries). The decision on where to focus the comparison can be assisted by considering the use of dense time-series comparisons of satellite sensor data and derived products (e.g., the NDVI). This approach was taken in the case of the mangrove dieback, where the NDVI time-series (and also knowledge) indicated the time periods to compare LCCS component codes and EVs before and after the event (i.e., 2014 and 2016). Further comparisons between two longer time-separated periods can also be used to track the recovery (or otherwise) of mangroves over time. Consideration generally should also be given to the temporal analysis of data, such as the NDVI, over time using algorithms such as the BFAST [12]. These can allow the targeting of the comparisons between t_1 and t_2.

The approach to identifying change differs from more traditional change detection procedures that typically compare classifications or measures (e.g., indices) generated or derived from an entire scene (or combination of scenes). These tend not to consider that change is often specific to a variable within and between different land covers and that multiple changes are occurring at the same time but also over different time frames. EODESM is therefore particularly beneficial, as a wide range of changes can be captured but also combined. As an example, information on hydroperiod change between two years can be combined with that of water extent change mapped between two sub-annual periods (e.g., weeks, months or each time a satellite image is acquired).

The EODESM system also allows for the detection and description of change to generate historical and potentially near real-time change descriptions. In the case of the mangrove dieback along the coastline of the Gulf of Carpentaria, evidence within or accessible through DEA included the change from aquatic (semi) natural vegetation to water (overlying mudflats). Furthermore, in areas where mangroves had retained a canopy cover, a decrease from a closed to a sparser canopy-covered forest was observed. No other changes were detected, as all other layers were generated from a single year. The transition from vegetation to non-vegetation and the loss of canopy could indicate deforestation, but this is unlikely given that mangroves are protected in Australia and observations post-dieback indicated canopy height did not decrease significantly. Alternative changes would be defoliation or dieback, but the introduction of Advanced Land Observing Satellite (ALOS-2) Phased Arrayed L-band Synthetic Aperture Radar (PALSAR-2) data prior to and following the dieback indicated a decrease in L-band backscatter and hence a loss of moisture content within the woody biomass. In the case of the Burdekin catchment, the loss of water between 2009 and 2016 was indicated by a change from non-vegetated (i.e., water) to vegetated and also the transition (at the margins) from aquatic to 'terrestrial' vegetation as lake levels progressively decreased over time. This sequence indicated colonisation by vegetation of what was assumed to be relatively shallow water. In both selected case studies, the integration of additional layers acquired at appropriate time points would benefit the evidence-based approach to change detection.

7.2. Application within the DEA

Within this study, the DEA datasets were capable of classifying changes in the extent of vegetation cover and hydroperiod within the EODESM system. These datasets are available in yearly summaries (from 1987 to present), and potentially at finer temporal scales, and form a significant part of the system's ability to describe the state and change across the Australian landscape. However, to provide more detailed descriptions of land covers, additional datasets not currently part of DEA were accessed, with these including the cultivated and urban datasets provided by ABARES and TERN's CVH. It is not necessary for these to be are hosted within the DEA environment, but they are critical to the base functionality of the EODESM system, and consideration needs to be given to how best to achieve access in an operational manner. While they are freely distributed by their respective suppliers, a more formal engagement with the parties responsible may provide benefits to DEA and the suppliers in terms of the validation (and error estimation) of datasets, public visibility of the datasets, and the potential to affect how often these datasets are maintained and updated.

Within EODESM, the land cover and change classifications were generated entirely from EVs, which were both continuous and thematically coded (e.g., the ABARE Cultivated Layer). In several cases (e.g., for WOfS, TERN CVH and National Mangroves), uncertainty estimates were available, although these were based on, for example, a selection of sites for which higher resolution data (e.g., LIDAR CVHs) were available. The overall class accuracies are therefore accumulated from—and are therefore dependent upon—the accuracy of the input layers, noting that these will vary both spatially and temporally. A criterion for using these layers is that they were available nationally. However, future efforts are now focusing upon the generation of spatially explicit accuracy layers that can give estimates of uncertainty at the pixel level and confidence in the classifications generated, including at the national level. A step towards this also has been the development of the EarthTrack mobile application (earthtrack.aber.ac.uk), which facilitates the ground level collection and public dissemination of land cover and change data in near real-time and according to the FAO LCCS Version 2 taxonomy and the TERN mangrove portal [29], which disseminates a diverse range of ground and airborne datasets to support national mangrove mapping and monitoring efforts.

As the EODESM system is not tied to comparisons over a fixed period of time, change metrics generated by the system have the potential to highlight events on the landscape at a variety of temporal scales. Dramatic changes, such as defoliation due to pests and damage from bushfires or flooding, occur over potentially much shorter timescale than the effects of long-term reduction in rainfall, but both scales can be monitored simply by specifying the beginning and end of the comparison time period to before and after such events. In this case, DEA could implement official products classifying the state of vegetation (according to the LCCS) at the continental scale with annual products (which is the current preference for major DEA products) and provide users with the ability to focus on areas of interest to generate LCCS classifications and change metrics for different time-periods (across multiple years, or within year change). However, one of the challenges to be faced will be to rank the identified changes and separate true changes (where a real and meaningful event or change in land use has taken place) from the many natural, seasonal and land management changes which will also be identified by the system.

The EODESM system has been shown to work under the DEA and National Computational Infrastructure (NCI) environments, but while the base inputs provided to the system are already generating useful data on the state of the landscape and the state of change, there is significant potential to enhance the capability of the generated products through the incorporation of additional datasets. Vegetation cover, height, hydroperiod, urban and cultivated environments are the base requirements (i.e., for LCCS Level 3). However, additional datasets should be explored, such as the use of high-dimensional pixel composites/statistics of time series to distinguish urban and cultivated environments [25,30]

The inclusion of datasets that can be retrieved periodically, ideally derived from the Landsat archive, would enable the robust assessment of the state of the landscape. Moreover, the retrieval of additional variables such as AGB (e.g., as derived from combinations of ALOS-1/2 PALSAR-1/2, Landsat sensor and ICESAT-1/2 or GEDI data) and water turbidity and depth (from optical sensors) would further provide information to describe evidence-based change. Additional work is required to better inform the EODESM system with respect to species distribution of vegetation, road networks, bare surface classifications and more. The LCCS standard includes numerous physical descriptors of the landscape (see Figure 1), and all of them can be incorporated into EODESM to form a better understanding of the state of the landscape both past and present, and to better predict how changes may affect the landscape in the future. The results for Lake Ross in the Townsville–Ayr region and the mangrove dieback around the mouth of the Norman river in the Leichhardt region show that both the WOfS and vegetation fractional cover are already useful layers that can be exploited to detect multiple changes across the Australian landscape.

There are several advantages to using the EODESM approach within DEA. The national availability of datasets and the 30-year archive of data make it possible to generate land cover and land cover change maps for any spatial extents within Australia and for any time period, and to compare changes between periods that are separated by variable time intervals. DEA also has computing capacity to generate LCCS classifications at a national (continental) level. Layers that are currently available on an annual basis (from 1987 to the present) are hydroperiod and fractional vegetation cover, whilst ITEM has been generated from an interannual time-series of Landsat sensor data in order to maximize the capture of the landscape at the lowest tides. Other layers that can be used as direct input to EODESM are either under review for final publishing on DEA or are separately maintained products. As such, while EODESM has been successfully tested using data on DEA, the publication of more EV data layers at a national scale would facilitate comprehensive classifications of land cover and change using EODESM.

8. Conclusions

This research provides evidence of the capability of the EODESM approach to deliver consistent classifications of land covers and change dynamics over diverse Australian landscapes. The research has established that the EODESM system can be integrated within the framework of the ODC, and applied to diverse landscapes, as demonstrated for Australia through DEA. EODESM therefore provides an option for the mapping of evidence-based land covers and change at a national level and for multiple time instances. The maps generated are at a higher spatial resolution than most current Australian and global land cover classifications, and the use of national EVs results in seamless mapping between tiles. The information content can also be increased by generating or accessing additional EVs. Mapping accuracy can be assessed using higher-resolution datasets, but there is a potential for validation using mobile applications or higher-resolution thematic maps (e.g., vegetation height and cover, water extent and turbidity) generated from airborne LiDAR or drone imagery; additionally, the latter can also be used to generate higher-resolution LCCS classifications for validation.

The EODESM system allows for the classifications of land covers for two points in time (determined by the user), and it can also detect basic and more advanced evidenced-based change alerts, as demonstrated for the Townsville–Ayr region (hydroperiod and vegetation/water change) and the Gulf of Carpentaria (mangroves). Change events and processes can also be identified within DEA through the analysis of the dense time series of Landsat sensor data and/or derived products (e.g., NDVI), thereby informing the time steps with which to best describe these changes using EODESM. A benefit of the approach is that new algorithms for generating EVs can be introduced and the resulting layers (e.g., hydroperiods) then inserted into EODESM to produce revisions of the classifications, including historically.

As result of this use case, we recommend (a) a national demonstration of the approach to land cover and change classification using EVs retrieved at the continental level as inputs to EODESM and within the framework of DEA; (b) the increase of the capacity for generating consistent, nationally available and temporally variable data layers (e.g., urban extent, cultivated area) for improved classifications; and (c) the advancement of the use of airborne (including drone) and mobile applications for validating classifications and advancing the retrieval of EVs.

Author Contributions: Conceptualisation, G.M.; Formal analysis, N.M., A.S. and C.O.; Investigation, C.K. and B.T.; Methodology, L.L.; Project administration, N.M.; Software, D.C., P.B., B.T. and B.L.; Visualization, R.L.

Funding: EODESM represents an advancement of the EODHaM system, which was developed originally through through the European Community's Seventh Framework Programme, within the FP7/SPA.2010.1.1-04: "Stimulating the development of GMES services in specific area", under grant agreement 263435 for the project 'Biodiversity Multi-Source Monitoring System: From Space To Species' (BIO SOS). The European Union's Horizon 2020 research and innovation program under grant agreement no. 641762 supported the development of EODESM with specific application to protected areas. The European Research Development Fund (ERDF) Sêr Cymru II program award (80761-AU-108; Living Wales) to Richard Lucas allowed continued development for application at national levels and realisation of the EarthTrack mobile app. Geoscience Australia provided support to demonstrate application within the Open Data Cube environment.

Acknowledgments: This paper was published with the permission of the CEO, Geoscience Australia. The authors would like to thank the considerable contributions made through the BIOSOS, ECOPOTENTIAL and Living Wales projects and thank PlanetLab for allowing access to the RapidEye data for Lake Ross.

Conflicts of Interest: The authors declare no conflict of interest.

References

1. Geoscience Australia. 2011. Available online: https://ecat.ga.gov.au/geonetwork/srv/eng/catalog.search#/metadata/71069 (accessed on 16 October 2019).
2. International Organization for Standardization. 2018. Available online: https://www.iso.org/standard/44342.html (accessed on 16 October 2019).
3. Atyeo, C.; Thackway, R. *A Field Manual for Describing and Mapping Revegetation Activities in Australia*; Bureau of Rural Sciences: Canberra, Australia, 2009.
4. Lewis, A.; Oliver, S.; Lymburner, L.; Evans, B.; Wyborn, L.; Mueller, N.; Raevksi, G.; Hooke, J.; Woodcock, R.; Sixsmith, J.; et al. The Australian geoscience data cube-Foundations and lessons learned. *Remote Sens. Environ.* **2017**, *202*, 276–292. [CrossRef]
5. Lucas, R.M.; Blonda, P.; Bunting, P.; Jones, G.; Inglada, J.; Aria, M.; Kosmidou, V.; Petrou, Z.I.; Manakos, I.; Adamo, M.; et al. The Earth Observation Data for Habitat Monitoring (EODHAM) System. *Int. J. Appl. Earth Obs. Geoinform.* **2014**, *37*, 17–28. [CrossRef]
6. Lucas, R.M.; Medcalf, K.; Brown, A.; Bunting, P.; Breyer, J.; Clewley, D.; Keyworth, S.; Blackmore, P. Updating the Phase 1 habitat map of Wales, UK, using satellite sensor data. *ISPRS J. Photogramm. Remote Sens.* **2011**, *66*, 81–102. [CrossRef]
7. EcoPotential NGO. Available online: https://www.ecopotential-project.eu (accessed on 16 October 2019).
8. Lucas, R.M.; Mitchell, A. Integrated Land Cover and Change Classifications. In *The Roles of Remote Sensing in Nature Conservation: A Practical Guide and Case Studies*; Díaz-Delgado, Lucas, R., Hurford, C., Eds.; Springer: Cham, Switzerland, 2017; pp. 295–308.
9. Lucas, R.M. The Earth Observation Data for Ecosystem Monitoring (EODESM). Available online: https://essilab.wixsite.com/eodesm (accessed on 16 October 2019).
10. Bunting, P.; Clewley, D.; Lucas, R.M.; Gillingham, S. The Remote Sensing and GIS Software Library (RSGISLib). *Comput. Geosci.* **2014**, *62*, 206–226. [CrossRef]
11. Bunting, P.; Gillingham, S. The KEA image file format. *Comput. Geosci.* **2013**, *57*, 54–58. [CrossRef]
12. Verbesselt, J.; Hyndman, R.; Zeileis, A.; Culvenor, D. Phenological change detection while accounting for abrupt and gradual trends in satellite image time series. *Remote Sens. Environ.* **2010**, *114*, 2970–2980. [CrossRef]
13. Tarantino, C.; Adamo, M.; Lucas, R.; Blonda, P. Detection of changes in semi-natural grasslands by Cross Correlation Analysis with WorldView-2 images and new Landsat 8 data. *Remote Sens. Environ.* **2016**, *175*, 65–72. [CrossRef] [PubMed]
14. Gill, T.; Johansen, K.; Phinn, S.; Trevithick, R.; Scarth, P.; Armston, J. A method for mapping Australian 584 woody vegetation cover by linking continental-scale field data and long-term Landsat time series. *Int. J. Remote Sens.* **2017**, *38*, 679–705. [CrossRef]
15. Sagar, S.; Roberts, D.; Bala, B.; Lymburner, L. Extracting the intertidal extent and topography of the Australian coastline from a 28 year time series of Landsat observations. *Remote Sens. Environ.* **2017**, *195*, 153–169. [CrossRef]
16. Lymburner, L.; Bunting, P.; Lucas, R.; Scarth, P.; Alam, I.; Phillips, C.; Ticehurst, C.; Held, A. Mapping the multi-decadal mangrove dynamics of the Australian coastline. *Remote Sens. Environ.* **2019**. [CrossRef]
17. Mueller, N.; Lewis, A.; Roberts, D.; Ring, S.; Melrose, R.; Sixsmith, J.; Lymburner, L.; McIntyre, A.; Tan, P.; Curnow, S.; et al. Water observations from space: Mapping surface water from 25 years of Landsat imagery across Australia. *Remote Sens. Environ.* **2016**, *174*, 341–352. [CrossRef]
18. ABARES. *The Australian Land Use and Management Classification Version 8*; CC BY 3.0; Australian Bureau of Agricultural and Resource Economics and Sciences: Canberra, Australian, 2016.
19. Scarth, P.; Armston, J.; Lucas, R.; Bunting, P. A Structural Classification of Australian Vegetation Using ICESat/GLAS, ALOS PALSAR, and Landsat Sensor Data. *Remote Sens.* **2019**, *11*, 147. [CrossRef]
20. Australian DoEE. Available online: https://www.environment.gov.au/ (accessed on 16 October 2019).
21. Bunting, P.; Rosenqvist, A.; Lucas, R.; Rebelo, L.M.; Hilarides, L.; Thomas, N.; Hardy, A.; Itoh, T.; Shimada, M.; Finlayson, C. The Global Mangrove Watch-a New 2010 Baseline of Mangrove Extent. *Remote Sens.* **2018**, *10*, 1669. [CrossRef]

22. Duke, N.C.; Kovacs, J.M.; Griffiths, A.D.; Preece, L.; Hill, D.J.; Van Oosterzee, P.; Mackenzie, J.; Morning, H.S.; Burrows, D. Large-scale dieback of mangroves in Australia's Gulf of Carpentaria: A severe ecosystem response, coincidental with an unusually extreme weather event. *Mar. Freshw. Res.* **2017**, *68*, 1816–1829. [CrossRef]

23. Clewley, D.; Bunting, P.; Shepherd, J.; Gillingham, S.; Flood, N.; Dymond, J.; Lucas, R.; Armston, J.; Moghaddam, M. A Python-Based Open Source System for Geographic Object-Based Image Analysis (GEOBIA) Utilizing Raster Attribute Tables. *Remote Sens.* **2014**, *6*, 6111–6135. [CrossRef]

24. Shepherd, J.D.; Bunting, P.; Dymond, J. Operational large-scale segmentation of imagery based on iterative elimination. *Remote Sens.* **2019**, *11*, 658. [CrossRef]

25. Roberts, D.; McIntyre, A.; Mueller, N. High-dimensional pixel composites from earth observation time series. *IEEE Trans. Geosci. Remote Sens.* **2017**, *55*, 6254–6264. [CrossRef]

26. Hacker, J. Targeted Airborne Data. Available online: http://wiki.auscover.net.au/wiki/Targeted_Airborne_Data. (accessed on 16 October 2019).

27. Lucas, R.; Finlayson, C.M.; Bartolo, R.; Rogers, K.; Mitchell, A.; Woodroffe, C.; Asbridge, E.; Ens, E. Historical perspectives on the mangroves of Kakadu National Park. *Mar. Freshw. Res.* **2017**, *69*, 1047–1063. [CrossRef]

28. PlanetLabs. Available online: https://www.planet.com/ (accessed on 16 October 2019).

29. The Terrestrial Environment Research Network (TERN) Mangrove Data Portal. Available online: http://wiki.auscover.net.au/wiki/TERN_Mangrove_Data_Portal_and_Monitoring_System (accessed on 25 June 2019).

30. Roberts, D.; Dunn, B.; Mueller, N. Open Data Cube Products Using High-Dimensional Statistics of Time Series. In Proceedings of the IGARSS 2018–2018 IEEE International Geoscience and Remote Sensing Symposium, Valencia, Spain, 30 June 2018; pp. 8647–8650.

Article

Paving the Way towards an Armenian Data Cube

Shushanik Asmaryan [1,*], Vahagn Muradyan [1], Garegin Tepanosyan [1], Azatuhi Hovsepyan [1], Armen Saghatelyan [1], Hrachya Astsatryan [2], Hayk Grigoryan [2], Rita Abrahamyan [2], Yaniss Guigoz [3,4] and Gregory Giuliani [3,4]

1   GIS and Remote Sensing Department, Center for Ecological-Noosphere Studies NAS RA, Yerevan 0025, Armenia
2   Institute of Informatics and Automation Problems NAS RA, Yerevan 0014, Armenia
3   Institute for Environmental Sciences, University of Geneva, 1205 Geneva, Switzerland
4   United Nations Environment Programme/Global Resource Information Database (UNEP/GRID-Geneva), 1219 Geneva, Châtelaine, Switzerland
*   Correspondence: shushanik.asmaryan@cens.am; Tel.: +374-9400-8214

Received: 14 June 2019; Accepted: 23 July 2019; Published: 2 August 2019

Abstract: Environmental issues become an increasing global concern because of the continuous pressure on natural resources. Earth observations (EO), which include both satellite/UAV and in-situ data, can provide robust monitoring for various environmental concerns. The realization of the full information potential of EO data requires innovative tools to minimize the time and scientific knowledge needed to access, prepare and analyze a large volume of data. EO Data Cube (DC) is a new paradigm aiming to realize it. The article presents the Swiss-Armenian joint initiative on the deployment of an Armenian DC, which is anchored on the best practices of the Swiss model. The Armenian DC is a complete and up-to-date archive of EO data (e.g., Landsat 5, 7, 8, Sentinel-2) by benefiting from Switzerland's expertise in implementing the Swiss DC. The use-case of confirm delineation of Lake Sevan using McFeeters band ratio algorithm is discussed. The validation shows that the results are sufficiently reliable. The transfer of the necessary knowledge from Switzerland to Armenia for developing and implementing the first version of an Armenian DC should be considered as a first step of a permanent collaboration for paving the way towards continuous remote environmental monitoring in Armenia.

Keywords: big earth data; sustainable development goals; swiss DC; Armenian DC; Landsat; sentinel; analysis ready data

---

1. Introduction

Environmental problems become an increasing global concern continuously put stress on natural resources. Global challenges with environmental compartments dimensions such as fresh water, air quality, deforestation, land management or urbanization require improved and updated information, which acquired the dynamic nature of environmental conditions [1,2]. Earth observations (EO) data (satellite and in-situ), provide strong monitoring mechanisms for above mentioned environmental problems because of their geospatial consistency, accessibility, repeatability, and global coverage [3,4]. It proves that by providing a summarized view of a given spatial extent remotely sensed EO becomes an important element to monitor the ecological state of the different environmental compartments (water, soil, plants, etc.). So, precise and reliable data are an important component of the environmental monitoring systems [5]. There are several open remote sensing (RS) data repositories that provide highly valuable, timely and precise remotely sensed EO information. However, there is a strong need of a set of geoprocessing tools, which would allow to retrieve the full information potential of EO data [6–8]. This is mainly because of EO data complexity, large-volume, and deficiency of good processing capacities [8–10].

Considering EO data as part of Big Data, because of their volume (e.g., Landsat archive is 7.5PB), variety (e.g., optical, radar), and velocity (e.g., Sentinel data temporal resolution is every 5 days), innovative tools are required to reduce the time and knowledge needed to access, prepare and analyze large volumes of EO data having steady and spatially adjusted calibrated observations [5].

EO Data Cube (DC) is a new paradigm aiming to meet Big Earth Data challenge as a new approach to store, organize, manage and analyze EO data [11,12].

Hence, Data-Cube is now considered as a promising technology to perform time-series analyses of large satellite Analysis Ready data-sets like Landsat and Sentinel [13].

There are several operational DC initiatives, covering different spatial scales and storing different data, using different infrastructures and software implementations (e.g., Earth Observation DC (EODC—http://eodatacube.eu), Earth on Amazon Web Services (EAWS—https://aws.amazon.com/earth/), Google Earth Engine (GEE—https://earthengine.google.com), Earth System DC (ESDC—http://earthsystemdatacube.net) [5].

As of end 2018, three countries (Australia, Switzerland, and Colombia) have DC on a national-scale (https://www.opendatacube.org/ceos).

Australian Geoscience DC (AGDC—http://www.datacube.org.au), renamed as Digital Earth Australia, was the first successful attempt, making entire continent's geographical datasets available to researchers and policy-makers [12,14]. Lessons learned from design and implementation of AGDC underpin Chinese DC (CDC) based on the new Open Geospatial Consortium (OGC) Discrete Global Grids System (DGGS) standard and cloud computing technologies and Colombian DC [15,16].

However, Switzerland is the second country in the world, which claimed to have a national-scale EODC. The Federal Office for the Environment supports the Swiss DC (http://www.swissdatacube.ch). It is developed, implemented and operated by the UN Environment (UNEP)/GRID-Geneva in partnership with the University of Geneva [5]. Currently, the Swiss DC contains 35 years of Landsat 5,7,8 (1984–2019), four years of Sentinel-2 (2015–2019), and 5 years of Sentinel-1 (2014–2019) Analysis Ready Data over Switzerland (total volume: 6TB; 200 billion observations) [17].

The Committee of Earth Observation Systems (CEOS) has vision, that more over 20 countries will be developing and realizing their Data-Cube infrastructure by 2022 [18].

Armenia is among these countries, aiming to gain the knowledge and to exchange experience from Switzerland implementing its own DC for several reasons: (i) Armenia still faces numerous environmental challenges as one of the most industrialized post-soviet countries; (ii) since the 90s, the economic policy moved towards supporting industrial development mainly ignoring environmental interests; (iii) in 2016, Armenia had initiated the Sustainable Development Goals (SDG) nationalization process and still face-off various problems caused by the lack of sufficient data hindering efficient national environmental monitoring; (iv) alternative ways need to be developed and realized to fill this gap and EODC represents a promising solution.

The paper aims to present the Swiss—Armenian joint initiative on the deployment of an Armenian DC, which is anchored on the best practices of the Swiss model.

2. Building the Armenian DC

Armenia was among the selected countries to contribute towards the shaping of the global development agenda, which was both a privilege and recognition of the country's unique perspective on development [19,20]. However, when monitoring the process of attaining several SDG targets (e.g., SDG target 6.6; SDG target 15.3; SDG target 15.4) an important problem of data disaggregation was encountered. EO can support the data aggregation process by providing policy makers with repeatable, continuous and multi-annual series of quantitative and qualitative data. The integration of EO technologies into decision making process is still to be improved in Armenia. So far, Armenian "decision makers" rely on the data provided by a few research or international organizations, which are experienced in working with EO data and technologies [21–23]. Taking into account the fact that reliable remotely sensed monitoring of the identified SDGs requires EO systems allowing systematic acquisitions, free and open-access data and high quality imagery, the Landsat and ESA's Sentinel missions are the main data sources used. But high performances computational resources are needed to maintain process, visualize and share the EO-based monitoring data. It could be done by creating linkages with new platforms such as DCs that empower data visualization by providing an easier way to visualize environmental changes.

Thus, Swiss-Armenian cooperation initiated the establishment of the Armenian DC as a full and updated archive of EO data (e.g., Landsat, Sentinel), benefiting from the experience of the University of Geneva in implementing the Swiss DC.

In order, to transfer the necessary knowledge, it is vital to develop new capacities. This helps to reach adoption, acceptance and commitment to this new technology for increasing the capacity to access and use Earth Observations [24]. Capacity development can be defined as "human, scientific, technological, organizational, and institutional resources and capabilities" to "enhance the abilities of stakeholders to evaluate and address crucial questions related to policy choices and different options for development" (GEO Secretariat 2006). Three levels of capacity building can be defined: (1) human (e.g., education and training); (2) institutional (e.g., improving the comprehension of the value of geospatial data for decision-making); and (3) infrastructure (e.g., installing/configuring/managing of the technology). This should help demonstrating the benefits of EODC through appropriate examples and best practices to strengthen: (1) existing observation systems; (2) capacities of decision-makers to use it; and (3) capacities of the general public to understand important environmental, social and economic issues at stake. Such initiatives can also be beneficial for providers to increase their visibility and reliability nationally and internationally by participating in the approach to build such systems [25,26].

Recognizing these needs and based on the experience acquired in developing capacity building material for implementing Spatial Data Infrastructure [24] similar to the Bringing GEOSS Services into Practice, the Swiss team started to develop an integrated set of teaching material and software to give the necessary knowledge to efficiently install, manage and use an EODC based on the Open DC software stack.

The successfully installed Armenian DC is already available via http://datacube.sci.am (Figure 1) and the "Bringing Open Data Cube into practice" material is available at: http://www.swssidatacube.ch/products.

As in the case of the Swiss DC, a fundamental aspect when building a DC is to have Analysis Ready Data products, ingested, stored and available in the database. Analysis Ready Data (ARD) are concerned by the four first steps (data acquisition, radiometric calibration, conversion to top of atmosphere(TOA) reflectance and Surface reflectance) allowing then to analyze data and generate time-series [5]. All procedures of discovering, downloading from different repositories (e.g., ESPA, Sentinel Data Hub) and preprocessing were planned to be automated as much as possible and should be interoperable.

Thus, the Armenian DC contains 3 years (2016–2019) of Landsat 7 and Sentinel-2 analysis ready data over Armenia.

The full coverage of Armenia includes 11 Sentinel-2 (38TLL, 38TML, 38TNL, 38TLK, 38TMK, 38TNK, 38SMJ, 38SNJ, 38SPJ, 38SNH, 38SPH) and 9 Landsat 7 (171031, 170031,169031, 171032, 170032, 169032, 168032, 169033, 168033) scenes. It requires around 30–40 min to download and process a single Sentinel-2 image. The system deployment environment is Ubuntu server version 18.04 with 64-bit virtual machine, 64GB of RAM, 8 cores and a storage space of 2 TB. For downloading the correct scenes of our region, the boundary and projection conditions are provided, after which the datacube platform allows to download the available satellite images from global databases and translate data from the Earth observation satellites into ready-to-use insights about the continent's environmental conditions. Armenia is located inside a rectangle with the upper left (38.32335165219022, 42.98858178626198) and lower right (41.551890393271684, 47.320774961261485) points in the Earth coordinate system.

The Armenian DC uses the National e-infrastructure, which is a complex IT infrastructure consisting of both communication and distributed computing infrastructures [16].

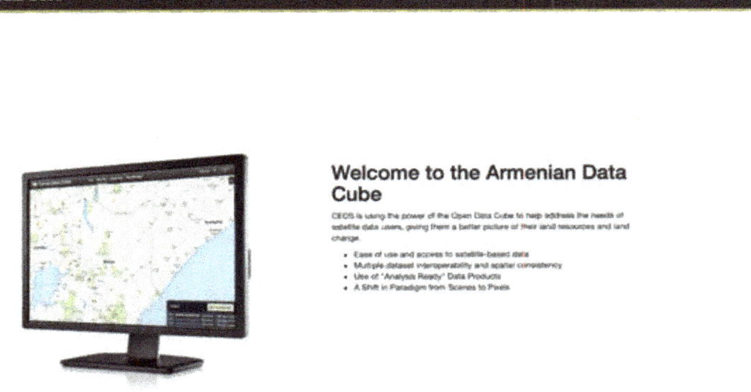

Figure 1. Interface of the Armenian DC.

3. Discussion

3.1. Lake Sevan as a Case Study: A Problem Statement

Among the issues where Armenian DC could provide a set of excellent tools and being demonstrated as a disruptive technology is the monitoring of the shoreline changes of Lake Sevan.

Lake Sevan is one of the most ecologically sensitive areas in Armenia. Since the beginning of the last century, the shoreline of this biggest freshwater lake in Armenia and South Caucasus has been changing continuously with different intensity causing many ecological problems: eutrophication of the lake, activation of erosion processes and so forth [27]. This makes it urgent to study the shoreline changes in order to understand the effects these produce on the near-shore belt [28,29].

Mapping and detection of coastline changes from satellite images have become increasingly important over recent decades, especially because satellites capture and provide data in visible and infrared spectral bands where the land and water can be easily distinguished [30–32]. These make optical satellite images containing visible and infrared bands of the electromagnetic spectrum widely used for coastline mapping especially when these images are easily obtainable [33].

There are several studies where the satellite optical imagery was used to assess Lake Sevan water quality [34–36]. However, there is no direct study on detecting changes of the Lake coastline using time-series analysis of satellite EO data and it is easy to perform if the data is openly available. The satellite image analysis enables to study the water boundary changes using the water detection service provided in the Armenian DC platform (Figure 2).

Exploration of the full potential of EO data requires huge computing resources enriched by specialized algorithms and tools [5]. So following Australian and Swiss experience on DC Swiss-Armenian research group decided to develop an automatized tool for shoreline delineation in ADC.

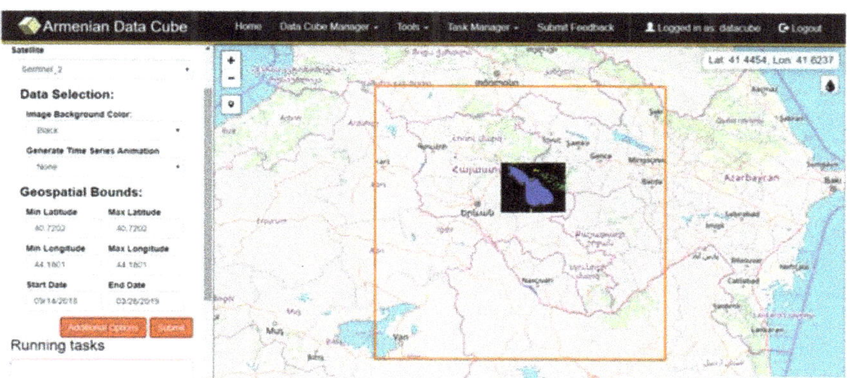

Figure 2. Sentinel-2 ingestion of Lake Sevan.

3.2. Analysis Ready Data Production (Data Availability, Access, Ingestion Preprocessing)

The main phase when developing DC is the preparation of ARD allowing to analyze data and generate time-series [5].

Satellite data scenes (Landsat 5,7,8 and Sentinel-2) were accessible via gsutil: a Python application, which gives an access to Google Cloud Storage from Command lines (https://cloud.google.com/storage/docs/gsutil).

The Live Monitoring of Earth Surface (LiMES) framework has been used for ARD preparation, which is a framework that helps to automate EO data discovery and (pre-) processing using interoperable set of tools transforming observations into the information products applicable for monitoring environmental changes. This framework is developed using a system of large storage capacities, high performance distributed computers, and interoperable standards to develop a scalable, coherent, flexible, and efficient analysis system, which can be used on various domains through decades of data for monitoring [5].

3.3. Image Processing

There are several methods of water object identification and shoreline delineation, which include classification and spectral signature feature analysis, which divided into single-band and multi-band methods [37].

Single band and multi-band threshold methods are widely used in optical RS to extract water bodies [38].

Single band method is a simple approach allowing to extract water surface information. Multi-band threshold methods are based on comprehensive consideration of each band and are widely used in water body extraction.

McFeeters [39] Normalized Difference Water Index (NDWI), which is well-known band-ratio method, which has been studied and used in the experiments via Python scripting with Sentinel-2 and Landsat scenes (Figure 3).

$$NDWI = (G - NIR)/(G + NIR) \tag{1}$$

It uses green (G) and near-infrared (NIR) spectral bands to maximize water feature identification (1). McFeeters proposed a zero threshold to separate water other land.

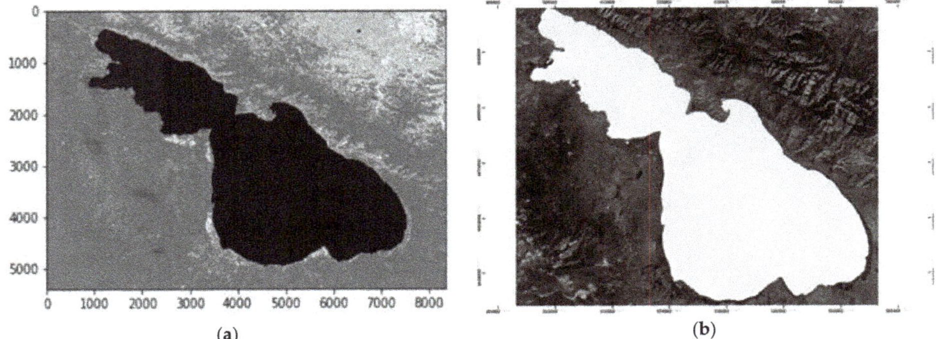

Figure 3. McFeeters band-ratio algorithm for Lake Sevan calculated from (**a**) Sentinel-2 and (**b**) Landsat sensors.

3.4. Validation of the Results

The validation step is an important component and not a straightforward task as DC has number of limitations, among which unknown quality of the automatized geoprocessing results using different verification algorithms. It comprises multiple components ranging from in situ measurements collection, modeling and retrieval of land surface variables to scale related analysis [40]. All these factors complicate the validation issue.

The validation step is an essential component and not a straightforward task, as the DC platform may generate unknown quality of the automatized geoprocessing results using different verification algorithms. It comprises multiple components ranging from in situ measurements collection, modeling and retrieval of land surface variables to scale related analysis [40]. All these factors complicate the validation issue.

The experimental verification of McFeeters band math (NDWI) calculation results was performed integrating the results of diverse observation, such as high-resolution remote-sensing products (UAV imagery) received during field campaign held in 2018 using Sensefly eBee and the hydrological data provided by the Service of Hydrometeorology and Active Influence on Atmospheric Phenomena SNCO, Ministry of Emergency Situation of Armenia (hereafter Service).

The shorelines derived via NDWI from Sentinel-2 and Landsat 8 were compared with the shoreline received from UAV for the small portion of the north-east shore (2 km).

To a first approximation the visual comparison of shorelines derived via NDWI from Sentinel-2 and Landsat 8 and UAV image shows that they match quite well despite the differences of spatial resolutions Landsat 8 (30 m), Sentinel-2 (10 m), UAV image (30 cm) (Figure 4).

The other approach was to compare the surface areas derived using NDWI from Sentinel-2 (12 January 2015) and Landsat 8 (29 December.2015) with the surface areas measured and calculated by the Service on 1 January.2015 and 1 January.2016 respectively (Table 1). The Table 1 shows that the differences between provided surface areas are 6.38 sq.km and 9.16 sq.km for Sentinel -2 and Landsat 8 respectively. It should be stressed that the images selected for comparison were acquired near the time of the hydrological data measurements.

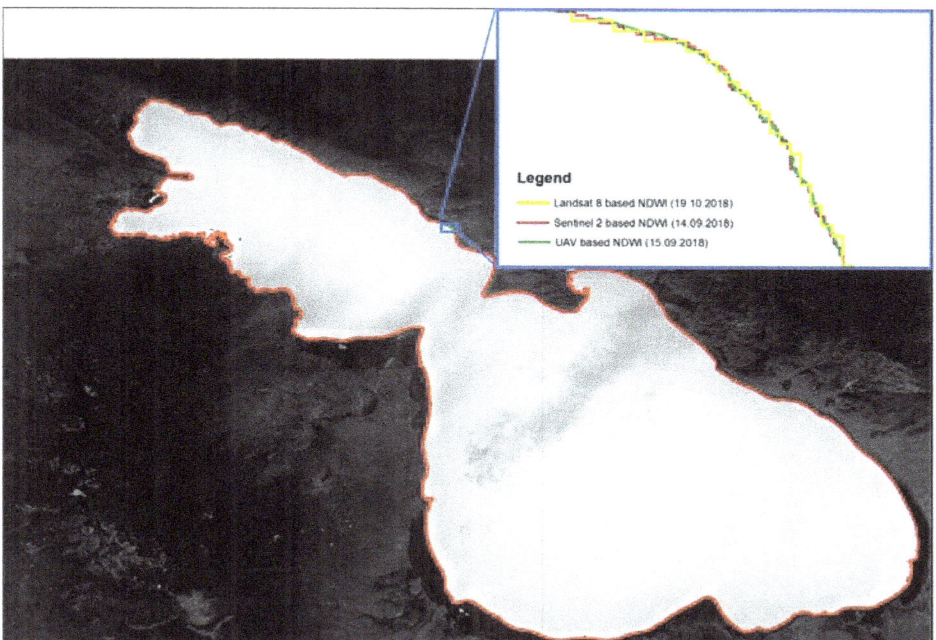

Figure 4. The comparison of shorelines derived from UAV, Sentinel-2 and Landsat imageries.

Table 1. The area of Lake Sevan according to Service and NDWI.

Sensor	Total Area (sq.km)		
	Satellite Image	Service	The Difference
Sentinel-2A (12 January 2015)	1269.18	1275.56	6.38
Landsat 8 OLI (29 December 2015)	1265.83	1274.99	9.16

4. Conclusions

This paper aimed to present the international Swiss—Armenian joint initiative to deploy next national DC in Armenia, which becomes the fourth national DC in the world after Australia, Switzerland and Cambodia. ADC is one of the best applications of the Armenian national e-infrastructure, which should be updated and empowered continuously in order to reveal the full potential of this innovative technology.

Thus far, the ADC is enriched with complete and up-to-date archive of EO data and successfully works for the simplest issues such as delineation of Lake Sevan.

Landsat and Sentinel image-based delineation of shorelines using NDWI spectral index gives sufficiently reliable results for Lake Sevan.

It should be added that the web-based User Interface has been developed by CEOS [41] to allow users exploring the mains functionalities of the data cube. However, for developing more advanced/tailored applications or services, the Python Application Programming Interface (API) is the preferred choice.

Once the Armenian Data Cube will be fully operational and will generate "official" products, they will be complied with the FAIR (Findable, Accessible, Interoperable, Re-usable) data principles [42,43], which will include adding a license such as Creative Commons and having a Digital Object Indentifier (DOI) for each generated datasets/products.

It could be stressed that Armenian DC has a potential to transform the EO into useful information for users and represents a prospective solution for remotely sensed environmental monitoring in Armenia. So the analysis between Armenian and Swiss DC and the transfer the necessary knowledge from Switzerland to Armenia for developing and implementing the first version of an ADC should be continued paving a way towards continuous remote environmental monitoring in Armenia.

Author Contributions: Conceptualization, S.A., A.S., H.A., Y.G. and G.G.; Data curation, V.M., G.T., H.G. and R.A.; Funding acquisition, S.A., H.A. and G.G.; Investigation, S.A., V.M., G.T., A.H., H.G. and R.A.; Project administration, S.A., H.A. and G.G.; Supervision, G.G.; Validation, V.M., G.T. and A.H.; Visualization, Y.G.; Writing—original draft, S.A., H.A., H.G., Y.G. and G.G.; Writing—review & editing, S.A., V.M., G.T., A.H., A.S., H.A., R.A., Y.G. and G.G.

Funding: This research was funded by the UNIVERSITÉ DE GENÈVE as the Leading House (hereinafter referred to as "LH") for the bilateral Science and Technology cooperation program with Russia and the CIS Region, Grant number "SFG 163".

Acknowledgments: The research was supported by the RA MES State Committee of Science and Russian Foundation for Basic Research (RF) in the frames of the joint research project "SCS 18RF-140" and "RFBR 18-55-05015 Arm-a" accordingly. The authors would like to also thank the Swiss Federal Office for the Environment (FOEN) for their financial support to the Swiss Data Cube. Methodologies used in this publication partly rely on the Swiss Data Cube (http://www.swissdatacube.org) material and methodologies, operated and maintained by UN Environment/GRID-Geneva, the University of Geneva, the University of Zurich and the Swiss Federal Institute for Forest, Snow and Landscape Research WSL.

References

1. Friedl, L. Benefits Assessment of Applied Earth Science. In *Satellite Earth Observations and Their Impact on Society and Policy*; Onoda, M., Young, O.R., Eds.; Springer: Singapore, Singapore, 2017; pp. 73–79, ISBN 978-981-10-3713-9. [CrossRef]

2. Durrieu, S.; Nelson, R.F. Earth observation from space-The issue of environmental sustainability. *Space Policy* **2013**, *29*, 238–250. [CrossRef]

3. Anderson, K.; Ryan, B.; Sonntag, W.; Kavvada, A.; Friedl, L. Earth observation in service of the 2030 Agenda for Sustainable Development. *Geo-Spat. Inf. Sci.* **2017**, *20*, 77–96. [CrossRef]

4. Yin, H.; Udelhoven, T.; Fensholt, R.; Pflugmacher, D.; Hostert, P. How Normalized Difference Vegetation Index (NDVI) Trendsfrom Advanced Very High Resolution Radiometer (AVHRR) and Système Probatoire d'Observation de la Terre VEGETATION (SPOT VGT) Time Series Differ in Agricultural Areas: An Inner Mongolian Case Study. *Remote Sens.* **2012**, *4*, 3364–3389. [CrossRef]

5. Giuliani, G.; Chatenoux, B.; Bono, A.D.; Rodila, D.; Richard, J.-P.; Allenbach, K.; Dao, H.; Peduzzi, P. Building an Earth Observations Data Cube: Lessons learned from the Swiss Data Cube (SDC) on generating Analysis Ready Data (ARD). *Big Earth Data* **2017**, *1*, 100–117. [CrossRef]

6. Earth Observation: Copernicus Sentinel Satellite Data-Open Access at ESA. Available online: http://open.esa.int/copernicus-sentinel-satellite-data/ (accessed on 14 June 2019).

7. USGS.gov | Science for a Changing World. Available online: https://www.usgs.gov/ (accessed on 14 June 2019).

8. Lewis, A.; Lymburner, L.; Purss, M.B.J.; Brooke, B.; Evans, B.; Ip, A.; Dekker, A.G.; Irons, J.R.; Minchin, S.; Mueller, N.; et al. Rapid, high-resolution detection of environmental change over continental scales from satellite data—The Earth Observation Data Cube. *Int. J. Digit. Earth* **2016**, *9*, 106–111. [CrossRef]

9. Gore, A. The Digital Earth. *Aust. Surv.* **1998**, *43*, 89–91. [CrossRef]

10. Lehmann, A.; Chaplin-Kramer, R.; Lacayo, M.; Giuliani, G.; Thau, D.; Koy, K.; Goldberg, G.; Richard, S., Jr. Lifting the Information Barriers to Address Sustainability Challenges with Data from Physical Geography and Earth Observation. *Sustainability* **2017**, *9*, 858. [CrossRef]

11. Baumann, P.; Mazzetti, P.; Ungar, J.; Barbera, R.; Barboni, D.; Beccati, A.; Bigagli, L.; Boldrini, E.; Bruno, R.; Calanducci, A.; et al. Big Data Analytics for Earth Sciences: The EarthServer approach. *Int. J. Digit. Earth* **2016**, *9*, 3–29. [CrossRef]

12. Purss, M.B.J.; Lewis, A.; Oliver, S.; Ip, A.; Sixsmith, J.; Evans, B.; Edberg, R.; Frankish, G.; Hurst, L.; Chan, T. Unlocking the Australian Landsat Archive—From dark data to High Performance Data infrastructures. *GeoResJ* **2015**, *6*, 135–140. [CrossRef]

13. Nativi, S.; Mazzetti, P.; Craglia, M. A view-based model of data-cube to support big earth data systems interoperability. *Big Earth Data* **2017**, *1*, 75–99. [CrossRef]

14. Open Data Cube. Available online: https://www.opendatacube.org (accessed on 14 June 2019).

15. Ariza-Porras, C.; Bravo, G.; Villamizar, M.; Moreno, A.; Castro, H.; Galindo, G.; Cabera, E.; Valbuena, S.; Lozano, P. CDCol: A Geoscience Data Cube that Meets Colombian Needs. In *Proceedings of the Advances in Computing*; Solano, A., Ordoñez, H., Eds.; Springer International Publishing: Cham, Switzerland, 2017; pp. 87–99. [CrossRef]

16. Yao, X.; Liu, Y.; Cao, Q.; Li, J.; Huang, R.; Woodcock, R.; Paget, M.; Wang, J.; Li, G. China Data Cube (CDC) for Big Earth Observation Data: Lessons Learned from the Design and Implementation. In Proceedings of the 2018 International Workshop on Big Geospatial Data and Data Science (BGDDS), Wuhan, China, 22–23 September 2018; pp. 1–3. [CrossRef]

17. Swiss Data Cube (SDC). Available online: https://www.swissdatacube.org/ (accessed on 14 June 2019).

18. Open Data Cube Manual—Open Data Cube 1.7+6.gd0ec48ca Documentation. Available online: https://datacube-core.readthedocs.io/en/latest/ (accessed on 14 June 2019).

19. *SDG Implementation Voluntary National Review (VNR) Armenia*; Report for the UN High-level Political Forum on Sustainable Development (9–18 July 2018). Yerevan, Armenia, 2018. Available online: https://sustainabledevelopment.un.org/content/documents/19586Armenia_VNR_2018.pdf (accessed on 2 August 2019).

20. UN in Armenia: News: Armenia's Roadmap for Sustainable Development Goals to Come Forth Soon. Available online: http://www.un.am/en/news/611?fbclid=IwAR1LmdmfQ0Ob6nNkE7lkP2wHJzr0mkT_3mn3TLCdQz27ASsneyjCJRfJ40w (accessed on 14 June 2019).

21. Muradyan, V.; Tepanosyan, G.; Asmaryan, S.; Saghatelyan, A.; Dell'Acqua, F. Relationships between NDVI and climatic factors in mountain ecosystems: A case study of Armenia. *Remote Sens. Appl. Soc. Environ.* **2019**, *14*, 158–169. [CrossRef]

22. Saghatelyan, A.; Asmaryan, S.; Muradyan, V.; Tepanosyan, G. The Utility of GIS for Assessing the Ecological State and Managing Armenian's Farmlands. *J. Geol. Resour. Eng.* **2014**, *2*. [CrossRef]

23. Asmaryan, S.; Saghatelyan, A.; Astsatryan, H.; Bigagli, L.; Mazzetti, P.; Nativi, S.; Lacroix, P.; Giuliani, G.; Ray, N. Leading the way toward an environmental National Spatial Data Infrastructure in Armenia. *South-East. Eur. J. Issue Earth Obs. Geomat.* **2014**, *3*, 53–62.

24. Giuliani, G.; Lacroix, P.; Guigoz, Y.; Roncella, R.; Bigagli, L.; Santoro, M.; Mazzetti, P.; Nativi, S.; Ray, N.; Lehmann, A. Bringing GEOSS Services into Practice: A Capacity Building Resource on Spatial Data Infrastructures (SDI). *Trans. GIS* **2017**, *21*, 811–824. [CrossRef]

25. Lehmann, A.; Nativi, S.; Mazzetti, P.; Maso, J.; Serral, I.; Spengler, D.; Niamir, A.; McCallum, I.; Lacroix, P.; Patias, P.; et al. GEOEssential—Mainstreaming workflows from data sources to environment policy indicators with essential variables. *Int. J. Digit. Earth* **2019**, 1–17. [CrossRef]

26. Nativi, S.; Santoro, M.; Giuliani, G.; Mazzetti, P. Towards a knowledge base to support global change policy goals. *Int. J. Digit. Earth* **2019**, *0*, 1–29. [CrossRef]

27. Pavlov, D.S.; Kopylov, A.I.; Poddubny, S.A.; Gabrielyan, B.K.; Chilingaryan, L.A.; Mnatsakanyan, B.P.; Bobrov, A.A.; Yepremyan, E.V.; Romanenko, A.V.; Hovsepyan, A.A.; et al. *Ecology of Lake Sevan during the Period of Water Level Rise*; Nauka DSC: Makhachkala, Russia, 2010; p. 348, ISBN 978-5-94434-162-4. (In Russian)

28. Babayan, A.; Hakobyan, S.; Jenderedjian, K.; Muradyan, S.; Voskanov, M. *Experience and Lessons Learned Brief*. 2005, pp. 347–362. Available online: https://iwlearn.net/iw-projects/1665 (accessed on 2 August 2019).

29. Baghdasaryan, A.B.; Abrahamyan, S.B.; Aleksandryan, G.A. *Physical Geography of Armenian SSR*; AS ArmSSR: Yerevan, Armenia; NAN of RA: Yerevan, Armenia, 1971.

30. Louati, M.; Saïdi, H.; Zargouni, F. Shoreline change assessment using remote sensing and GIS techniques: A case study of the Medjerda delta coast, Tunisia. *Arab. J. Geosci.* **2015**, *6*, 4239–4255. [CrossRef]

31. Alesheikh, A.A.; Ghorbanali, A.; Nouri, N. Coastline change detection using remote sensing. *Int. J. Environ. Sci. Technol.* **2007**, *4*, 61–66. [CrossRef]

32. Durduran, S.S. Coastline change assessment on water reservoirs located in the Konya Basin Area, Turkey, using multitemporal landsat imagery. *Environ. Monit. Assess.* **2010**, *164*, 453–461. [CrossRef]
33. Toure, S.; Diop, O.; Kpalma, K.; Maiga, A.S. Shoreline Detection using Optical Remote Sensing: A Review. *ISPRS Int. J. Geo-Inf.* **2019**, *8*, 75. [CrossRef]
34. Tepanosayn, G.; Muradyan, V.; Hovsepyan, A.; Minasyan, L.; Asmaryan, S. A Landsat 8 OLI Satellite Data-Based Assessment of Spatio-Temporal Variations of Lake Sevan Phytoplankton Biomass. *Ann. Valahia Univ. Targoviste Geogr. Ser.* **2017**, *17*, 83–89. [CrossRef]
35. Heblinski, J.; Schmieder, K.; Heege, T.; Agyemang, T.K.; Sayadyan, H.; Vardanyan, L. High-resolution satellite remote sensing of littoral vegetation of Lake Sevan (Armenia) as a basis for monitoring and assessment. *Hydrobiologia* **2011**, *661*, 97–111. [CrossRef]
36. Hovsepyan, A.; Muradyan, V.; Tepanosyan, G.; Minasyan, L.; Asmaryan, S. Studying the Dynamics of Lake Sevan Water Surface Temperature Using Landsat8 Sateliite Imagery. *Ann. Valahia Univ. Targoviste Geogr. Ser.* **2018**, *18*, 68–73. [CrossRef]
37. Li, W.; Du, Z.; Ling, F.; Zhou, D.; Wang, H.; Gui, Y.; Sun, B.; Zhang, X. A Comparison of Land Surface Water Mapping Using the Normalized Difference Water Index from TM, ETM+ and ALI. *Remote Sens.* **2013**, *5*, 5530–5549. [CrossRef]
38. Haibo, Y.; Zongmin, W.; Hongling, Z.; Yu, G. Water Body Extraction Methods Study Based on RS and GIS. *Procedia Environ. Sci.* **2011**, *10*, 2619–2624. [CrossRef]
39. McFeeters, S.K. The use of the Normalized Difference Water Index (NDWI) in the delineation of open water features. *Int. J. Remote Sens.* **1996**, *17*, 1425–1432. [CrossRef]
40. Zhang, R.; Tian, J.; Li, Z.; Su, H.; Chen, S.; Tang, X. Principles and methods for the validation of quantitative remote sensing products. *Sci. China Earth Sci.* **2010**, *53*, 741–751. [CrossRef]
41. Rizvi, S.R.; Killough, B.; Cherry, A.; Gowda, S. The Ceos Data Cube Portal: A User-Friendly, Open Source Software Solution for the Distribution, Exploration, Analysis, and Visualization of Analysis Ready Data. In Proceedings of the IGARSS 2018–2018 IEEE International Geoscience and Remote Sensing Symposium, Valencia, Spain, 22–27 July 2018; pp. 8639–8642. [CrossRef]
42. Wilkinson, M.D.; Dumontier, M.; Aalbersberg, I.J.; Appleton, G.; Axton, M.; Baak, A.; Blomberg, N.; Boiten, J.W.; Da Silva Santos, L.B.; Bourne, P.E.; et al. The FAIR Guiding Principles for scientific data management and stewardship. *Sci. Data* **2016**, *3*, 160018. [CrossRef]
43. Stall, S.; Yarmey, L.; Cutcher-Gershenfeld, J.; Hanson, B.; Lehnert, K.; Nosek, B.; Parsons, M.; Robinson, E.; Wyborn, L. Make scientific data FAIR. *Nature* **2019**, *570*, 27. [CrossRef]

Article

National Open Data Cubes and Their Contribution to Country-Level Development Policies and Practices

Trevor Dhu [1,*], Gregory Giuliani [2], Jimena Juárez [3], Argyro Kavvada [4,5], Brian Killough [6], Paloma Merodio [3], Stuart Minchin [1] and Steven Ramage [7]

[1] Geoscience Australia, Canberra GPO Box 378, Australia; stuart.minchin@ga.gov.au
[2] Institute for Environmental Sciences/GRID-Geneva, University of Geneva, 1205 Geneva, Switzerland; gregory.giuliani@unige.ch
[3] Instituto Nacional de Estadística y Geografía (INEGI), 20276 Aguascalientes, Mexico; jimena.juarez@inegi.org.mx (J.J.); paloma.merodio@inegi.org.mx (P.M.)
[4] National Aeronautics and Space Administration, Earth Science Division, 300 E St. SW, Washington, DC 20546, USA; argyro.kavvada@nasa.gov
[5] Booz Allen Hamilton, 8283 Greensboro Dr, McLean, VA 22102, USA
[6] National Aeronautics and Space Administration, Hampton, VA 23666, USA; brian.d.killough@nasa.gov
[7] GEO Secretariat, 2300 CH-1211 Geneva, Switzerland; sramage@geosec.org
* Correspondence: trevor.dhu@ga.gov.au

Received: 15 June 2019; Accepted: 25 October 2019; Published: 5 November 2019

Abstract: The emerging global trend of satellite operators producing analysis-ready data combined with open source tools for managing and exploiting these data are leading to more and more countries using Earth observation data to drive progress against key national and international development agendas. This paper provides examples from Australia, Mexico, Switzerland, and Tanzania on how the Open Data Cube technology has been combined with analysis-ready data to provide new insights and support better policy making across issues as diverse as water resource management through to urbanization and environmental–economic accounting.

Keywords: Open Data Cube; UN 2030 Agenda for Sustainable Development; UN System of Environmental Economic Accounting; Earth observation data

---

1. Introduction

Earth observation (EO) data from ground, airborne, and space platforms and associated applications have the potential to, and already, provide insights into global policy frameworks including: the United Nations (UN) 2030 Agenda for Sustainable Development [1], the UN System of Environmental Economic Accounting [2], the Sendai Framework for Disaster Risk Reduction [3] and the Paris Climate Agreement [4]. EO data can support, validate, and augment traditional data inputs, including national statistics, administrative data, household survey data and census information. In addition, EO data contribute as a direct indicator to inform relevant goals and targets; help optimize surveys and other traditional data collection efforts; and support disaggregation of targets and indicators, where relevant, to ensure that no one is left behind. Today, petabytes of EO data and geospatial information, coupled with analytical methods and innovation in technology, and enabled by free and open data policies, are applied widely around the world to derive useful information about the drivers, pace, and associated impacts of change on Earth, as well as to inform policies and support decision making.

Significant work is still needed, however, to ensure that different types of end-users are harnessing the full potential of EO to address local challenges and assist with the monitoring and implementation of global agendas, such as the UN 2030 Agenda for Sustainable Development. Improvements are needed to overcome challenges such as: EO data accessibility and handling; EO data validity and fitness for purpose; integration of information from different data streams; and data continuity [5]. Organizations such as the Group on Earth Observations (GEO) and the Committee on Earth Observation Satellites (CEOS) are working to reduce the barriers that are faced by end-users across multiple sectors and regions in accessing, analyzing, and integrating satellite-based and other sources of EO data into national processes and decision support systems. More and more, there is a recognized need for new ways of managing and providing easy access to the vast amounts of EO that is increasingly available, as well as for raising awareness about the value of the data and translating science into policy.

In recent years, there has been a global move towards satellite operators producing analysis-ready data, to reduce the work needed by users prior to exploiting and analyzing satellite data. For example, CEOS has led the creation of the CEOS Analysis Ready Data for Land (CARD4L) framework. This framework defines CARD4L data as " … satellite data that have been processed to a minimum set of requirements and organized into a form that allows immediate analysis with a minimum of additional user effort and interoperability both through time and with other datasets" [6].

The CARD4L framework has provided a set of product family specifications (PFS) for surface reflectance [7], surface temperature [8], and radar backscatter [9]. These specifications, while not prescriptive, provide both minimum and target thresholds for general metadata, per-pixel metadata, radiometric and atmospheric corrections, and geometric corrections.

Satellite operators such as the United States Geological Survey (USGS) are now in the process of using these PFS to produce global collections of CARD4L compliant EO data. This transition to easily accessible CARD4L compliant data provides an incredible opportunity for EO data to be more impactful across a wide range of global challenges. However, the sheer amount of data that is now, or will soon be, available for use demands that we move away from the historical approach of users downloading data and local processing toward "processing into high performance computing data centers (e.g., Google Earth Engine, planet-API, National Computing Infrastructure in Australia, DigitalGlobe DGBX platform) using Big Data processing tools … along the lines of moving the algorithms to the data not the data to the algorithms" [10]. Bringing together data, analytical methods, infrastructure, and application insights is essential to promote and accelerate social, economic, and environmental sustainability.

A range of software (open source and proprietary), tools, and analysis platforms exist for accessing, storing, processing, and facilitating the use of EO to derive insights and for societal applications. Cloud computing has had a tremendous effect on the emergence of computational infrastructure designed to provide EO analysis capabilities such as Google Earth Engine [11], the Copernicus Data and Information Access Services [12] and a range of other platforms provided via Amazon Web Services, Microsoft Azure (Layerscape), etc. Before choosing a software product, tool or analysis platform, end-users—including national statistical agencies, line ministries, and national mapping agencies, among other stakeholders—need to take into account their local needs including governance requirements, institutional capacity, geospatial analytics expertise, associated costs, and sustainability of the respective tool(s) or EO analysis platforms.

Countries such as Australia, Mexico, Switzerland, and Tanzania either have adopted or are in the process of adopting an open source solution, the Open Data Cube (ODC) [13,14], to enable them to integrate insights from Earth observation data into their national policy and information systems. The ODC builds on the work of the Australian Geoscience Data Cube [15] and seeks to increase the value and impact of global EO data by providing an open and freely accessible exploitation architecture, while fostering a community of cooperation that promotes open EO data, reuse of algorithms, and related information usage and sharing for the benefit of society.

The open source nature of the ODC was an important factor in this tool being selected by Australia, Mexico, Switzerland, and Tanzania. However, the other critical factor is the ODC's ability to be implemented on diverse computational infrastructures ranging from national supercomputing facilities such as Australia's National Computational Infrastructure, through to numerous commercial cloud infrastructures. Together, this combination of open source software and infrastructure flexibility has enabled the establishment of sovereign, operational capabilities that can be controlled and managed in-country. This is critical in order to build trust that the information products being generated can be both sustained and relied upon for use in a wide variety of policy problems.

2. Open Data Cube Examples

This paper provides examples of how the provision of CARD4L compliant EO data together with operational ODC implementations is enabling countries to better tackle challenges ranging from water resource management through to urbanization. The examples vary in terms of the maturity of their implementation and their success to date in influencing national development agendas. However, all of these deployments have the common aim of trying to ensure that there is a better connection from global EO data to national level development practices.

2.1. Water Resource Management in the Murray-Darling Basin, Australia

2.1.1. Overview of Digital Earth Australia

In May 2018, the Australian Government announced an ongoing investment of approximately 9.1 M USD per year in Digital Earth Australia (DE Australia) to deliver new and innovative satellite-based applications and services to the Australian Government and businesses. DE Australia is the world's first operational, continental scale implementation of the ODC technology. The technical details of DE Australia are outlined in [15,16].

DE Australia currently produces its own analysis-ready data for the Landsat and Sentinel-2 satellites, which include corrections for position, terrain, radiometry, atmosphere, and sun-sensor geometry [17,18]. This approach is currently being reviewed by CEOS to ensure that it is compliant with the CARD4L standards.

DE Australia is helping the Australian Government to understand environmental changes such as water availability, crop growth and urban expansion, supporting improved decision making and planning. DE Australia is also driving Australia's economic growth by enabling small businesses and industry to more readily access satellite data to innovate and create new products. This will present new opportunities and increase the profitability and productivity of businesses in sectors such as land planning, construction, agriculture, and mineral exploration [19,20]. For example, information drawn from satellites is vital to help grazers increase the capacity of paddocks and make their farms more viable and sustainable [21].

2.1.2. Managing Water in the Murray-Darling Basin, Australia

The Murray-Darling Basin (the Basin) is one of Australia's most important agricultural regions. It covers 14% of Australia's land mass, produces more than one-third of the nation's food and $22 billion in agriculture on average each year [22]. The Basin is also home to more than 30,000 wetlands, 16 of international importance under the Convention on Wetlands of International Importance (Ramsar Convention), which are fundamental to the health and viability of the whole basin [22].

Managing water across an area as large as the Basin with as many different competing uses for the water presents numerous challenges that must be addressed to ensure that Australia achieves progress against a wide range of Sustainable Development Goals (e.g., SDG 2 and 6). In 2012, the Murray-Darling Basin Plan passed into law in Australia as a major step forward in managing this complex system. This Murray-Darling Basin Plan was developed " ... to bring the Basin back to a healthier and sustainable level, while continuing to support farming and other industries for the benefit of the Australian community" [23].

These challenges are further compounded given that Australia is currently in the midst of a drought that is covering vast tracks of the Basin. In order to help manage the current drought, Geoscience Australia and the New South Wales government are exploiting one of DE Australia's first decision-ready products, Water Observations from Space (WOfS). DE Australia's current implementation of WOfS detects water on the land surface from Landsat imagery [24]. WOfS provides an understanding of where water is usually present, where it is seldom observed, and where inundation of the surface has been occasionally observed by satellite.

DE Australia has used the WoFS algorithm, updated for every Landsat pass, and a map of the locations and spatial extent of over 60,000 water storages, referred to as farm dams, to estimate how full these farm dams are on a monthly basis. The product does not provide an estimate of the volume of water available, but rather an estimate of how full the dam is as a percentage of the maximum of the horizontal extent of the farm dam (Figure 1).

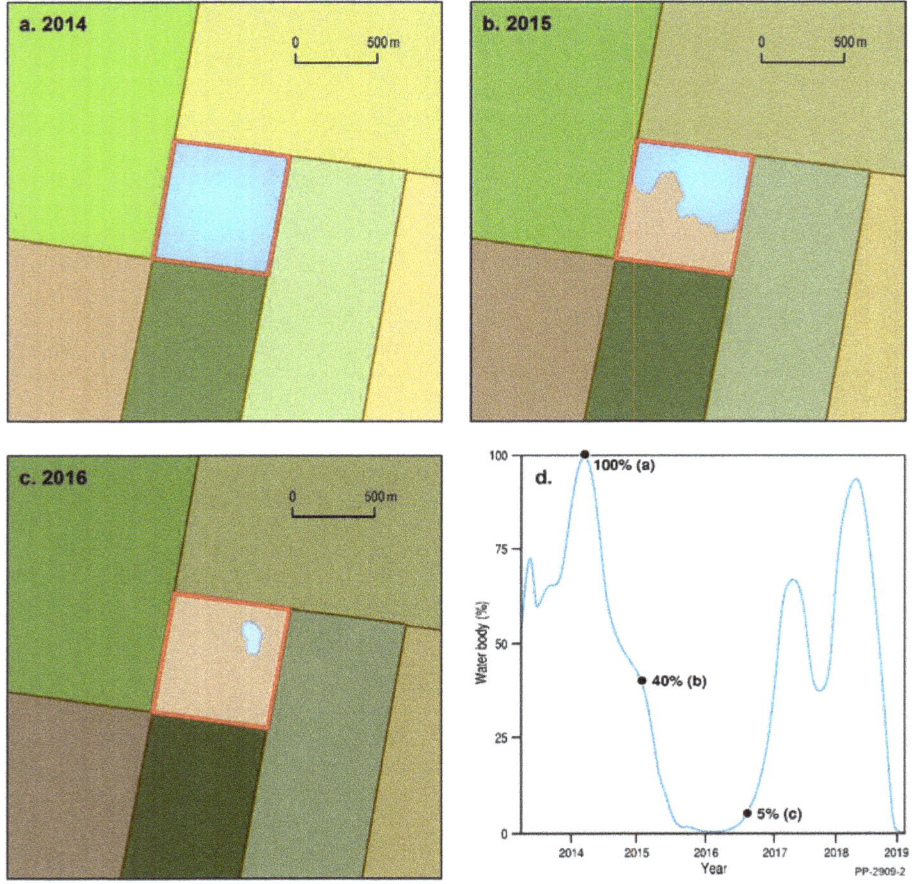

Figure 1. Idealized example demonstrating how DE Australia describes the relative 'fullness' of a farm dam. (**a**) In this panel the dam (identified by the red square) is 100% full as there is water across the entire spatial extent of the dam; (**b**) Here the dam is classified as 40% full as the water covers 40% of the spatial extent of the dam; (**c**). Here the dam is classified as being 5% full as the water only covers 5% of the spatial extent of the dam; (**d**) A time history of the 'fullness' of the dam with the fullness of the dam from panels a, b and c plotted.

The farm dam fullness estimated by this method is not a precise estimate and is meant to be a qualitative indicator of water availability. For example, this method does not work well for steep sided farm dams nor for farm dams with a spatial extent smaller than ~2500 m². Nonetheless, initial qualitative attempts to validate this data have demonstrated its utility in providing governments with a broad-scale understanding of the availability of water. The New South Wales (NSW) government is taking DE Australia's monthly estimates of dam fullness, aggregating it spatially across local regions called parishes, to produce and publish a map of Farm Dam Water Status on a monthly basis (Figure 2). This is the first ever comprehensive audit of farm dam levels across NSW and it is providing the NSW government with new insights into how much water is available on farms and where the current drought is posing the largest risks to water supply for agriculture. This insight is, in turn, informing the prioritization of a wide range of drought management and response programs of the NSW Government.

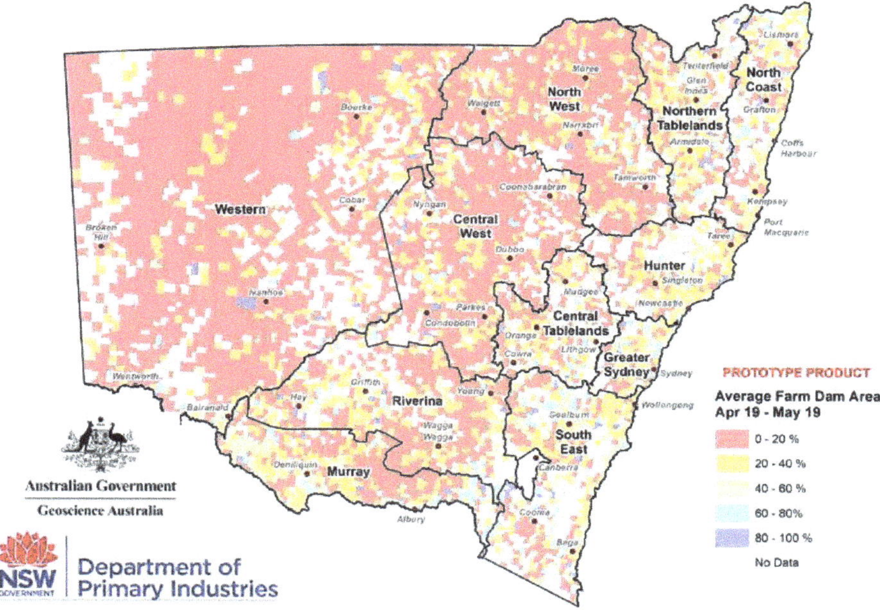

Figure 2. Farm Dam Water Status for New South Wales showing how full farm dams are across New South Wales for the period of 19 April–19 May, 2019 [25].

The NSW Farm Dam Assessment forms part of the NSW Government's State Seasonal Update that is used by the NSW Regional Assistance Advisory Committee in making recommendations on potential support for farm businesses, families, and communities. The Farm Dam Assessment is now an integral part of the data used by this committee in making recommendations to the NSW Minister for Primary Industries and NSW Government on relative priority and timing of introducing and withdrawing drought response programs and initiatives identified in the NSW Drought Framework.

At the same time, given the scarcity of water in the Basin and the various competing uses, it is critical that the Australian Government is able to ensure that when water has been embargoed for the environment, that it is not illegally used for other purposes such as irrigation. The Australian Government is now using Digital Earth Australia (DEA) to ensure compliance with the Murray-Darling Basin Plan and to help make sure that no one is stealing water that has been earmarked for environmental purposes.

The Murray-Darling Basin Authority (MDBA) is the Australian government agency responsible for developing and implementing a plan to ensure the sustainable use of the Basin's water resources. In 2018, the MDBA undertook the " . . . first large-scale use of satellite imagery for tracking the progress of an environmental flow event covering a large fraction of the Basin … " [26]. This pilot study was undertaken with the explicit intent of using satellite data to help prevent and detect water theft. The MDBA demonstrated that the imagery from DE Australia is able to detect water in irrigation channels and on-farm storages as well as detect changes to crops [26]. It is also capable of detecting water in the wetlands and river reaches that are being targeted with a specific water release. The MDBA has identified that DE Australia is providing them with " … a valuable new tool to ensure water is delivered to where it is needed and is not diverted for unauthorised use" [27].

DE Australia was subsequently used to provide the MDBA with EO data that they are using to identify waterbodies that have filled during periods when farmers were not allowed to extract water from the river [28]. EO data on its own is not enough to determine whether or not these waterbodies have been filled legally or illegally. However, this data does offer a tool for prioritizing where more expensive on-ground compliance investigations are undertaken. This approach to using satellite data to protect water is a new capability that has been enabled by DE Australia and will transform Australia's approach to managing this precious resource into the future.

2.2. Snow Cover Monitoring, Switzerland

2.2.1. Overview of the Swiss Data Cube

Following the work done in Australia, the Swiss Data Cube (SDC) [29] is an initiative supported by the Federal Office for the Environment (FOEN) and developed, implemented, and operated by the United Environment Program (UNEP)/GRID-Geneva in partnership with the University of Geneva (UNIGE), the University of Zurich (UZH), and the Swiss Federal Institute for Forest, Snow and Landscape Research (WSL). The objective of the SDC is to support the Swiss government in environmental monitoring and reporting, as well as enable Swiss scientific institutions to fully benefit from EO data for research and innovation.

Currently, the SDC contains 35 years of analysis-ready data of both optical (e.g., Landsat 5, 7, 8; Sentinel-2) and radar (e.g., Sentinel-1) satellite data over the entire country (total volume: 6TB; 200 billion observations) [30]. The SDC is an innovative analytical framework, based on the Open Data Cube software stack [14,31], allowing users to benefit from this new generation of EO data, and in particular minimizing the time and scientific knowledge required to access, prepare, and make it possible to analyze a large volume of data with consistent and spatially-aligned, calibrated observations.

The SDC is aiming at contributing to the national Digital Switzerland strategy by (1) supporting innovation and growth, (2) improving efficiency and effectiveness of government investments, (3) improving management of natural resources, (4) stimulating research, (5) generating decision-ready information products, and (6) improving data access and use and enabling new products/services that can transform everyday life. Ultimately, the SDC will deliver a unique capability to track changes in unprecedented detail using Earth observation satellite data and enable more effective responses to problems of national significance [32].

2.2.2. Monitoring Snow Cover Evolution, Switzerland

Like many other countries in the world, Switzerland faces challenges (e.g., land management, environmental degradation) caused by increasing pressures on its natural resources [33]. These challenges need to be overcome to meet the needs of a growing population. Switzerland is acknowledged as the water reservoir of Europe. While its territory represents four thousandths of the continent's total area, 6% of Europe's freshwater reserves are stored in Switzerland [34,35].

Monitoring snow cover and its variability is an indicator of climate change and identification of snowmelt processes is essential for effective water-resource management. Indeed, it is expected that by 2085, the proportion of snow contributing to water bodies will decrease by 25%, strongly affecting the water regime of major European rivers like the Rhône, Rhine, and Danube [36–38].

Earth observation (EO) data acquired by satellites are helpful to monitor snow conditions through time. Synthetic-aperture radar (SAR) images are effective and robust measures to identify melting snow, whereas optical data are able to identify snow cover extension [39,40]. Detailed knowledge of snow cover and its evolution in Switzerland is an essential tool for public policies and decision-making. Beyond the economic issues related to tourism, other questions arise such as flood risk management or water supply, given the storage role that snow plays, retaining water in winter to release it in spring and summer.

Consequently, to better understand the spatial distribution and evolution over time of snow cover nationwide, the UN Environment/GRID-Geneva and the University of Geneva have developed a snow detection algorithm benefiting from the analysis-ready data archive and analytical framework offered by the Swiss Data Cube. The Snow Observations from Space (SOfS) algorithm is an integrated and innovative solution for monitoring snow cover and its variability across the entire country and ultimately will allow generation of a decision-ready product that can be readily used as a basis for the design, implementation, and evaluation of policies, programs, and regulation, and for developing policy advice [41]. Preliminary results have shown a clear decrease of snow cover over Switzerland in the last 20 years. The perennial snow zone, where the probability of snowfall varies between 80% and 100%, still covered 27% of the Swiss territory in the decade 1995–2005. Ten years later, it has fallen to 23%, a loss of 2100 km^2 equal to seven times the size of the canton of Geneva. While areas with little or no snow covered 36% of the territory during the decade 1995–2005, the area lacking snow cover increased to 44% between 2005 and 2017, corresponding to an increase of approximately 5200 km^2 (Figure 3).

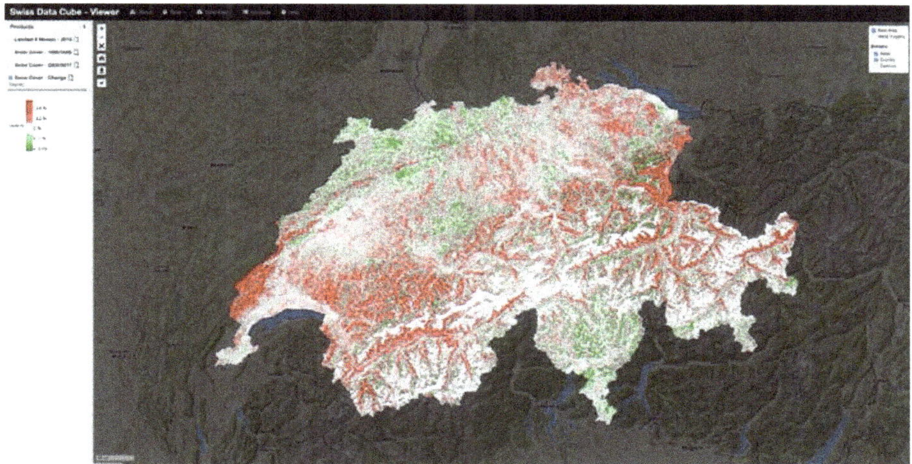

Figure 3. Snow cover change between periods 1995–2005 and 2005–2017 visualized in the Swiss Data Cube Viewer [42]. Red areas show a decline in snow cover over the two decades while green areas show an increase.

SOfS offers interesting potential for environmental monitoring and can serve as a pilot example, which can be of interest for other countries/regions. To test its robustness, SOfS is currently being applied in two protected areas in Italy and France (Gran Paradiso and La Vanoise National Parks) where in-situ measurements will help validate satellite observations [43], as well as in Armenia to provide insights for water management of Lake Sevan, the largest lake in the country that has significant economic, cultural, and recreational value [44].

To turn this algorithm into an integrated and effective mechanism to monitor snow cover and its variability, providing actionable information to decision makers, we are aiming to (1) consolidate the algorithm "Snow Observations from Space" (SOfS) to get the best results possible for monitoring snow cover conditions using both optical and radar imagery allowing the identification of dry and wet snow, and (2) automate the algorithm and transform it into a service that allows measurement of different periods and scales (e.g., a value every month, trimester, year at national, cantonal, and communal levels). Results will be made available through an interface targeted at decision and policy makers (at the national, cantonal, and communal levels) so they can access the latest values and trends on the snow cover indicator. These two activities are currently under development and a first prototype will made available in 2019–2020.

In a broader context, given its small territory and dense population, effective land management has become a national priority, as exemplified by a recent vote on land use (e.g., "Stopper le mitage – pour un développement durable du milieu bâti, initiative contre le mitage/Zersiedelung stoppen – für eine nachhaltige Siedlungsentwicklung, Zersiedelungsinitiative"). The SDC has the potential to support the Swiss government to monitor environmental changes nationwide in near real-time. Currently, land cover and land use data are generated from visual interpretation of aerial photography over a 6-year period to cover the entire surface of Switzerland [45]. Therefore, following the seasonal dynamics of vegetation, water, and snow is almost impossible. The SDC can contribute to overcome this limitation, providing effective monitoring services of snow coverage, drought conditions, water quality, urban development, agricultural activities, or health of vegetation.

2.3. Testing the UN System for Economic-Environmental Accounting, Mexico

2.3.1. Overview of the Mexican Geospatial Data Cube

Through a collaboration with Geoscience Australia, the Mexican Geospatial Data Cube (MGDC) is being developed at the National Institute of Statistics and Geography of Mexico (INEGI). Initially, the MGDC will contain over 109,000 images of Landsat analysis-ready data provided by the USGS and NASA, including data since 1984. Figure 4 shows the number of images by year that are currently contained within the MGDC. This archive is already loaded into the institutional infrastructure of INEGI and is estimated to increase its volume from 30 TB to 90 TB when decompressed.

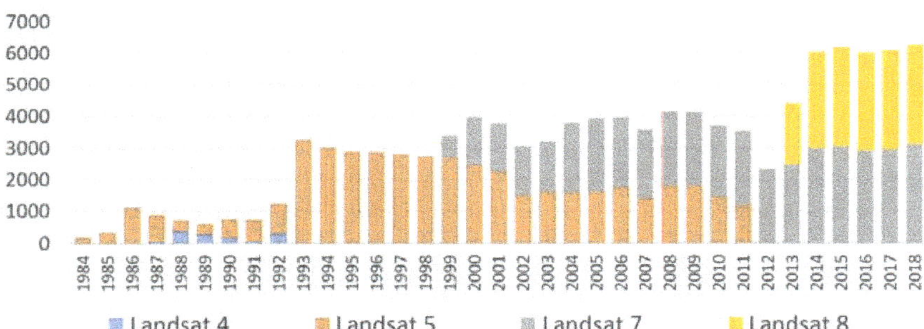

Figure 4. Overview of the number of images contained within the MGDC broken down by year and Landsat mission.

In order to facilitate future interoperability between sensors, INEGI has chosen to implement a defined grid onto which all data in the MGDC will be projected. The grid was defined using Albers Equal Area projection, with 150 × 150 km tiles and 30 m resolution (5000 × 5000 pixels); this design will allow future interoperation with 10 m resolution from the data provided by Sentinel-2. This 416 tile grid (144 of which cover the territory) also ensures covering Mexico's exclusive economic zone (Figure 5).

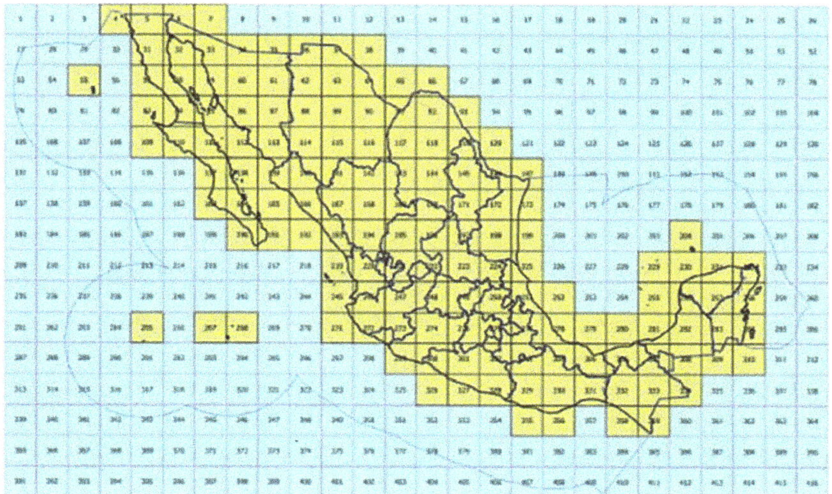

Figure 5. Overview of the standardized grid that is used for the MGDC (each tile represents a 150 × 150 km region).

As with the DE Australia infrastructure described in Section 2.1, the MGDC is designed to produce an ever expanding range of derived, decision-ready products (Figure 6). INEGI is currently testing a diverse set of products that range from common indexes such as the Normalized Difference Vegetation Index and Modified Normalized Difference Water Index, through to more complex classification tools such as the Water Observations from Space (WOFS) product described in [24]. More important, INEGI is working to ensure that the EO data is supplemented and integrated with a wide range of in-situ validation data to ensure that the derived MGDC products are robust and tangibly improving the design and monitoring of public policies and internationally agreed objectives, such as the 2030 Agenda for Sustainable Development [1] and the System of Environmental-Economic Accounting [2].

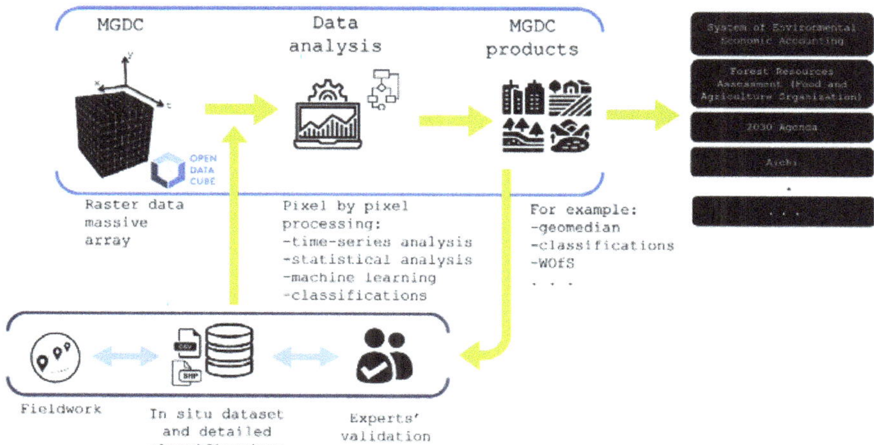

Figure 6. Notional workflow demonstrating how the MGDC intends to combine in-situ data with EO data to produce products that can drive better decision making and drive progress against a range of national and international policy priorities.

The ability to exploit the large amount of time series data contained within the MGDC is allowing the use of high-dimensional statistical methods [46] to generate robust composite images for all of Mexico (e.g., Figure 7). These annual, cloud-free summaries that were generated using the geometric median (or geomedian) algorithm, have already been used to provide insights into matters of national significance in Mexico like Natural Resources and Agriculture Statistics. A series of initial product validations by INEGI's thematic specialists is being carried out, and it is planned that the MGDC system will be a transversal service platform in INEGI.

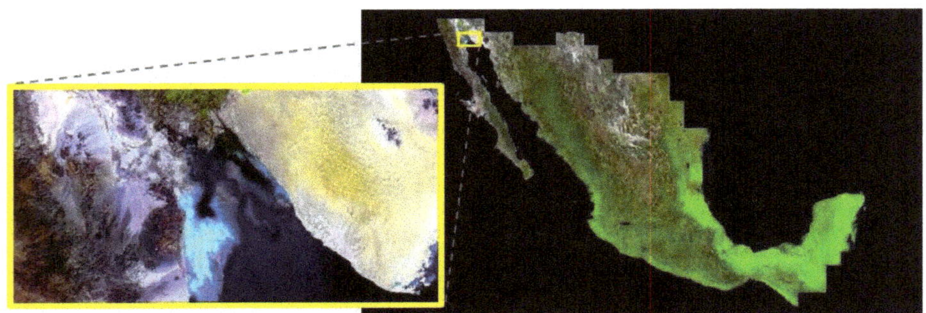

Figure 7. Example of an annual cloud free image for Mexico, using the available Landsat data from 2015 (January to December) in the MGDC. The image was produced using some 6074 Landsat images in MGDC and the high-dimensional statistic methods of [46]. The inset shows the Colorado River Delta region.

2.3.2. Using the MGDC for Natural Capital Accounting and Valuation of Ecosystem Services

One of the priority uses for the MGDC is to support the Natural Capital Accounting and Valuation of Ecosystem Services (NCAVES) project. This project was launched jointly by the European Union and implemented by the United Nations Statistics Division (UNSD), the United Nations Environment (UN Environment), and the Secretariat of the Convention on Biological Diversity. The objective of the project is to advance the knowledge agenda on environmental–economic accounting, ecosystem accounting, and by initiating pilot testing of the System of Environmental Economic Accounting (SEEA). The results will improve the measurement of ecosystems and their services (both in physical and monetary terms) at the subnational level and develop an internationally agreed methodology.

This project integrates information from the Economic Statistics and the Natural Resources-Environment Departments. As producer of geospatial information, INEGI generates the data for the extent (quantity) and condition (quality) of ecosystems. It is also in charge of the monetary valuation of ecosystem services, also based on the SEEA EEA methodology.

INEGI is currently evaluating the ability of the MGDC to overcome some of the constraints that the NCAVES project faces while routinely and efficiently using EO data. For example, in the initial phase of the project, geospatial data from different themes are used in a pilot study developed for Aguascalientes, a state in the centre of Mexico. This requires the mapping of features of interest such as rivers, other water bodies and elements of infrastructure, such as roads. These maps and geospatial data are currently derived from remote sensing imagery, i.e., satellite images from several sensors and platforms. However, this is based mainly on visual procedures that impose limits to the spatial details included and are time-consuming if we are to keep those details updated.

However, the MGDC's ability to do rapid, national-scale analyses with the Landsat archive, combined with the ability to undertake time series analysis, offers the potential for new, scalable and, hopefully, automatable image analysis techniques to support these accounts. For example, vegetation evolution in time can be easily monitored. Figure 8 shows visual results of a MGDC test consisting of comparing two false-colour images from the areas of Montes Azules and Marqués de Comillas from

1986 and 2017; to the right of the Lacantún River, the deforestation (Marqués de Comillas) during these 3 decades is observed. This analysis better supports public policy implementation and monitoring, as this river observed in the area acts as the limit for the natural protected area of Montes Azules.

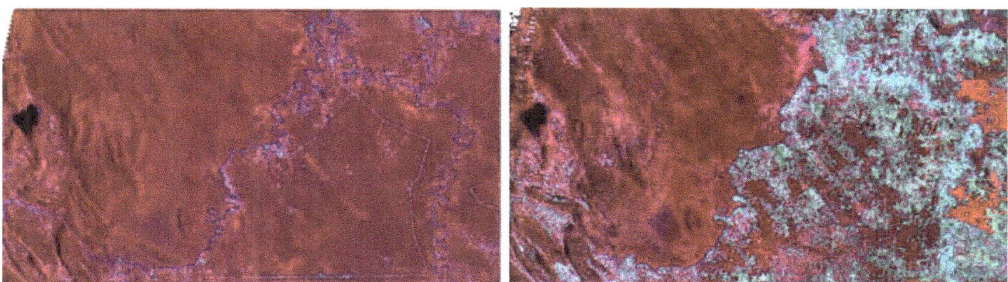

Figure 8. MGDC images from the areas of Montes Azules and Marqués de Comillas from 1986 (left) and 2017 (right). Rendering (RGB): Red (near infrared band), Green (red band), Blue (green band).

The image from 1986 is a single observation (Landsat 5), while the 2017 image is a pixel-level statistical summary (geomedian) from the 2017 time series, using Landsat 7 and 8 data. With pixel-level time-series analysis, there is also huge potential for the development and application of new machine learning techniques. Collectively, these advances have the potential to dramatically reduce the time it takes to identify the required features and to improve the accuracy and spatial scale at which they can be identified.

Based on the observed results, current testing of the MGDC is focused on two stages of the ecosystem accounting process (Figure 9). The MGDC is demonstrating potential to improve our ability to determine the ecosystems extent across all of Mexico and to provide insights into the condition of some of those ecosystems.

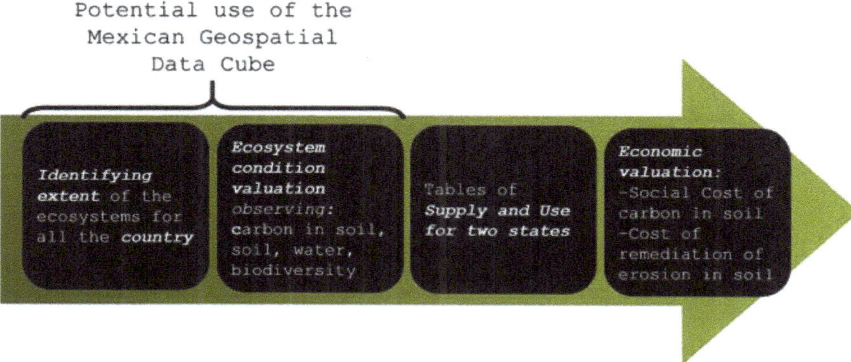

Figure 9. Notional workflow demonstrating the sequence of accounts and the components where the MGDC has the most potential to offer improvements.

The MGDC has the potential to support the identification of ecosystems, which can be improved by classification algorithms applied to an already segmented image. Furthermore, the classification process can be improved by using new products such as the geomedian images discussed in Section 2.3.1 and shown in Figure 7. These images may be used to generate the input for the current classification processes in Land Use Land Cover maps.

Similarly, the MGDC has implemented the same water classification algorithm (WOfS) used in Australia to support the detection and analysis of water bodies as described in Section 2.1.2. INEGI is currently testing the accuracy of this product in the Mexican context. However, assuming it has similar levels of accuracy in Mexico as it does in Australia, it will provide a rich source of information on the extent of these critical ecosystems in Mexico.

At this stage, the time-series analysis enabled by the MGDC represents the main advantage of having a massive analysis-ready raster data array. In particular, addressing the ecosystem condition can be eased with certain algorithms for detecting change over time. For example, results of the ecosystem extension accounts for the state of Aguascalientes show continuous expansion of urban areas (particularly in private and communal lands) and no change in the extension of coniferous forest from 2007 to 2015.

Results from ecosystem services valuation studies will be useful to inform and strengthen key public policies such as land-use planning, land-use changes, and nature conservation measures. Over time, it is expected that, with the MGDC, data access and use will become easier and faster, enhancing the timeliness of valuation procedures. The MGDC will improve the implementation of policy instruments such as payment for ecosystem service schemes, environmental responsibility and liability and environmental impact assessment.

2.4. Using the African Regional Data Cube to Manage Urbanization in Tanzania

2.4.1. Overview of the African Regional Data Cube

The Africa Regional Data Cube (ARDC) was launched in May 2018 by the Global Partnership for Sustainable Development Data (GPSDD), the Committee on Earth Observation Satellite (CEOS), and Amazon, to support five countries: Kenya, Senegal, Sierra Leone, Ghana, and Tanzania. The ARDC is focused on building the capacity of users in this region to apply EO satellite data to address local and national needs, as well as the objectives of the Group on Earth Observations (GEO) and the 2030 Agenda for Sustainable Development [1]. The ARDC will support a number of key users, including government ministries, national statistical agencies, geographic institutes, and research scientists.

The satellite data in the ARDC (approximately 11 TB volume) is considered analysis-ready data in that it is processed to a minimum set of requirements and organized into a form that allows immediate analysis through time and is interoperable with other datasets. To date, the ARDC only includes Landsat analysis-ready data since the year 2000, but the ARDC is working to add Copernicus data from the Sentinel-1 and Sentinel-2 missions. These pre-processed analysis-ready datasets were compiled in time series stacks to allow valuable assessments of changing land and water resources, a task that would be very time consuming and quite difficult using traditional, scene-based analysis methods.

2.4.2. Urbanization in Tanzania

Many governments, such as Tanzania, are interested in tracking urbanization to understand the changes in land resources and corresponding population growth rates. These government agencies include national statistical offices, urban planning managers, and the ministries of agriculture and environment. It is known that increases in urbanization have an impact on the environment and the health of a population. With urbanization products from the ARDC, government decision-makers can measure the extent and location of urban growth to help planning of water and land use. In addition, these data and information products can be used to directly address Sustainable Development Goal (SDG) 11.3.1 (ratio of land consumption rate to population growth rate).

An urbanization analysis was conducted over the city of Dar es Salaam, Tanzania using the ARDC. The results of this study were compared to a report from [47] and the European Space Agency (ESA) Urban Thematic Exploration Platform [48].The analysis used a fractional cover [49] threshold approach (bare soil fraction > 25%) to identify urban regions. This threshold range was selected using visual interpretation with an attempt to match the results of [47] (Figure 10) and [48] (Figure 11). Though this is not a complex scientific method, it is a good first-order approximation for urban extent and can be used to assess long-term urban change. As suggested in [47], a random forest classification approach should be used to obtain more accurate results in order to validate the measured urban extent.

According to [47], the urban growth was 100% (or 8% per year) from 2007 to 2016. The analysis results from the ARDC showed an urban growth of 123% (or 9% per year). This urban growth rate can be compared with the population growth rate (~5% according to several online sources) for the same time period. Using the equation from the SDG 11.3.1 guidelines, the ratio of land consumption rate (urban growth) to population growth rate is 1.8. This suggests that the population is rapidly moving out of the city and expanding its urban footprint at a rate significantly faster than the population is growing. Though these ARDC results are promising, a more accurate supervised land classification analysis is needed in the future to validate the results with ground-based data and to remove false-positive urban areas.

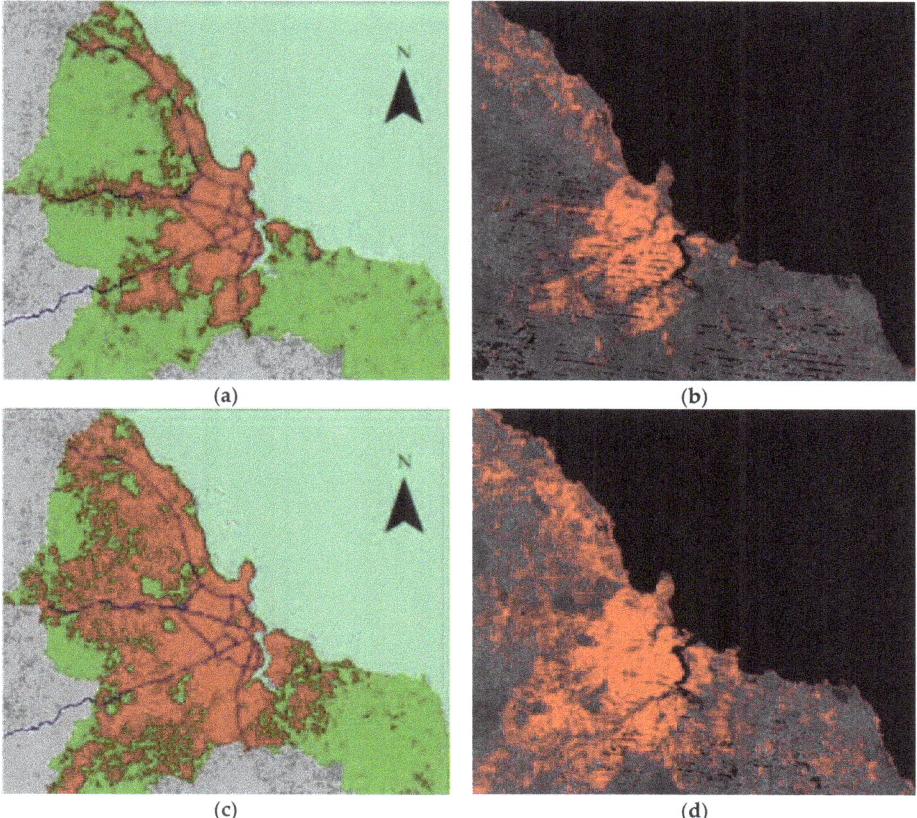

Figure 10. Urbanization results are compared for two time periods, 2007 (**a**,**b**) and 2016 (**c**,**d**), and two analysis methods, [47] (**a**,**c**) and ARDC (**b**,**d**), over Dar es Salaam, Tanzania. Though these results are quite similar, the ARDC results show more urban pixels that are likely associated with non-urban rural areas that lack vegetation and are classified as urban using the threshold approach. In addition, there are some common Landsat-7 "banding" issues visible on the background greyscale image and in the urban results. These artifacts could be removed with further compositing and post-processing.

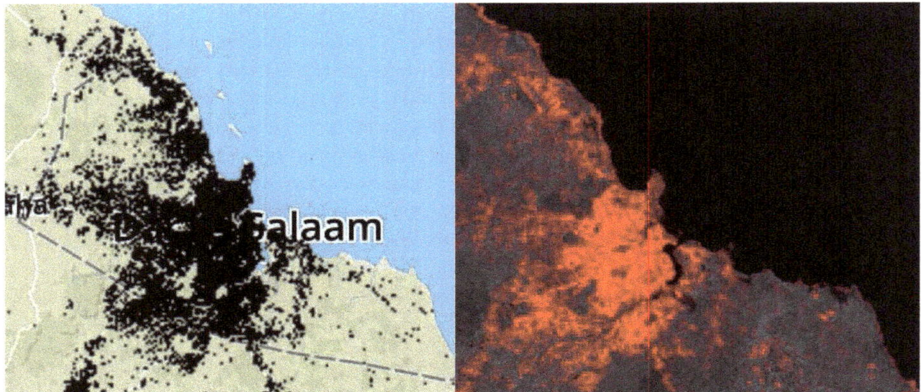

Figure 11. Urbanization results are compared for 2015 from [44] (left) and the ARDC (right) to gain confidence that the urban footprint results were reasonably accurate. The [44] results (left) are based on proprietary X-band radar and the ARDC results (right) are based on Landsat-8 optical data. Overall, the primary urbanization results are quite similar and suggest the fractional cover threshold approach yields sufficiently accurate results to apply to other years in the time series.

3. Conclusions

Satellite missions will continue to provide increasingly larger volumes of free and open analysis-ready data for global users. With recent advances in the global provision of analysis-ready data and proven and innovative open source data technology solutions such as the Open Data Cube (ODC), global users now have the unprecedented ability to routinely utilize satellite data for national policy and decision-making needs.

The examples shown in this paper have demonstrated that governments have the need and desire to use satellite data to tangibly improve their management of natural resources and policies that support sustainable development. One of the reasons that ODC has been successful at enabling this is that it is an open source and scalable architecture; it allows countries to establish and operate their own sovereign analysis capability. Countries are able to control the quality and timeliness of their analyses and rely upon their own operational capability to underpin regulatory and official reporting processes.

However, great effort is needed for deploying an operational ODC at a national level. The vast amounts of data that need to be integrated and managed means that operational deployments need to have access to a wide range of system engineering skills and high-performance computational infrastructure. Consequently, there is an emerging trend to move from sovereign ODC deployments to larger scale regional centres that support sovereign ODCs. For example, by the end of 2019, the ARDC will merge into a larger regional cube including data from an ever growing range of satellites, including Landsat and Sentinel-2, which will cover the entire continent of Africa. This initiative is called Digital Earth Africa and serves as an example of a vision to create many regional data cubes using the ODC infrastructure.

Regional-scale data cubes are more manageable in terms governance and institutional arrangements, whereas a full global data cube would be impractical to implement and manage effectively. In addition, these regional cubes allow users to address transboundary topics that otherwise would not be possible with individual country-level data cubes. In this matter, standardization is key; it is essential that data and processes are consistent and measurements regard similar criteria. In the future, a set of regional data cubes could share technical approaches and application algorithms, while maintaining local management of data and products relevant to regional decision-making needs.

With a vision toward a global set of regional data cubes, it will be possible to take advantage of consistent time series satellite data and different, yet interoperable, datasets. Such data cubes and their corresponding open source application algorithms can be enhanced and shared across the world to address national and transboundary issues while maintaining data sovereignty and political separation through a regional implementation. In addition, to realise the full potential of the ODC products to address local and regional decision-making and policies, it is important to increase research and gather in-situ ground data for proper algorithm and product validation. Over time, it is expected that open data products will increase, their accuracy will improve, and data access and use will become easier and faster for everyone.

Author Contributions: Conceptualization, all authors.; methodology, T.D., J.J., P.M., B.K., G.G., and A.K.; writing—original draft preparation, T.D., J.J., P.M., B.K., G.G., and A.K.; writing—review and editing, All Authors.

Funding: This research received no external funding.

Acknowledgments: Results of this publication partly or fully rely on the Swiss Data Cube (http://www.swissdatacube.org), operated and maintained by UN Environment/GRID-Geneva, the University of Geneva, the University of Zurich and the Swiss Federal Institute for Forest, Snow and Landscape Research WSL. This paper is published with the permission of the CEO, Geoscience Australia.

Conflicts of Interest: The authors declare no conflict of interest.

References

1. United Nations. Transforming Our World: The 2030 Agenda for Sustainable Development. Available online: https://www.un.org/ga/search/view_doc.asp?symbol=A/RES/70/1&Lang=E (accessed on 15 June 2019).
2. United Nations. System of Environmental-Economic Accounting. Available online: https://seea.un.org/ (accessed on 15 June 2019).
3. United Nations Office for Disaster Risk Reduction. Sendai Framework for Disaster Risk Reduction 2015–2030. Available online: https://www.unisdr.org/files/43291_sendaiframeworkfordrren.pdf (accessed on 15 June 2019).
4. United Nations. Paris Agreement. Available online: https://unfccc.int/sites/default/files/english_paris_agreement.pdf (accessed on 15 June 2019).
5. Committee on Earth Observation Satellites. Satellite Earth Observations in Support of the Sustainable Development Goals. Available online: http://eohandbook.com/sdg/files/CEOS_EOHB_2018_SDG.pdf (accessed on 15 June 2019).
6. Committee on Earth Observation Satellites. CEOS Analysis Ready Data. Available online: http://ceos.org/ard/ (accessed on 15 June 2019).
7. Committee on Earth Observation Satellites. Product Family Specification, Optical Surface Reflectance (CARD4L-OSR). Available online: http://ceos.org/ard/files/CARD4L_Product_Specification_Surface_Reflectance_v4.0.pdf (accessed on 15 June 2019).
8. Committee on Earth Observation Satellites. Product Family Specification, Land Surface Temperature. Available online: http://ceos.org/ard/files/CARD4L_Product_Specification_Land_Surface_Temperature_v4.0.pdf (accessed on 15 June 2019).
9. Committee on Earth Observation Satellites. Product Family Specification, Normalised Radar Backscatter. Available online: http://ceos.org/ard/files/CARD4L_Product_Specification_Backscatter_v4.0.pdf (accessed on 15 June 2019).
10. United Nations Statistical Division. Earth Observations for Official Statistics Satellite Imagery and Geospatial Data Task Team Report. Available online: https://unstats.un.org/bigdata/taskteams/satellite/UNGWG_Satellite_Task_Team_Report_WhiteCover.pdf (accessed on 15 June 2019).
11. Gorelick, N.; Hancher, M.; Dixon, M.; Ilyushchenko, S.; Thau, D.; Moore, R. Google Earth Engine: Planetary-scale geospatial analysis for everyone. *Remote Sens. Environ.* **2017**, *202*, 18–27. [CrossRef]
12. Copernicus. DIAS. Available online: https://www.copernicus.eu/en/access-data/dias (accessed on 15 June 2019).
13. Open Data Cube. Available online: https://www.opendatacube.org (accessed on 15 June 2019).
14. Killough, B. Overview of the Open Data Cube Initiative. In Proceedings of the IGARSS 2018 - 2018 IEEE International Geoscience and Remote Sensing Symposium, Valencia, Spain, 22–27 July 2018; pp. 8629–8632.

15. Lewis, A.; Oliver, S.; Lymburner, L.; Evans, B.; Wyborn, L.; Mueller, N.; Raevksi, G.; Hooke, J.; Woodcock, R.; Sixsmith, J.; et al. The Australian Geoscience Data Cube—Foundations and lessons learned. *Remote Sens. Environ.* **2017**, *202*, 276–292. [CrossRef]

16. Dhu, T.; Dunn, B.; Lewis, B.; Lymburner, L.; Mueller, N.; Telfer, E.; Lewis, A.; McIntyre, A.; Minchin, S.; Phillips, C. Digital earth Australia—unlocking new value from earth observation data. *Big Earth Data* **2017**, *1*, 64–74. [CrossRef]

17. Li, F.; Jupp, D.L.B.; Reddy, S.; Lymburner, L.; Mueller, N.; Tan, P.; Islam, A. An Evaluation of the Use of Atmospheric and BRDF Correction to Standardize Landsat Data. *IEEE J. Sel. Top. Appl. Earth Obs. Remote Sens.* **2010**, *3*, 257–270. [CrossRef]

18. Li, F.; Jupp, D.L.B.; Thankappan, M.; Lymburner, L.; Mueller, N.; Lewis, A.; Held, A. A physics-based atmospheric and BRDF correction for Landsat data over mountainous terrain. *Remote Sens. Environ.* **2012**, *124*, 756–770. [CrossRef]

19. Frontier SI. Digital Earth Australia Industry Strategy. Available online: https://frontiersi.com.au/wp-content/uploads/2019/04/FrontierSI_Digital_Earth_Industry_Strategy_March_2019-v2.pdf (accessed on 20 October 2019).

20. Frontier SI. Community Focus Key to APSEA Award Success. Available online: https://frontiersi.com.au/community-focus-key-to-apsea-award-success/ (accessed on 20 October 2019).

21. Salleh, A. How Satellites and Machine Learning Algorithms Are Helping Farmers to Be More Sustainable. Available online: https://www.abc.net.au/news/science/2019-09-27/machine-learning-and-satellites-help-farms-sustainability/11500214 (accessed on 20 October 2019).

22. Murray-Darling Basin Authority. Murray-Darling Basin Authority Annual Report 2017–2018. Available online: https://www.mdba.gov.au/sites/default/files/pubs/MDBA-Annual-Report-2017-18.pdf (accessed on 15 June 2019).

23. Murray-Darling Basin Authority. A Plan for the Murray-Darling Basin. Available online: https://www.mdba.gov.au/basin-plan/plan-murray-darling-basin (accessed on 15 June 2019).

24. Mueller, N.; Lewis, A.; Roberts, D.; Ring, S.; Melrose, R.; Sixsmith, J.; Lymburner, L.; McIntyre, A.; Tan, P.; Curnow, S.; et al. Water observations from space: Mapping surface water from 25years of Landsat imagery across Australia. *Remote Sens. Environ.* **2016**, *174*, 341–352. [CrossRef]

25. New South Wales Department of Primary Industries. NSW State Seasonal Update—May 2019. Available online: https://www.dpi.nsw.gov.au/climate-and-emergencies/droughthub/information-and-resources/seasonal-conditions/ssu/may-2019 (accessed on 15 June 2019).

26. Murray-Darling Basin Authority. A Case Study for Compliance Monitoring Using Satellite Imagery. Available online: https://www.mdba.gov.au/sites/default/files/pubs/Compliance-monitoring-using-remote-sensing.pdf (accessed on 15 June 2019).

27. Murray-Darling Basin Authority. Satellites helping to guard against water theft. Available online: https://www.mdba.gov.au/media/mr/satellites-helping-guard-against-basin-water-theft (accessed on 15 June 2019).

28. Murray-Darling Basin Authority. *Monitoring 'First Flush' Flows in the Namoi, Macquarie and Warrego Rivers—Remote Sensing for Compliance and Ecohydrology*; Murray-Darling Basin Authority: Canberra, Australia, 2019.

29. Swiss Data Cube (SDC) - EO for monitoring the environment of Switzerland in space and time. Available online: https://www.swissdatacube.org (accessed on 15 June 2019).

30. Giuliani, G.; Chatenoux, B.; De Bono, A.; Rodila, D.; Richard, J.-P.; Allenbach, K.; Dao, H.; Peduzzi, P. Building an Earth Observations Data Cube: Lessons learned from the Swiss Data Cube (SDC) on generating Analysis Ready Data (ARD). *Big Earth Data* **2017**, *1*, 100–117. [CrossRef]

31. Rizvi, S.R.; Killough, B.; Cherry, A.; Gowda, S. The Ceos Data Cube Portal: A User-Friendly, Open Source Software Solution for the Distribution, Exploration, Analysis, and Visualization of Analysis Ready Data. In Proceedings of the IGARSS 2018–2018 IEEE International Geoscience and Remote Sensing Symposium, Valencia, Spain, 22–27 July 2018; pp. 8639–8642.

32. Giuliani, G.; Chatenoux, B.; Honeck, E.; Richard, J. Towards Sentinel-2 Analysis Ready Data: A Swiss Data Cube Perspective. In Proceedings of the IGARSS 2018–2018 IEEE International Geoscience and Remote Sensing Symposium, Valencia, Spain, 22–27 July 2018; pp. 8659–8662.

33. Giuliani, G.; Dao, H.; De Bono, A.; Chatenoux, B.; Allenbach, K.; De Laborie, P.; Rodila, D.; Alexandris, N.; Peduzzi, P. Live Monitoring of Earth Surface (LiMES): A framework for monitoring environmental changes from Earth Observations. *Remote Sens. Environ.* **2017**, *202*, 222–233. [CrossRef]
34. Beniston, M.; Farinotti, D.; Stoffel, M.; Andreassen, L.M.; Coppola, E.; Eckert, N.; Fantini, A.; Giacona, F.; Hauck, C.; Huss, M.; et al. The European mountain cryosphere: A review of its current state, trends, and future challenges. *Cryosphere* **2018**, *12*, 759–794. [CrossRef]
35. Beniston, M.; Stoffel, M. Assessing the impacts of climatic change on mountain water resources. *Sci. Total Environ.* **2014**, *493*, 1129–1137. [CrossRef] [PubMed]
36. Changement climatique et hydrologie. Available online: https://sciencesnaturelles.ch/topics/water/climate_change_and_hydrology (accessed on 1 April 2019).
37. Beniston, M.; Uhlmann, B.; Goyette, S.; Lopez-Moreno, J.I. Will snow-abundant winters still exist in the Swiss Alps in an enhanced greenhouse climate? *Int. J. Climatol.* **2011**, *31*, 1257–1263. [CrossRef]
38. Lehmann, A.; Guigoz, Y.; Ray, N.; Mancosu, E.; Abbaspour, K.C.; Rouholahnejad Freund, E.; Allenbach, K.; De Bono, A.; Fasel, M.; Gago-Silva, A.; et al. A web platform for landuse, climate, demography, hydrology and beach erosion in the Black Sea catchment. *Sci. Data* **2017**, *4*, 170087. [CrossRef] [PubMed]
39. Small, D. Flattening Gamma: Radiometric Terrain Correction for SAR Imagery. *IEEE Trans. Geosci. Remote Sens.* **2011**, *49*, 3081–3093. [CrossRef]
40. Small, D.; Miranda, N.; Ewen, T.; Jonas, T. Reliably flattened radar backscatter for wet snow mapping from wide-swath sensors. In Proceedings of the ESA Living Planet Symposium, Edinburgh, Scotland, 9–13 September 2013.
41. Frau, L.; Rizvi, S.R.; Chatenoux, B.; Poussin, C.; Richard, J.; Giuliani, G. Snow Observations from Space: An Approach to Map Snow Cover from Three Decades of Landsat Imagery Across Switzerland. In Proceedings of the IGARSS 2018–2018 IEEE International Geoscience and Remote Sensing Symposium, Valencia, Spain, 22–27 July 2018; pp. 8663–8666.
42. Swiss Data Cube Viewer. Available online: http://www.swissdatacube.org/viewer (accessed on 15 June 2019).
43. Guigoz, Y.; Palazzi, E.; Terzago, S.; Chatenoux, B.; Poussin, C.; Giuliani, G. Snow cover evolution in Gran Paradiso and Vanoise protected areas using Earth Observation Data Cube. *Data* **2019**, *4*, 138.
44. Asmaryan, S.; Asastryan, H.; Guigoz, Y.; Giuliani, G. Paving the way towards an Aremnian Data Cube. *Data* **2019**, *4*, 117. [CrossRef]
45. *Land Use in Switzerland–Results of the Swiss Land Use Statistics*; Swiss Federal Statistical Office (FSO): Neuchâtel, Switzerland, 2013.
46. Roberts, D.; Mueller, N.; Mcintyre, A. High-Dimensional Pixel Composites From Earth Observation Time Series. *IEEE Trans. Geosci. Remote Sens.* **2017**, *55*, 6254–6264. [CrossRef]
47. Gombe, K.E.; Asanuma, I.; Park, J.-G. Quantification of Annual Urban Growth of Dar es Salaam Tanzania from Landsat Time Series Data. *Adv. Remote Sens.* **2017**, *6*, 175–191. [CrossRef]
48. Thematic Exploration Product, Urban Footprint. Available online: https://urban-tep.eu (accessed on 1 May 2019).
49. Guerschman, J.P.; Scarth, P.F.; McVicar, T.R.; Renzullo, L.J.; Malthus, T.J.; Stewart, J.B.; Rickards, J.E.; Trevithick, R. Assessing the effects of site heterogeneity and soil properties when unmixing photosynthetic vegetation, non-photosynthetic vegetation and bare soil fractions from Landsat and MODIS data. *Remote Sens. Environ.* **2015**, *161*, 12–26. [CrossRef]

MDPI
St. Alban-Anlage 66
4052 Basel
Switzerland
Tel. +41 61 683 77 34
Fax +41 61 302 89 18
www.mdpi.com

Data Editorial Office
E-mail: data@mdpi.com
www.mdpi.com/journal/data